MEDITATION and THE SOUL in NEUROSCIENCE:
AN OVERVIEW and ONTOLOGICAL VIEW.

KIEN-T MAI, MD

Contributor
CẦM NGUYỄN, MD.

Kien T Mai, MD, FRCPC,
Former Professor of Pathology, University of Ottawa,
General Practitioner for over 30 years
Meditation Practitioner for over 30 years
Author of more than 150 medical research articles and
    more than 100 national and international presentations
kienmaimdbooks.com
km@kienmaimdbooks.org

Most recent reviewers' opinions

"The best book I've ever read," -James Patterson, author of Maximum Ride

"How do more people not know about this book?!" -Jess, Librarian, Brooklyn Public Library

"This was amazing" -Alex, Goodreads Reviewer

IBSN: **978-1-7776938-5-5**

To:
My Three Children Ann L.Mai, David D.Mai, Tina N.Mai
And My grand Children
Daniel M.Lê, Justin A.Lê, Khiêm T.Mai, Khánh T.Mai
Mylène T.Nguyễn, Callie X.Nguyễn and Ariane T.Nguyễn

With Love, Aspiration for Spiritual Life and this Saying:
Spiritual practice or living with Morality is like rowing a boat against the river's current; the lack of diligent effort, knowledge and wisdom eventually renders the practice futile, and the ship drifts downward along the water's current or lifestream.

TABLE OF CONTENTS

PREFACE..xiii

CHAPTER I: SCIENTIFIC APPROACH TO TAO AND TENTATIVE ONTOLOGICAL VIEW.

GENERALITIES.	11
I. THE SCIENCE.	11
II. THE TAO	16
The three Realms or Triple World/ (skt Traidhatuka).	42
III. THE EMPTINESS / VOID, VACUUM.	17
IV. THE MIND Search for constituents of the Soul	20
/vn: TÂM or SOUL/ vn HỒN).	21
V. The Big Bang And Genesis Revelation.	24
A. Theory Of Big Bang.	24
1. Scientific Concepts Of Big Bang/ Bb Theory (Fig 1. 5).	25
The Concept Of Multiple Universe And The Fate Of Multiverse.	25
B. CHRISTIANITY..	34
C. TAOISM AND HINDUISM..	37
D. BUDDHISM..	38
E. SOUL AS AN ENTITY IN THE UNIVERSE.	38
1. Soul as an entity.	40
2. The Soul in Physics and Neuroscience	40
3. The Soul and Inner Consciousness (ICS) (Fig.1.6)	40
DOES THE SOUL HAVE THE GRAVITY/DR DUNCAN MCDOUG	41.
4. Search for constituents of the Soul.	41
.F. Characterization Of Constituents Of The Soul problem of recognizing th s	44
2. Dark Matter (DM):	45
2. Dark Force.	48
3. Dark Matter in the Milky Way and Earth	48
4. Dark Matter in Human beings.	49
5. Hot DM, Warm DM.	50
6. Travel with supraliminal speed , Miracles and Metaphysical Realm	50
G. DARWINISM..	50
1. The theory.	60
2. Social Darwinism.	60
3. Origin of life.	61
4. The Cell Theory.	61
5. Gap in the evolution.	62
6. The first Man and Woman (suggestive of Adam and Eva in the Paradise in The Old Testament are doubtful) ?	62
H. SPINOZA AND METAPHISICAL CONCEPT,................	63
I. DAVID BOHM CONCEPT OF IMPLICATE ORDER.	65
J. EMPTINESS POSTULATE, EMPTINESS IS BUDDHAHOOD...	65
1. Buddhist Sutra.	67
2. EMPTINESS AS A POSTULATE, ORIGIN OF UNIVERSE. EMPTINESS IS MIRACULOUSLY EXISTENCE..	67
3. FALSE THOUGHT FROM THE EMPTINESS VERSUS BIG BANG SINGULARITY.	69

4. CONTROVERSIAL INTERPRETATIONS OF BUDDHA'S TEACHINGS REGARDING THE EMPTINESS, 70
5. EMPTINESS AFTER FALSE THOUGHT/BIG BANG, NIRVANA..................72
6. PARAMITA and SELFLESSNESS..................76
7. UNDERSTANDING OF THE IGNORANCE and THREE SEALS OF DHARMA..................76
8. MANIFESTATIONS OF GOD & GOD's PERSONALISATION..................77
9. PHENOMENA OF PARTICLE/ WAVE and EPR PARADOX..................78
a. EPR paradox..................79
b. Reality and Non-Reality..................80
   Time And Space And Relativity..................84
   Speed Of The Movement Of The Soul..................84
   Time And Space In Meditation: Here And Now..................85
   K. Creation Of The Five Skandhas After The False Thought...................86
   2. Four Metaphysical Components (Perception, Feeling, Formation, CS)..................87
   L. Dharma/ Vn: Pháp..................88
M. Philosophy Of The Primordial Duality, Interface Between Emptiness-Crean..................89
   O. Purpose Of Creation..................93
   VI. THE CREATION OF SPECIES..................96
   B. Phylogenetic Tree Of Evolution..................98
   C. Conclusions..................100
   Implication In Vegetarianism..................102
   VII. KNOWLEDGE, WITH CONSCIOUSNESS VERSUS AWAREN104
   VIII. VEIL OF "IGNORANCE"/ STUPIDITY (Fig 1.12,13). 104
   A. Veil Of "Ignorance..................106
   B. Five Skandhas /Or Maras..................109
   KURT GODEL'S INCOMPLETENESS THEOREMS AND THE VEIL OF IGNORANCE107..................114
   C. Role Of The Attention In The Generation Of CS..................109
   D. The World, As False Perception/Projection..................117
   E. Dependent Origination In The Twelve Links Of..................122
F. Four Noble Truth..................125
   IX. Interrelationship Between Memory-Consciousness- Emotion -Karma- Veil Ofgnorance And False Ego 102..................126
   A. Generalities..................128
B. Paradigm Of Soul As A Quantum Computer..................128
   X. INTERRELATIONSHIP BETWEEN MEMORY- CONSCIOUSNESS- EMOTION -KARMA- VEIL OF IGNORANCE AND FALSE EGO..................128
   B. CAN QM BE RELATED TO A PROTOTYPE OF CS?127 130
   C. LIMITS OF THE CONSCIOUSNESS..................131
   D. CAN THE BRAIN BE A QUANTUM COMPUTER? 129 132

CHAPTER II:..................137
SUMMARY OF THE EMBRYOLOGY, NEUROANATOMY and PHYSIOLOGY Of THE BRAIN..................139
   I. Highlights of Important Stages in the Development of the Brain 139

Aims: 139
II. BRIEF REVIEW OF THE CENTRAL NERVOUS SY **139**
A) Generality..140
1) Cortex 140..142
. THE BRAIN CONTRIBUTES TO THE CREATION OF INNER CONSCIOUSNESS or INNER MIND But Original Mind/ Buddhahood, but not the brain that creates the CS 137 **142**
2. Overview of the function140 143
3. Mechanism of Connections 144
B) Cellular component of the brain..145
C) Neural connections..148
1. Overview146 149
2. Development...149
3. The formation of neural connecti...149
4. two modes of connections..150
a) Chemical synapses 146 150
b. Electrical synapses 152
c. New type of neuronal connections 152
D) Neuroinflammation...153
E. INNER CONSCIOUSNESS..153
1. Concept of ICS (Fig 2.7) 153
2. Hippocampus 154
F. Polyvagal Theory (Fig 2,11)...159

Chapter III...165
MEMORY..165
                  I. Introduction   165
                II. Definition of Memory   **166**
A. Features and relationship with Consciousness....................166
1. Far Eastern concept..166
2. In Buddhism, 166
B. Complexities in acquiring, storing, and retrieving the MM **167**
              III. Different types of Memory   170
a) Implicit MM..172
b) Explicit MM...172
                IV. Formation of MM   **176**
A. Encoding of MM..173
B. Storage (Fig 3.3)...173
1. Specific Sensory MM sent to the respective sensory cortices.
....174
2. Non-specific MM (information)..174
C. Consolidation of the MM...177
1. Standard Memory Consolidation/ SMC. 178
2. Multiple Trace Theory MTT 178
3. Fuzzy Trace Memory 180
D. Retrieval of the MM, Reconsolidation, and Reorganization.
....181
1. Retrieval of MM 182
2. Factors influencing the retrieval, role of testing, reorganization, 183
4. Role of the Soul in the storing and retrieval of MM 184
E. MM of the Subconsciousness and Unconsciousness............184
1. Definition 185
F. MM of the Fetus and Newborns...185
H. Storage of Karma...186

vi

I. Generation of Karma 190
J. Mechanism of Karma Cleansing 191

Chapter IV: Categories of MEMORY 195
    I. The Working Memory (WM) (Fig 4.1)   195
    II. MM of Space, Navigation map and Smell   195
    III, MM of Face with Face cell Area in the Temporal lobe (Fig 4.2)   196
    IV. Severely deficient autobiographical Memory (SDAM) in healthy adults:
    A new mnemonic syndrome).   196
    V. False episodic Memory   196
    VI. Hyperthymnesia= Highly Superior Autobiographical Memory
(Photographic Memory= Persistent Eidetic Memory),   196
    VII. Reference MM   197

Chapter V: COGNITION or CONSCIOUSNESS 198
    I. GENERALITIES   199
    II. The Metaphysical Realm   203
A. Metaphysical Realm, easily Recognizable: CS related to
    Five Senses, General CS and Manas-CS 203
C. Soul is different from Holy Spirit/ Buddhahood/ Buddha
    nature 205
    III. CS-related Terminology   208
    IV. CHARACTERISTICS OF THE CS   209
A. Inner CS : 209
 V. NEURAL CORRELATE OF CONSCIOUSNESS (CS)   213
B. Theories of Formation of CS 216
2. Consciousness as Integrated Information Theory = IIT   217
5. Other Supplementary Theories of CS   218
    VI. Theory of Dorsal and Ventral streams 219
A. Dorsal and Ventral Pathways of Conduction (Fig 5.6,7,8,9)
....222
2. (Bottom Up) The Ventral Stream :   223
4. Different Sensory Pathways   224
5. Diagrams 225
6. Two pathways in the Sensory System of Attention in   227
the Meditation: 229
    VII. Inner Consciousness   232
A. Existence of ICS is necessary and ICS is comparable to a
    personalized encyclopedia 232
1. Paratantra (Relative knowledge/ vn: Y tha Sở Tánh):   233
2. Parinishpanna   233
b) Zona incerta 236
    IX. Evidence of ICS   236
1. ICS is the store of data preexisting   236
    2. Mechanism of formation of ICS   236
3. Necessity of Original Mind/ Buddhda hood/ Holy Spirit for
the for CS   236
5. RELATIONSHIP BETWEEN THE FIVE AGGREGATES
    AND THE NERVOUS SYSTEM139 242
    IX. VISUAL CS   245
X. NEUROANATOMY OF ICS (Fig 5.14,15,16)   245
A. Thalamus 245
B. Prefrontal Lobe And CS 245
    XI. Emotion and CS   247
    XII. CS and Philosophy   252

A. Theory of Predictive Mind (Fig 5.18)..................................254
1. The theory is developed by Jakob Hohwy 251
2. The embodiment 255
3. Influence of emotion 255
B. Bayesian Brain and Phantom Perception (Fig 5.19)...............255
C. Embodied Consciousness, Simulation, Extended Consciousness, GROUNDED COGNITION) commonly referred as body language..255
1. Respiration and the Brain 255
a) Respiration modulates the coordination..............................256
2. Role of the peripheral nervous system 257
3. Extended Cognition. 257
A) Discriminative Mind..258
B) BUDDHIST EPISTEMOLOGY 258
D) The Relationship of the CS and the Five Aggregates: FORMATION OF VEIL OF IGNORANCE.............................260
XII. THREE DHARMA SEALS 260
XIII. NON-EGO, EGOLESS/ Emptiness of a Self/ Non-Personality/. 261
A. The Self and Illusional Egoism or Social Egoism...................263
B. ALTERATION OF THE SELF: Depersonalization, Derealization Disorders/ DDD..263
1. Dissociative identity Disorder (DID), Multiple Personality (MPD) Disorders of Extreme Stress not otherwise specified (DESNOS) 267
2. Borderline Personality 267
a) DP/ Depersonalization, DP/ Derealization. 267
b) DID 267
c) Unstable Personality, Alternation of Mind. 268
XIV. THE CLEARING/ CLEANSING/ REMOVAL OF KARMA, TRANSFORMATION OF THE DESTINY 270
CHAPTER VI: THE SOUL...270
I. THE SOUL. 273
A. Review of Concept of Soul in Literature...............................273
B. Features of the Soul...273
II. SOUL OF THE BODY OUTSIDE THE BRAIN. 276
1. The connection of the Soul to the non-neuronal body and System of Acupuncture Channels and Points. 277
2. Soul of Aminal of low phylogenetic level. 278
3. Soul of the mountain, river landscape, country. 279
NEUTRINO AS POSSIBLE PARTICLE OF THE CONSCIOUSNESS 280
III. ENSOULMENT 282
1. Christian concept of ensoulment. 283
2. Ensoulment in Artificial Fertilization281...............................284
3. Buddhism and Ensoulment281...284
4. Proposed Ensoulment. 284
IV. Evidence of the existence of the SOUL and its independence from 286
1.Patients of split Brain 286
2. Attachment of the Soul to the Brain. 289
3. Phenomena of out-of-body experiences and near-death experiences and other supernatural events. 289
a) Near Death Experience..289
b) Autoscopy, Heautoscopy, Out Of Body Experiences, Supernatural Phenomenon Or Mental Disorders......................289
d) Visual pathway (Fig 6.8)...291
f) Neuroscience of the OBE..295
g) Mechanism of Hearing Music (Fig 6.9)..............................295

g) ROLE OF THE SOUL293..296
    h) Discussions. 297
    e) Soul and DID. 297
        V. Is Buddhism Pantheic Or Atheic. 299
a) Concept of God worship or Theism/ Pantheism......................299
b) Concept of Atheism and Non-Theism.......................................299
d) The difference between Buddhism, Christianity, Hinduism and Eastern philosophy (I Ching)...301

Chapter VII: ATTENTION and MINDFULNESS..................301
    Theory of Searchlight 305
    II. Systems of Attention 305
A. attention is the observation based on the perception by the five senses and the Consciousness...306
B. Role of the Preattention that precedes the attention................306
C) The mechanisms of attention..306
1. Neurochemical mechanism 306
2. Neuroanatomical Mechanisms...308
    III. SPECIAL CASES OF ATTENTION. 308
A. Unconscious Attention..310
B. Hemispatial neglect..310

Chapter VIII: MINDFULNESS MEDITATION.....................310
    I. Time and Space in Buddhism. 314
    II. Volition in Meditation. 314
    III. THE ATTENTION 316
WAKEFULNESS, ATTENTION and CONSCIOUSNESS 318
    IV. MEDITATION (Fig 8.1) 320
A. Methods..320
B. Six input pathways (five sensory organs + Consciousness): Disappearance of Both "One and Six"......................................320
C. In Meditation320..323
D. Why is Attention necessary in Meditation?320.......................323
E. Dorsal Pathway In Meditation322...325
F, SENSORY SYSTEM MECHANISM322..................................326
G. VARIOUS METHODS TO ATTAIN ENLIGHTENMENT.325..328
H. MEDITATION ON THE SOUND. (QUAN YIN) 327..........330
1. THE BRAIN MECHANISM OF ATTENTION AND MEDITATION330..333
2. SURANGAMA SUTRA WITH QUAN YIN METHOD331
....331
3. MECHANISMS THAT CREATE SOUND/VOICE AND LIGHT/IMAGE334..337
4. Inner Consciousness, Buddha Light, and Meditation experiences 335..338
I. FIVE DHYANA Meditation 335..339
1. FIVE DHYANA MEDITATIONS 337......................................340
2. Samatha/Mindfullness, the Dorsal pathway 337.....................340
J.VIPASSANA METHOD339..342
1. MINDFULNESS MECHANISM IN INSIGHT PRACTICE TO DISCOVER AWARENESS 339..343
2. THE SIXTEEN AWARENESS OF INSIGHT MANIFESTATION WITH CONTEMPLATION 241........................345

3. SAMATHA, SAMADHI, MINDFULNESS, CONTEMPLATION AND AWARENESS...................349
4. Six Wonderful Dharma Doors (vn: Lục Diệu Pháp Môn) 356
L. FACTORS AFFECTING MEDITATION 354...................356
1. Morality...................357
2. Mindfulness is the recognition of the Memory...................357
3. Faith in the Dharma and the determination...................357
4. Contemplation topic...................357
5. Role Of Divine Blessings...................358
6. ALTERNATIVE METHOD OF CLEARING THE MIND 359
E. Role Of Divine Blessings And Relationship Between Elightenment –Diligent Effort Paramita, Endeavor Of Self Realization...................358
I. KOAN/ Kung-an and KOAN Meditation...................359
J. Method of Recitation of Namo Amitabha Buddha of Pure Western Land (of Ultimate Bliss) Doctrine...................360
Different Disfavored and Favored Opinions.   361
Discussion on methods of Meditation. 362
K. Recommended Duration and the Posture.   363
L. The Third Eye...................363
M. Kenosis.   366
N. Phenomenon Of Illusions/ False States In The Meditation. 366
O. Dreams In Meditation...................370
P. Hindrances in Meditation.   370
1. Illness, Pain.   370
2. Somnolence and Sleep...................370
V. MEDITATION EXPERIENCES.   372
VI. MEDITATION and KARMA CLEASNING   374

Chapter IX: THE SLEEP...................377
I.   INTRODUCTION   379
II. CHARACTERISTICS OF SLEEP   379
A. Electroenceplogram EEG and the Sleep .   379
1. NREM-REM...................380
B. Suprachiasmatic Nucleus SCN/ and Molecular mechanism 380
C. Melatonin and Pineal gland   382
III. Mechanism of Sleep and Wakefulness 382
- WAKEFULNESS...................383
- SLEEP...................387
IV. OTHER MECHANISMS   388
V. PHENOMENON IN SLEEP.   388
A. Muscle atonia: (under influence of GABA and Glycine)
B. REM Intrusions.   389
Further reading   390
D1. REMSD=REMS deprivation   390
E. Sleepiness.   390
H. Insomnia.   392
I. Restless leg syndrome...................394
J. Mouth Dryness...................394

Chapter X: THE DREAMS...................394

x

      I. Generalities   395
      II. MECHANISM.   395
A. Cortices of Dreaming...396
B. Default Mode Network and Dreaming...................................396
C. Dreams and Neurohormones...396
E. Weak, Non-coordinated Connections and Insufficient Activation of Consciousness center..397
F. Dream Contents...398
G. Attention and Acetylcholine in dreaming...........................399

Chapter XI: MORALITY, FOUR IMMEASURABLE MINDS OF VIRTUES, EMPATHY, ALTRUISM, THREE POISONS (GREED/ ANGER/ IGNORANCE)..407

CHAPTER XII:...431
CURIOSITY, IMAGINATION AND CREATIVITY 431......................431
Curiosity..432
Imagination..432
Creativity..433
Neuroscientific Wiring..434
Creativity and Aging..435

SUMMARY
Since innumerable books and excerpts regarding the phenomenology of Tao can not adequately express the essence, this book investigates the ontological aspect of the Tao to understand the mechanism of metaphysical phenomena. This is to keep and follow the right path as Buddha said: *Be a lamp unto yourself.*

1. Everything exists in its wholeness. An eventual analysis will create two parts: the physical and metaphysical parts.
 Buddhahood possesses immeasurable power, Ultimate Omniscience/ UO, and other attributes like ingenuity, infinite light, sound, and non-discrimination, which are the origin of the Creation of the Universe. As the corollary, the phenomenon of entanglement and interconnectivity in quantum mechanics represent the noumenon of the Emptiness.
2. The Soul or Consciousness/ CS is imperceptible to the sensory organs, like sound and gravity. Synaptic connections represent memory. As a result, Soul must be attached to the Default Mode Network/DMN where Memory/ MM is stored. Since the Soul is the product of the Big Bang, it is likely constituted by the Dark Matter/ DM, and Neutrinos sharing its properties. Since the Soul represents Consciousness, Neutrino may be an elementary particle of Consciousness.

3. The UO recognizes the data as a whole package, as it is, without splitting it into information specific to sensory organs. Although the UO can form knowledge, the data must be integrated into the nervous system to coordinate with any motor activity. This process constitutes the filtering system or *Veil of ignorance* (or incomplete and deviated CS) that deviates and distorts the data.

4. The brain is a box of prediction because the brain contains all the information for predicting future events. In consciousness formation, the error detection system will trigger the MM in the DMN for comparison and then label the incoming data. In Meditation, the Intra-Parietal Sulcus /IPS aims to minimize incoming information; due to the absence of external input, MM from the DMN is retrieved

and will render the synaptic unstable, susceptible to neuroplasticity phenomenon.

5. After the sentient's death, the ensoulment into a new body is the imprint of the Karma the Soul carries. Supernatural phenomena like Out of Body Experience/ OBE and Near Death Experience/ NDE share many characteristics with psychedelic-drug effects, Dreams, and biochemical cerebral hypoxic changes. The natural pathway of most NDE and OBE represents not only the neuroscience pathway but the *additional* "non-local" extension of the Soul. After death or in some specific condition like Meditation, the brain and the body are detached or become "transparent," enabling the Soul to be non-local.

6. The universe is bound to the rule of duality and plurality. As a result, one body may harbor more than one Soul. , In *some* circumstances, double or triple Identity/ ID can account for disorders like Dissociated Identity Disorders (DID) or Sleepwalking.

7. Veil of ignorance (or incomplete and deviated CS) is not the obliteration of intelligence but the deviation and restriction of the CS to different areas like denial of the metaphysical world is ignorance. CS is deep in insight but narrow in extent. Awareness that does not require much attention; is broad in extent but superficial in insight. Enlightenment develops as the extension of Awareness.

8. Karma is the cause of the loss of free will. The Soul is predestined to a mandate for the entire life. Humans only have the free will to act according to Morality: Four immeasurable Minds. Otherwise, free will eventually create more karma.

9. The spiritual path of Meditation often consists of Samantha (calm state), Samadhi (mindfulness), Contemplation, and Realization of Awareness and Buddhahood, but meditators often lack understanding mechanisms of Meditation.

10. Despite the illusion, the noumenon of this nature and the reality of the metaphysical realms are not deniable.
KM.

PREFACE

This book aims to contribute to understanding the Buddhist Dharmas and Meditation in light of neuroscience and science in general, as well as book reviews of prominent Far Eastern philosophers and experiences of colleagues and meditation practitioners. Neuroscience is a branch of science with recently renewed interests. It studies the development, structure, functions, and particularly the mechanisms of these functions in creating the impact on behaviors and Consciousness, including learning and Memory. The recent rapid advancement is due to the development of Psychology, Psychiatry, Neurology, Neurosurgery, and Neuropathology and the invention of many sophisticated technologies of brain stimulation and brain imaging, exceptionally functional Magnetic Resonance Imaging/ fMRI. Understanding brain functioning is colossal but has remained in the revelation of I Ching, especially in major religions. Many Neuroscience researchers have been awarded Nobel Prizes in Physiology or Medicine, like Golgi, and Cajal in neuronal connections, Spermann about Embryonic organizer in the development of the central nervous system, Kandel regarding the basis of Memory storage in neurons, Sperry in the study of split brains revealing different functions of the Right and Left hemispheres, David H. Hubel and Torsten N. Wiesel regarding visual neurology, Carlsson about neuronal mediator DOPAmine in the application to the treatment of Parkinson's disease O'Keefe, Edward Moser and May-Britt Moser about the map of navigation in Hippocampus, Jeffrey C. Hall, Michael Rosbash and Michael W. Young for their discoveries of molecular mechanisms controlling the circadian clock, Richard Axel, Linda Buck discoveries of odorant receptors and the organization of the olfactory system. In addition, it is worth citing Francis Crick's discovery of DNA (Nobel Award in 1962) and Gerald Eldelman's discovery of antibody molecules. (Nobel Award 1972). These researchers finally turned their interest to Neuroscience.

Besides the motor function, in the last three decades, Neuroscience has successively discovered the mechanism of the formation of non-physical components of sentients like Memories, emotions/ behaviors, and Consciousness by integrating information from peripheral inputs and different brain centers. This information displays electrical potentials on the external membranes of neurons, transmitted along the axons (long cytoplasmic extensions) through synaptic connections aided by neurohormones. Other phenomena, like sleep and dreams, have been unlocked.

Despite the enormous progress of science with the development of sophisticated technology helping many aspects of human life, understanding of the Soul and many related metaphysical phenomena have remained elusive for complete understanding. These metaphysical phenomena and entities are Out of Body Experiences/OBE, Near Death Experiences/NDE, experience in mindfulness Meditation, mediumship, somnambulism, Cotard's syndrome, conversion disorder with unexplained blindness, deafness, paralysis, and seizures...

This book is based on a crucial preamble/ concept: The Big Bang/ False Thought arising from the Emptiness/ Original Mind/ God's Will /the Oneness is the primary event creating this Universe. The Oneness is non-created and imperishable/ permanent/ having an original nature (vn:Tự Tánh). All phenomena following the Oneness are secondary, represent the Duality/ Plurality, and are perishable / impermanent. As a result, all entities in the Universe are composed of at least two components: the visible part, called the Form or Physical aspect, accessible to the five sensory organs, and the invisible part, called the Soul/ Metaphysical part, not accessible to the five sensory organs. For example, all sentients, including human beings, consist of the body and the Soul; the Universe itself comprises baryonic matter, all particles, and the invisible component: the Dark Force and Dark Matter. The author believes that the above Preamble is essential in discussing and studying the Soul. This is essential because Consciousness/CS, even with the most sophisticated devices to investigate the Universe of its wholeness is impossible to locate or identify the invisible part physically. However, CS can identify the secondary impacts of the Soul on the physical part.

Accepting the above concept, the information in this book is from the review, logistic deduction, and ontological study of the Buddhist sutras in the light of current scientific findings of neuroscience with the integration of general knowledge from the ancient Far East and Western philosophy, which are most consistent with Buddhist Dharma and neuroscience.

Unlike most current Buddhist books in English and Vietnamese studying Buddhist phenomenology, this book is aimed at the ontology (the self-nature) of Buddhism. For this purpose, the Creation of the Universe/ Genesis is reviewed, and the concept of the Big Bang is discussed. Since the Soul is also the product of the Universe/ Genesis/ Big Bang and can be considered as an entity such as gravity. Both entities are not perceptible to sensory organs. Gravity is related to a physical mass, while the Soul is related to the brain and is responsible for the essential activity: Memory and Consciousness.

The Soul of the non–neuronal part, the extremities, and the body account for the strength of muscles or Chi Gong in martial arts and Chi in Acupuncture.

The Soul, known as the metaphysical component, harbors the MM, CS, and previous lives CS/ the Karma. The Soul resides in the brain at the site of MM, particularly the implicit Memory (MM) and semantic MM, corresponding to the Default Mode Network. With understanding the Soul and its role in the mechanism of formation of Memory and CS (acquisition of knowledge of the surrounding nature), it is feasible to unlock mysteries of karma formation, the law of Samsara (reincarnation), the problem of Free Will, mystic phenomena like OBE, NDE, Sonambulism, Mediumship, Cotard's syndrome...

At last, this book is to reveal the neuroscientific mechanism of mindfulness Meditation. Meditation is the way of Tao, for the perfection of life with purification of Karma, clearing the veil of ignorance/ stupidity, and being ready to be back home, the Nirvana. The tributary effects are happiness and healthy

life. The Meditation practitioner usually depends on the Meditation master for the practice. However, it is challenging to distinguish authentic masters from false masters, as Buddha said:
Be a lamp unto yourself, be a refuge to yourself. Take yourself to no external shelter (vn: *Tự thắp đuốc mà đi*).

The author wishes to acknowledge the Master for her instruction and initiation into the method of Quan Yin/ Sound Meditation, our friends for sharing their experience of Meditation, my teachers in Vietnam at different levels of education, professors of Medicine at Saigon, my friends, particularly Dr. Cẩm Nguyễn, Mỹ Quang Trần, Điệp Ngọc Bùi ...and Late Dr. Hưng Kim Nguyễn, for their support and shared knowledge, my colleagues at the Department of Pathology, University of Ottawa. This book writing can not be completed without my understanding of I ching, Far Eastern philosophy, and spiritual excerpts from Vietnamese scholars: Nguyễn Hiến Lê, Ngô Tất Tố, Phan Bội Châu, Dr. Nguyễn văn Thọ.... and other Scholars. The author owes so much knowledge from many Buddhist Venerables who make important Buddhist sutras available: Nikaya Surangama, Flower Adornment, Lotus, Diamond sutras... This book is for the memoirs of our parents for their dedication to fostering our education and morality, particularly my mother, a typical Vietnamese woman devoted to her children, who took me to the village school during wartime and showed me the path to medicine. Lastly and to share with readers.

Respectfully especially, this book can not be finished without my wife, my co-meditator on the spiritual path, and my lifelong friend with devotion, care, love, and sharing experiences in life and Meditation.

As Lao Tzu said '*Tao that can be told is not the eternal Tao, Name that can be named is not the eternal name*". This book only represents what the author would like
KM.

IMPORTANT GRAY NUCLEI AND CORTICES.

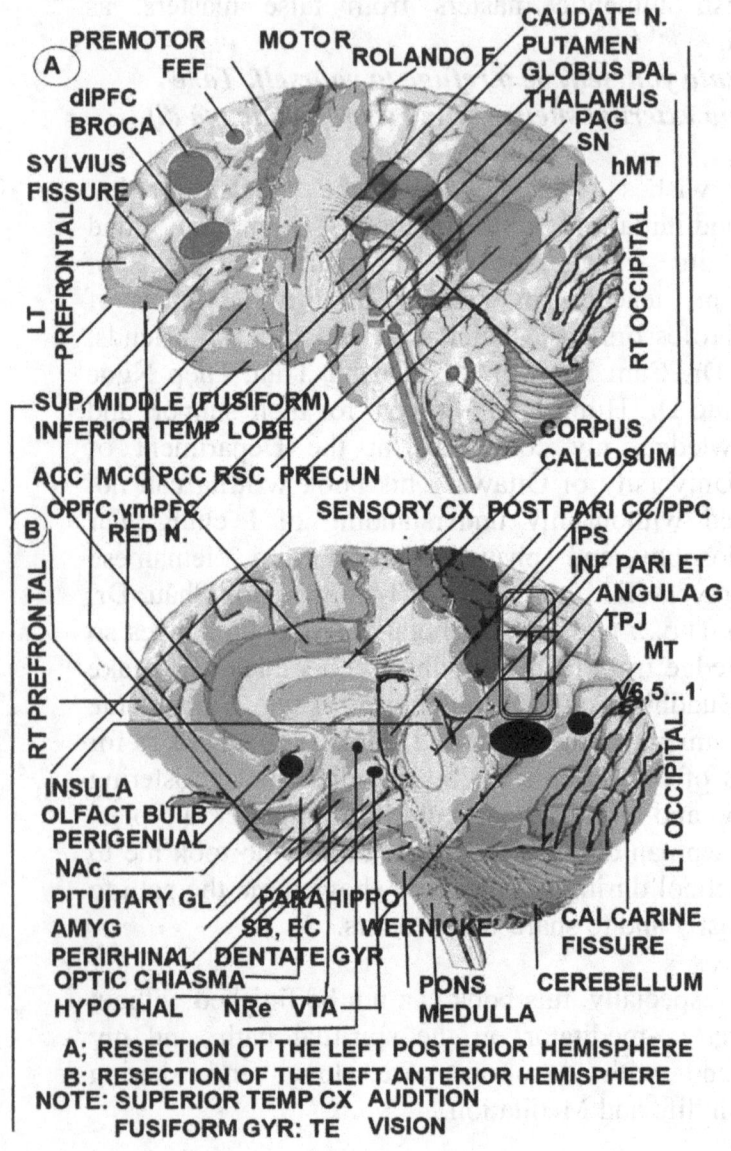

A: RESECTION OT THE LEFT POSTERIOR HEMISPHERE
B: RESECTION OF THE LEFT ANTERIOR HEMISPHERE
NOTE: SUPERIOR TEMP CX AUDITION
      FUSIFORM GYR: TE VISION

Fig 1

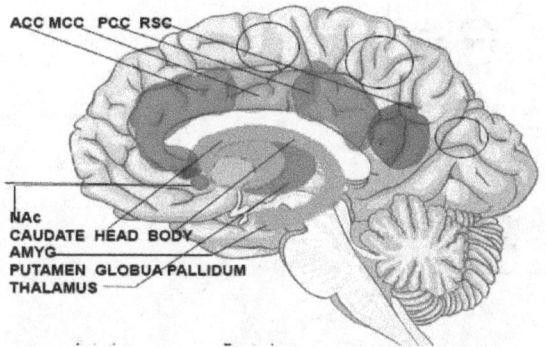

Fig 2: Medial view Cortices and large nuclei

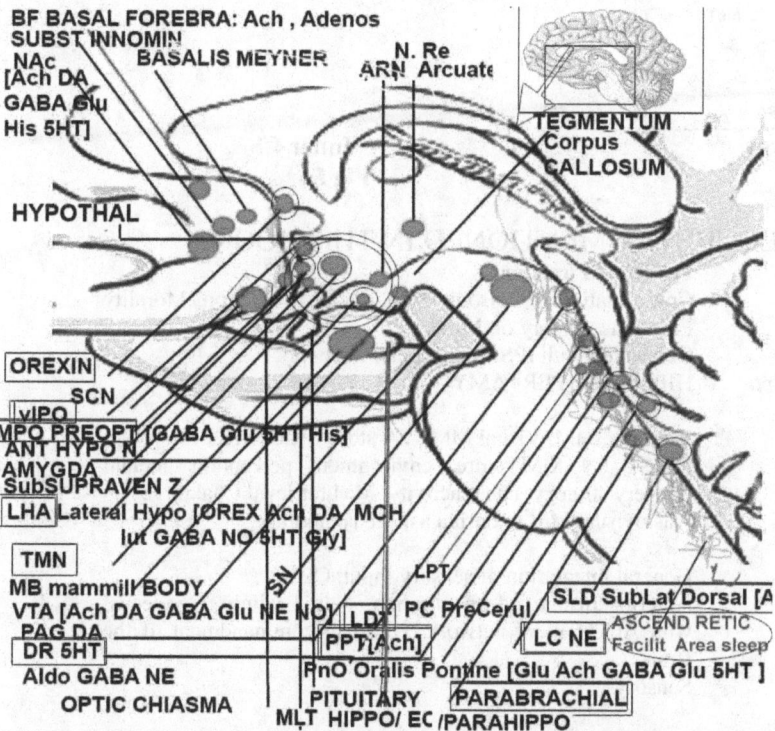

ACC: Anterior Cingular Cx, ARAS :Ascending Reticular Network, dlPFC: dorsolateral PFC, DR: Dorsal Raphe, EC Entotrhinal cortex, FEF: Frontal Eye Field, Fusi: Fusiform gyrus, LC: Locus Ceruleus, LDT: LateroDorsal Temental, LHA : Lateral HypoT Area, LPT: Lateral Pontine N. M: Motor, MTL: Medial Temporal Lobe, MT Special visual Cx, N.Re: N Reuniens, NAc: N. Acumbens , OCC: Occipital, PAG: PeriAqueductal Gray, PFC: PreFrontal Cortex, PM: Premotor, PPC: Posterior Parietal Cx, RSC RetroSplenial Cx, SCN: Suprachismati N. SN: Substantia Nigra, MCC: Middlle CC, PCC Posterior CC, Striat: Striatm, TEMP: Temporal, Thalam:Thalamus/DN, TMN:TuberoMammillary N. TPJ:

Fig 3: Medial view: Small Nuclei

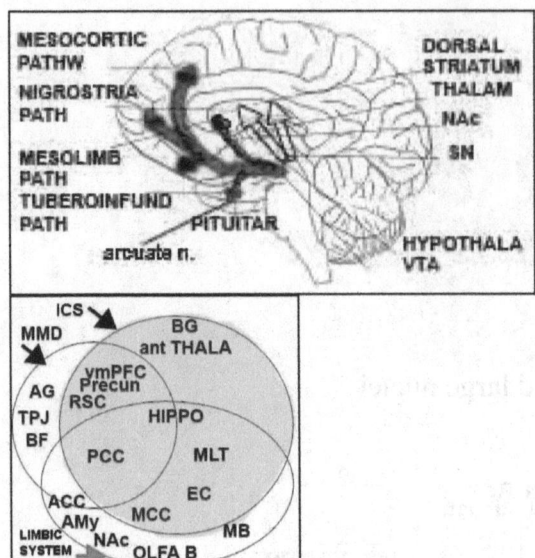

DOPA Pathway
Fig 4

Inner CS
Fig 5

AREAS FREQUENTLY MENTIONED IN THIS BOOK.

Areas	Features/ Function
vmPFC, OPFC	Consolidation of episodic MM, social interaction, Morality Emotions, Theory on Mind.
dlPFC	Connected with IPS: Management network
MedialTemporal L	HIPPO+EC+PER+AMYG
PCC	Semantic MM, Visual MM. CS store (PCC+MCC)
Precuneus	DMN, TN MM store, environment perception, mental imagery strategy, cue reactivity, MM retrieval Chakra 12
Temporal Lobe	Ausiovisual MM (external aspect Temporal)
TPJ	General integration of sensory input, OBE
Insula	Anterior Insula: deep emotion. Pain Feeling Connected with ACC:Salient network, Ultimate commandment of the Brain
Anterior Thalam	Sensory
Amygdala	Fear, Anxiety, Agressions, Pavlov reflex, Learning
NAc (Ventral Stria=NAc+Olfact)	Joy, Reward, connected with Ventral Tegmental Area, Ventral Striatum
PAG:Periaqueduc	Automatic reactional emotion expression, related to nerve X
VTA	Joy, impulses
ACC	Error detection, critical role in CS formation by comparing new input to store MM
Basal Gangl	-Ventral Striatum=**NAc**+Olfact Bulb: CS, Rewards -Dorsal **Striatum=Caudate** (GABA movt, (jerky in Hungtinton), attention,working Memory, cognitive , emotion)+Putamen decision making, -Globus Pallidus /DOPA or Dorsal **Pallidium**, SN,

| Composed of: | SubThal Movements with or without volition, hyperactivity <u>Cerebral palsy</u>:, <u>Chorea</u>, <u>Dystonia</u>, <u>Epilepsy</u> Ventral Pallid (below NAc) reward |

- **Diencephalon** = EpiThalamus (Pneal gland and Habenular nuclei), Thalamus, SubThalamus (Zona Incerta and Reticular nucleus), MetaThalamus (geniculate bodies); HYPO, MB
- **Midbrain=mesencephalon** colliculi, Tegmentum, (ARAS)

NEUROTRANSMITTER SYSTEMS.

Acetylcholinergic System: Consciousness, MM.
Serotonergic system (DR of midbrain, Pons, Medulla): pleasure, depression, MM, new CS.
Noradrenergic System: (LC): alertness, wakefulness, fight-flight, MM, CS.
DOPAnergic system:
  Nigro-Striatal: SN→Striatum: manner of movements
   Mesolimbic : VTA→ limbic areas (HIPPO, AMYG, Nac, Ventral Striatum/ reward)
   Mesocortical: VTA→ PFC: MM, CS in psychosis

ABREVIATIONS

ACC: Anterior Cingulate Cortex
ADHD: Attention Deficit Hyperactivity Disorder.
AI: Artificial Intelligence
AG: Angular Gyrus.
AHN: Anterior Hypothalamic Nucleus 365.
AMGD/AMYD/ AMYG/ Amyg: Amygdala
AMPAR: DL-alpha-amino-3-hydroxy-5methylisoxazole-propionic acid.
APRA: Anterior Paralimbic REM Activation Area
ARAS: Ascending Reticular Activating System.
AS: Autoscopy
ATL: Anterior Temporal Lobe.
ATN: Anterior Thalamus Nucleus
AUD: Audiotory Cortex.
BBB: Blood Brain Barrier
BDNF: Brain Derived Neurotrophic Factor.
BF: Basal Forebrain
BG: Basal Ganglia
BNST: Bed Nucleus of Striata Terminalis
BOLD: Blood Oxygenation-Level-Dependent
CCC: Conformal Cyclic Cosmology.
CMB: Cosmic Microwave Background.
CREB: cAMP Response Element-Binding Protein
CS: Consciousness
Cx: Cortex.
dACC: Dorsal Anterior Cingulate Cortex.
DESNOS: Disorders of Extreme Stress not otherwise specified DID: Dissociative Identity Disorder, Divided I D
DIDNOS: Dissociative Identity Disorder not otherwise specified dlPFC: dorso-lateral Prefrontal Cortex
DM: Dark Matter
DMN: Default Mode Network
DMH: Dorso-Medial Hypothalamus
DOPA: Dihydroxyphenylalanine
DP: Depersonalization
DR: Derealization.
DR: Dorsal Root (Ganglion)
dSPZ: Dorsal Sub-Paraventricular Zone.
DVC: Dorsal Vagal Complex.
EC: Entorhinal Cortex
EEG: Electro Encephalogram.
EOG: Electro-Oculogram
EP: Emotional Part of Personality..
EPR PAR: Einstein-Podolsky-Rosen Paradox
EPSP: Excitatory Post Synaptic Potential.
ESP: Extra Sensory Perception
FEF: Frontal Eye Field.
FEP: Free Energy Processing.
fMRI: Functional Magnetic Resonance Imaging.
FTT: False Tagging Theory.
FTT: Fuzzy Trace Theory
FG/ FUSI: Fusiform Gyrus
GABA: Gamma -Aminobutyric Acid.
GPe and GPi: Globus Pallidus External and Internal:
GWS: Global Work Space Theory of Consciousness.
HAA: Hypothalamus-Adrenal Axis
HI: Hjgh Order for Inhibition 202
HIPPO: Hippocampus
hMT: Special Visual Cortex
HOT: High Order Perception Theory
HPA: HIPPO-Pituitary-Adrenal Axe
IBD: Inflammatory Bowell Disease.
IC: Inner Consciousness, Inferior Colliculi
ICS: Inner Consciousness System
IEG: Immediate early Gene.
IFG:Inferior Frontal Gyrus
IIT: Integrated Information Theory.
IMC: Intermediolateral Column of spinal cord.
IPFC: Inferior Pre-Frontal Cortex IPS: Intra-Parietal Sulcus:
IPL: Inferior Parietal Lobule
IPS: Intra-Parietal Sulcus.
IPSP: Inhibitory Post Synaptic Potential.
ITC: Inferior Temporal Lobe.
LC: Locus ceruleus.
LD: Lucid Dream.
LDT: Latero-Dorsal Tegmental
LHA: Lateral Hypothalamus Area.
LGN: Lateral Geniculate Body.
LT: Left.
LTD: Long Term Depotentiation
LTM: Long Term Memory.
MACHO: Massive Compact Halo
MB: Mammillary Body
MB: Mid Brain
MC: 202
MCC: Middle Cingulate Cortex
MCH: Melanin Concentrating Hormone
MeV: Megaelectron Volt
MGN: Medial Geniculate Body
MLT
MM: Memory
MnPO: Median Preoptic Nucleus
MPAR:
MPOA: Median Preoptic Area
MPD: Multiple Personality.
MPFC:
MPO: Myeloperoxidase
MS: Multiple Sclerosis.
MST: Motor Simulation Theory.
MST: Medial Superior Temporal
MT: Melatonin.
MT: Special Visual Cortex
MTL: Medial Temporal Lobe.
MTT: Multiple Trace Theory
MUM: Minimal Unified Model.
MVPC: Multi-Voxel Pattern Classification
MWI: Many World Interpretation
N: Nucleus
NAc: Nucleus Accumbens.

NDE: Near Death Experience
NMDAR: N-Methyl-D-Aspartate Receptor
NCC: Neural Correlate of Consciousness.
NO: Nitric Oxide (Nitrogen Monoxide)
NRe: Nucleus Reunion
OBE: Out of Body Experience
OCC: Occipital
OFC: Orbital Frontal Cortex
OPFC: Orbital Pre-frontal Cortex
p: Pali
PAG: Peri-Aqueductal Grey.
PB: Parabrachial Nucleus
PC: Pre-Ceruleus
PCC: Post-Cingular Cortex.
PD: Parkinson Disease
PER:
PFC: Pre-frontal Cortex.
pgACC: pre genuos Anterior Cingular Cortex.
PGO: Ponto-Geniculo-Occipital (wave)
PKC: Protein-Kinase C.
PM: Premotor.
PnO: Nucleus Pontis Oralis
PPA: ParaHIPPO Place Area.
PPC: Posterior Parietal Cortex.
PPN: Pedunculopontine Nucleus.
PPT: Pedunculopontine Tegmentum
PULV" Pulvinar
PVH/ PVN/ PVA: Preventricular Nucleus

QM: Quantum Mechanics
RAS: Reticular Activating System
RBD: REM Sleep Behavior Disorder
REM: Rapid Eye Movement
REMSD: REMS Deprivation.
RM: Rostral Medial.
RMPFC: Rostral Medial Pre-Frontal Cortex
ROS: Reactive Oxygen Species.
RPFC: Rostral Pre-Frontal Cortex
RSA: Respiratory Sinusal Arrhythmia.
RSC: Retro-Splenial Cortex.
RT: Right.
SB; Subiculum
SB: Somnambulism.
SC Superior Colliculi.
SCG: Superior Cervical Sympathetic Gland

SCN: Suprachiasmatic Nucleus
SDAM: Severely Deficient Autobiography
sgACC: Subgenual Anterior Cingulate Cortex
sgPFC:: Subgenual Pre-Frontal Cortex
SI: Substantia Inominate
Skt: Sanskrit
SLD: Sub-Lateral Dorsal.
SMA: Supplementary Motor Area.
SMC: Standard Memory Consolidation..
SMH: Somatic Marker Hypothesis.
SN: Substancia Nigra.
SPL: Superior Parietal Lobe
SPZ: Sub-Paraventricular Zone
SSRI: Selective Serotonine Reuptake Inhibitor
STRIAT: Striatum.
SWR: Sharp Wave Ripple.
SWS: Slow Wave Sleep
TE, TEO: Special Visual Cortex.Areas of
Temporal Lobe for Form TEMP: Temporal.
Thalam: Thalamus.
TLE: Temporal Lobe Epilepsy.
TMN: Tubero-Mammillary Nucleus
TOE: Theory of Everything
TPJ: Temporo-Parietal Junction.
TRN: Thalamic Reticular Network
UO: Ultimate Omniscience
vBST: ventral Bed Nucleus of the stria
Terminalis 365
VIP: Vasoactive intestinal peptides
vlPAG: Ventrolateral PeriAqueductal Gray
vlPO:Ventro-lateral Pre-Optic (Nucleus).
VMH: Ventromedial Hypothalamus
vmPFC: ventro-medial Prefrontal Cortex
vOT: Ventral Occipito-temporal Cortex
VP: Ventral Pallidum.
VPC: Ventral Prefrontal Cortex
vSPZ: Ventral Sub-Paraventricular Zone
VTA: Ventral Tegmental Area
VVC: Ventral Vagal Complex.
VWFA: Visual Form Word Area.
WM: Working Memory.
ZI: Zona Incerta.

blood oxygenation-level-dependent (BOLD

AIMS AND SUMMARY OF SIGNIFICANT POINTS AND FINDINGS

Buddhist Dharma is inseparable from the Worldly Dharmas. The teachings from leaders of major religions are similar. This is also confirmed in the Connected sutra (vn: kinh Tương ưng) that all Buddha sayings are similar and that Buddha is aware of what human beings need:
*It is on account of the deeper meaning that the
eternally-abiding reality of self-realization is talked of by
me, and between myself and [all the other] Buddhas, in this
respect, there is no distinction whatever*,

In the Flower Adorment sutra (kinh Hoa Nghiêm):
*Son of Buddha! Of the past, future, And present guides,
None expounds just one method To become enlightened.
Buddhas know beings' minds. Their natures are different;
According to what they need to be freed, Thus do the Buddhas teach ....*

After forty-five years of proclamation of the Dharma,
in Lankavatara sutra (kinh Lăng Già Tâm ấn), Buddha concluded
*From the night of Enlightenment till that of Nirvana,
have not, and will not speak, in the meantime, made any proclamation whatsoever*. Buddha said: *"I make such a statement because the truth depends on personal realization that is beyond explanations or distinctions and beyond dualistic terms*. In another instance, Buddha said:
*To please the host of beings / they render figures
unfaithful with their art / but teachings are for truth isn't in the words.*
Saying that because the proclaimed Dharma is obvious and essential for the moral pathway of living, although unrecognized by human beings because of the veil of ignorance (incomplete and deviated CS, **according to Gödel's incompleteness theorems)**
. Teaching in words is not very helpful.

Buddhist Dharma consists of scriptures transcribing Buddha proclamations stored in sutras by both Theravada and Mahàyàna Buddhism.
Nowadays, after more than 2500 years, humans have progressed in worldly life, especially in science per se, in understanding the Universe at macroscopic and quantum levels, the origin of species and life, and the mechanisms of disease.

Buddha took up a few simsapa leaves in the simsapa grove overhead in his hand and addressed the bhikkhus thus: "Which is more numerous... The things I have directly known but have not taught you are numerous, while the things I have taught you are few, like the leaves in my hand."

Buddha's teaching is to save humans from suffering in this era of degenerative Dharma. Therefore, what Buddha proclaimed is limited compared to Buddha's Ultimate Omniscience (vn: Trí Huệ Bát Nhã). Furthermore, this perceptible material world is less than 5% of the Universe, according to astrophysicists. Given the boundless extent of the Universe and possibly other numerous universes, despite the progress in human Consciousness, the scientific knowledge of our physical world is still minimal compared to Buddha's Ultimate Omniscience.

As a result, scientific knowledge is negligible. This is evidenced by scientific discoveries staying within what was registered in the Buddhist sutras. Although science and philosophy are very much helpful in understanding Buddha's teaching, the will, steady effort/diligence/ Virya-paramita (skt) —Diligence-paramita (vn: Tinh tấn) is critical in the path of spiritual perfection.

Contrarily to some Buddhist Scholars, Buddha's revelation of the Creation was complete from Emptiness through the False Thoughts up to the creation of the Multiverse. The revelation is not illusional. When Junjuro Takakusu a Buddhist scholar, said: " Buddhism does not give importance to the idea of the Root-Principle or the First cause as other systems of philosophy often do; nor it discuss the idea of cosmology", the saying only reflects personal limited knowledge of Buddha's teachings. Since Buddhism is not a product of the brain, as in the case of consciousness, but it is the experience of Buddha in meditation. Furthermore, it is unreasonable to compare Buddhism as a religion with a philosophy which develops from human thinking and conception that are almost always incomplete or subjective to adjustment with time

Similarly, it is widespreadly believed in some Buddhist communities that Buddha is not interested in the ontological aspects of the Creation and life, such as: *"Whether the Universe is eternal, not eternal, both eternal and not eternal, neither eternal nor not eternal, and so on. Whether the Universe is finite, infinite, both finite and infinite, neither finite nor infinite, and so on. Whether Buddha exists after death, does not exist after death, either exists or does not exist after death, neither exists nor does not exist after* *death. Whether the soul/Mind is identical with the body, different with the body, ...(62 false views).* .Such questions are not problematic for Buddha to answer but may be impossible for those who pose the question to understand due to the veil of ignorance/ incomplete and deviated CS, according to Gödel's incompleteness theorems). Buddha is omniscient. The fact that some Buddhist devotees relate the silence of Buddha to these questions as Agnosia likely reflects their incomplete knowledge when reviewing Buddhist sutras. In some paragraphs in **the Surangama sutra** or in ***Mahàpadàna Sutta***, Buddha critically informs this ultimate truth

I. AIMS

The Buddhist scripture is voluminous, and discussions of Buddha teachings and different techniques of spiritual perfection, including Meditation, are extensive. This book investigates the ontologic aspect (the origin) of the revelation of the Universe and its Creation. This will light into other mechanisms accounting for various phenomenologic elements of the spiritual path, such as the unimaginable revelation of Buddha's land in scripture, the power represented by the Universe expansion, Buddha's word, sound, and light.... Other common problems in understanding metaphysical phenomena like Near Death experiences (NDE), Out of Body Experiences (OBE), ... Other phenomena like Dissociated Identity Disorders (DID or Multiple Personalities) and sleepwalking are very close to daily life but have remained elusive in pathophysiology (mechanism of the disease). Understanding the mechanism of phenomenon is essential in the spiritual path since spiritual masters are challenging to distinguish from fake <u>but</u> famous masters. This is to keep and to follow the right path as Buddha said: ***Be a lamp unto yourself, be a refuge to yourself. Take yourself to no external shelter,*** (vn: *Tự thắp đuốc mà đi).*

II. SIGNIFICANT POINTS AND FINDINGS

Because of a large amount of information from Buddhist scriptures, neuroscience, and Eastern philosophy intermingled in each chapter to reveal the mechanism of the phenomena in the spiritual path, this book reading is challenging and requires much attention and time.

As mentioned above, the essence of major religions is always consistent with correct scientific findings. Scientific knowledge only constitutes a very small portion of reality, but major religions possess the Ultimate Omniscience (UO) (vn: Trí Huệ Bát Nhã)

1. Everything in the Universe, without exception, exists in its wholeness. Any analysis or division for examination is irrational since the whole entity is unknown. An eventual analysis/division will create

two parts: the positive part represents the physical part of the body. The negative or metaphysical part represents the invisible part /the soul/ Memory and Consciousness. However, the division is never equal in any meaning and complete: in the positive part, there is a remnant of the negative part, and vice versa. The negative remnant part represents the memoir, and the positive remnant represents the synaptic junction/ the physical part of the MM. This is consistent with the Cartesian (Descartes) concept of duality.

Emptiness is believed to be the origin of the Universe. Emptiness/ Ultimate pole (Thái cực) / Original Mind/ are synonyms to designate Buddhahood/Holy Spirit, the ultimate origin of the Universe and all species. Buddhahood/Holy Spirit possesses immeasurable power, Ultimate Omniscience (UO, Trí Huệ Bát Nhã), and other attributes like ingenuity, infinite light, sound, being splendid, homogenous, non-discriminatory, not associated with birth or death. Because of the homogeneity and non-discrimination, Emptiness is characterized by the similarity between the microscopic and macroscopic structure in all parts. As the corollary, the phenomenon of entanglement and interconnectivity in quantum mechanics represent the noumenon of the Emptiness.

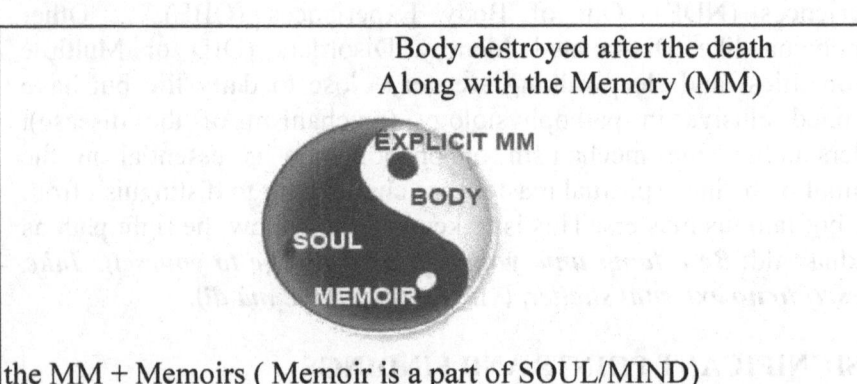

the MM + Memoirs (Memoir is a part of SOUL/MIND)

Lao Tsu said that *Tao is to take those who have too much to give to those who do not have enough to make the high lower; the low raised*. Therefore, Emptiness itself must be filled with content that is imperceptible to the sensory organs. Therefore, the void per se never exists.

2. The three self-natures characterize Emptiness: i) impermanence, therefore being non-created and imperishable; ii) the false self/ownership; and iii) Dukkha or Affliction, therefore being the

Creator. Emptiness is also characterized by unlimited power source, Ultimate Omniscience, homogeneity (Neither identity Nor difference), and being associated with no space and no time (Thusness or Thus-come and Thus-gone, vn: Nhu Lai). As a result, in Emptiness, separate particles share the similarity (as being interconnected /entangled). False thought/ Big Bang arising from Emptiness eventually causes immediate loss of homogeneity. As a result, the false thought instantaneously evolves the entire "Emptiness Space"; therefore, the Big Bang is seen in all directions of the Universe (Multiverse concept). This accounts for the phenomenon of galaxy recession with acceleration and speed faster than light.

3. The Soul is an entity. Soul, Sound, and Gravity are three entities very close to humans, but their structure has remained unknown. Elementary particles constituting sound and gravity have remained elusive to the research. The Soul is composed of Memory (MM) and Consciousness (CS) because of their imperceptibility to the sensory organs, but like sound and gravity, sentient beings can feel its secondary effects. Since CS is a product of the MM, MM is represented by synaptic connections (the connection between the arborizations of two neurons). As a result, Souls must be attached to the sites of the brain storing MM, particularly the Default Mode Network (DMN, a system of the cortex for semantic Memory (MM), storing the semantic MM and also representing the seat of the "False Ego," because false Ego is constituted by the CS of the present life.

DMN is the essential part of storing the MM and the CS. Since the information stored in the DMN becomes individualized and specific for each individual, it is called inner CS, distinct from external data.

Since the Soul is the product of the Big Bang and is associated with imperceptible, pervasive, and non-local/mobile features, it is likely constituted by the Dark Matter, and Neutrinos. Furthermore, Neutrinos can be attached to electrons of charge (-). As a result, the Soul can be associated with electromagnetic features, as in cases of the Soul attached to the synapses in MM connections, embodied Soul as Chi in acupuncture and in Chi Qong.

This book proposes a groundbreaking idea that the Soul is primarily composed of Dark Matter (DM) and Neutrinos. This proposal, if supported by facts or concepts, could revolutionize our understanding of the Soul and its composition.

4. External information/data, when in contact with the Ultimate Omniscience (UO), generates the knowledge. The UO recognizes the information as a whole package of information, as it is, without splitting it into information specific to sensory organs. However, living sentients need the nervous system, i.e., the brain, to move and express CS after receiving the information. Therefore, although the UO can form knowledge, the information must be integrated into the nervous system to coordinate with any motor activity like eating, moving, and talking.... As a result, sentients see, hear, move ... and understand after the nervous system processes the information. This integration process constitutes the filtering system that changes, deviates, and distorts the data. This filtering system is called *the veil of ignorance/False Ego* (incomplete and deviated CS, according to Gödel's incompleteness theorems)).

Since all entities have an imperceptible part, the non-nervous part of the body is also accompanied by the Soul. As the imperceptible component of the brain corresponds to the Soul that is related to MM and CS, the imperceptible part of the body is the embodied Soul, commonly known as Chi in acupuncture and Qi Gong in Martial Arts. The embodied Soul is responsible for the power and energy in acupuncture and part of the power and energy in Martial Arts. In this circumstance, the Soul is electronically charged since Chi and Chi Gong have been demonstrated to show some electrical activity.

5. Ensoulment. After the sentient's death, the Soul detaches from the body and carries with it all the significant semantic karma to be submitted to the law of karma.
(In the physical world, sentients obey the law of determinism of the world. Therefore, the Soul is not free but has to follow the predestined mandate). Eventually, the ensoulment occurs for the particular Soul. The essential feature of the ensoulment is the imprint of the Karma which the Soul carries to the new physical body. Therefore, good merit or karma acquired in this life is automatically transferred to the next lives, whereas the physical strength of the body is completely lost after the death.

Supernatural phenomena like Out of Body Experience (OBE) and Near Death Experience (NDE) share many characteristics with psychedelic-drug effects, Dreams, and biochemical cerebral hypoxic changes. However, neuroscience can not account for the whole spectrum of OBE and NDE because the natural pathway of most NDE and OBE represents not only the neuroscience pathway but the *additional* "non-local" extension of the Soul. For other supernatural phenomena like Sleepwwalking, Cotard's Syndrome, and Conversion Disorder Dissociated Identity Disorders, the Soul is also implicated in the pathophysiological mechanisms. In normal living sentients, the Soul is attached and enclosed within the five skandhas (body and brain). After death or in some specific condition like Meditation, the brain and the body are detached or become "transparent," enabling the Soul to be non-local.

6. After the Creation/Genesis, the Universe is bound to the rule of duality and plurality. As a result, one body may harbor more than one Soul, which will create a living sentient, in *some* circumstances, with double or triple ID, accounting for disorders like Dissociated Identity Disorders (DID) or Sleepwalking.

7. MM formation is similar to the installation/ connection of electrical devices: all devices must be connected to the in-house electrical panel that must be connected to the electrical source of the city. Turning on a device, i.e., a light bulb, is equivalent to reactivating the connection, similar to retrieving the MM to remember an event

8. The brain is a box of prediction because the brain contains all the information from the past, similar to a department of statistics collecting past information for the prediction of future events. In the case of the brain, each time, at the onset of any event, the preliminary information triggers the system of error detection (Anterior Cingulate Cortex/ACC) to compare and predict new events.

In the Meditation, voluntary attention (or mindfulness) under the control of the Intra-Parietal Sulcus/IPS is aimed to block and reduce, to a minimum, incoming information, The system of error detection (ACC) that is devised to work continuously in a wakeful state will trigger the MM in the inner CS for comparison with incoming data. As a result, the MM is reactivated. This is the mechanism of looking to the inner self in Meditation. Furthermore, the reactivation of synaptic connections in MM retrieval will render the synaptic unstable if not reconsolidated. The retrieved MM are eventually discarded by IPS since retrieved MM are inconsistent with the pre-selected focus of attention. The non-reconsolidation of the MM of bad karma is the mechanism of karma cleansing. The process of neuroplasticity (destruction and repair) removes the synapses involved.

9. The information input must traverse many barriers before becoming CS. Successively, the barriers are the five Skandhas (ngũ Âm) consisting of Form (vn: Sắc, environment), Perception (vn: Thọ, sensory organs), Feeling (vn: Tưởng, dorsal pathway of integration going through the motor cortex for the reflex of defense such as closing eyes when an object appears), Formation (vn: Hành, integration of information to be ready for CS formation and CS (vn: Thức). These steps constitute the filtering system, also called the *veil of ignorance* (or **incomplete and deviated CS, according to Gödel's incompleteness theorems**)). With the attention, the integration renders the CS to be focused and insight-rich in detail. Ignorance is not the obliteration of intelligence but the deviation and restriction of the concentrated area to another area of interest. For example, denial of the metaphysical world is an ignorance/stupidity Therefore, the CS is deep in insight but narrow in extent. In contrast, Awareness does not require much attention; therefore, it is broad in extent but superficial in insight.

10. Knowledge is composed of CS/understanding and Awareness. In contrast to the CS, depending on the attention and the dorsal pathway, Awareness follows the ventral pathway and only uses a low level of attention. These two components are mutually inhibitory. In Meditation, the attention will render the focus area small enough to uncover the Awareness. As a result, Awareness makes the information close to reality, as it is.

11. Karma is the cause of the loss of free will. As mentioned above, each individual is governed by the rule of determinism. The Soul is predestined to mandates for the entire life. Human beings are only free to act according to Morality. (Four immeasurable Minds, vn: Tứ vô lượng tâm). A free will to act <u>not</u> according to Morality will lead to bad karmas. Therefore in worldly lives, one can say that there is no Free Will. Reaction to an aggressor to the physical body or assets is only advised for self defense to <u>only protect the body</u>, if possible but not to cause harm to the aggressor as Jesus said:

"*anyone slaps you on the right cheek, turn to them the other cheek also*"
Neuroscience proves that humans are not completely free to act in this insane world. The bad behavior creates a new Karma. Man only has the freedom to act according to morality (Ten Commandments or Five Precepts: In Christianity, it is believed that God created humans in this physical world and gave man the right to act. It is true that human beings have the freedom to act, but they are recommended to only follow the Ten Commandments!

12. Science developed mainly depending on the CS more than on Awareness. CS and MM are the product of the brain and are stored as the Inner CS in the Default Mode Network that acts as the Veil of Ignorance in many instances. Awareness, which is close to Ultimate Omniscience, is likely the source of creativity and curiosity. The Meditation restricts the CS and enlarges the Awareness, therefore may enhance curiosity and creativity. However, the purpose of the Meditation, with continuous diligent effort, is to facilitate the path of going back home, to find the Original Mind with Emptiness, harmony, stillness, non-discrimination.

Despite the limited role of science in the understating of the metaphysical realm as outlined in the above paragraphs, Neuroscience is also helpful to clarify or avoid some serious misunderstandings of Buddha's teachings, such as:

Feeling (in five skandhas: Form, Perception, Feeling, Formation, and Consciousness) must be distinguished from Formation and Consciousness. Both Feeling and Formation represent the process of integration of data in the brain. Feeling represents the dorsal pathway for immediate/non-conditioned reflex reaction (without Consciousness) to coming data; Formation (in five skandhas or the twelve links of dependent causation) represents the complex integration of data for the generation of Consciousness. This process uses the Ventral pathway. Therefore, Formation is not a motor activity that activates the physical parts of the body.

The view of average humans is through the Veil of ignorance that renders all objects modified, distorted and illusional. Spiritual correction of this view is the solution. As living sentients of the Creation, it is seriously erroneous to have the view of the rejection of this world, despite its impermanence, because the living creature's lifespan is much shorter than that of nature. The Three characteristics of a Bodhisattsava are: Pity, Awareness, and Effort, vn:Bi Trí Dũng) with a positive attitude toward life. In spite of the illusion, the noumenon of this nature is not deniable.. The problem is the failure to recognize the wholeness of the nature, which includes the physical and the metaphysical realms. Chinese Zen Master Du Tin (Song dynasty) said:

" before meeting the intellect, I saw that mountains are mountains and rivers are rivers. After learning, I understood that mountains are not mountains and rivers are not rivers. Thirty years later, I realize that mountains are mountains and rivers are rivers."

CHAPTER I: SCIENTIFIC APPROACH TO TAO AND TENTATIVE ONTOLOGICAL VIEW.
GENERALITIES.
I. THE SCIENCE.

Science is defined, in Wikipedia, as a systematic endeavor that builds and organizes knowledge in the form of testable explanations and predictions about the Universe. In another instance, Science is the pursuit and application of knowledge and understanding of the natural and social world following a systematic methodology based on evidence.

Knowledge is a dynamic interplay of two crucial components: Consciousness (understanding or Cognition) and Awareness. As we delve deeper into this book, we'll explore how both these elements demand varying levels of attention to the subject, enabling the assimilation of information into the brain, thereby forming Consciousness or Awareness. It's important to note that without this attention, knowledge cannot be formed. Consciousness or Cognition (understanding) corresponds to the information with in-depth content (high level of details) but is narrow in extent. On the other hand, Awareness is superficial in content (fewer details) but broad in extent.

Consciousness is the product of the environment with sensory organs and the brain; these two components constitute an information-filtering system. Furthermore, it is gradually known that acquiring sophisticated devices improves the perception of the five sensory organs. Science has tremendously extended human knowledge beyond the limit of the macroscopic and microscopic world using only the five sensory organs. Humans can see the Universe as far as 300,000 million years after the Big Bang with the James Webb Space Telescope (**JWST** in 2021), giving the Universe an age of 13.7 trillion. (This means that, with JWST, it took at least 200.000 million years for the light from the early Big Bang phenomenon to reach us nowadays). For the microscopic level, a superstring length equivalent to Plank constant $\hbar$ of 10-34cm can be conceived. Nevertheless, there was no significant progress in the metaphysical/ nonphysical world, as more than 95% of the Universe remained in the domain of Emptiness. Emptiness, in this context, refers to the vast expanse of space that is not occupied by perceptible matter (baryonic matter), but is more related to the Dark force and Dark matter. These are theoretical concepts that are inferred from the observed gravitational effects on visible matter, and their nature and properties are still largely unknown. Furthermore, our Universe may be one entity of the multiverse.

This makes the scope of scientific knowledge much lower than the potential scope of human realization.

i. So far, **the understanding of some basic quantum phenomena is incomplete**:
- Entanglement/ interconnectedness/ non/ locality of particles.

Einstein, in a debate with Niels Bohr, brought attention to a phenomenon that would later be known as the Einstein Poldolsky Rosen paradox, challenging the completeness of the Copenhague quantum interpretation. This thought experiment, which became a cornerstone in the field, was later replicated in the laboratory by Alain Aspect, a feat that earned him a Nobel Prize in 2022. The experiment involved two photons of the same system moving in opposite directions. The spin of the photons (spin; rotation around itself) was measured separately and independently by two observers. Despite the distance between the two photons, they always rotated in the same direction. This raised the question of the supraluminal communication between the two photons and then violated the absolute cosmic speed limit principle: The speed of light in a vacuum is the upper limit. *Nothing can go faster than 3.0×10^8 meters per second.*

- The phenomenon of particles and waves of light.

The light displays the phenomenon of duality in the form of particles and waves. However, physicists need help understanding why when they try to visualize light as a particle, they only see the light in the form of waves and vice versa.

Bohr said:" If quantum mechanics hasn't profoundly shocked you, you haven't understood it yet. Perhaps, the paradoxical and irrational phenomenon in Quantum mechanics is even less disturbing than the phenomenon of "Emptiness is wonderfully Existing" in Buddhism (vn: Chân Không Diệu Hữu).

For nearly a century, the paradoxes and 'irrational' phenomena in quantum mechanics have remained unsolved. It is possible that the root of these issues does not lie in modern physics, but in the metaphysical aspects of religion.

Heisenberg's Uncertainty Principle is a fundamental concept in quantum mechanics. It states that the position and velocity of a particle, such as a photon or electron, can not be precisely measured at the same time. Instead, they can only be determined with a certain degree of uncertainty. This means that a particle can never be exactly pinned down at any given moment.

$\Delta x \Delta p = \hbar/4\pi$
Δx Momentum error
Δp Position error
$\hbar$ Planck constant

Heisenberg devised a thought microscope experiment using gamma rays (short-length waves) to measure with enhancing accuracy because long visible light waves rendered measuring vague. However, gamma rays are associated with high energy levels that may displace the particle. In the end, there is no means to determine the particle accurately. Furthermore, Bohr went deeper than Heisenberg when he rejected the Heisenberg microscope experiment, arguing that particles do not even have a precise path. *Their path is probably unknown to us humans due to the veil of ignorance* incomplete and deviated CS, according to Gödel's incompleteness theorems).

The above quantum interpretation, which refers to the quantum world as a 'veiled reality', suggests that the quantum world is fundamentally different from the rest of Nature. This concept implies that our understanding of the quantum world is limited by our human perception and that the true nature of quantum entities is inherently uncertain. Heisenberg eventually accepted Bohr's interpretation, despite his initial disappointment at the contradiction between his concept of a particle having an actual path and Bohr's concept of particle uncertainty forming a cloud of possibilities.

Given his concept of determinism, Einstein was unhappy with the above interpretation when he raised the question: *Is the moon really there when no one looks at it?*. With Alain Aspect's confirmation of the EPR paradox, the Entanglement/ interconnectedness/ non/ locality of particles is confirmed: there is an unlikely reality at the Quantum level. However, the EPR can not confirm or reject Bohr or Einstein's arguments; everybody agrees that Einstein's moon exists. However, quantum uncertainty has remained poorly understood. However, it was said that toward the end of his life, he even allowed quantum indeterminism and, hidden variables and local reality to be retained with something new to its support.

It is critical to mention that scientists and Buddhists conceive of reality (the truth) differently. To physicists, reality is accessible to the five sensory organs. In contrast, Buddhists view reality as only represented by 'Emptiness', a concept that is not accessible to the five sensory organs. In the Buddhist perspective, Emptiness is not a void, but a state of potentiality from which all phenomena arise. This view of reality as emptiness creates a universe with everything that is impermanent, perishable, and therefore unreal, according to Buddhist philosophy.

ii. The Big Bang theory, a cornerstone of cosmology, attempts to explain the creation of the Universe. However, the mechanism behind the formation of different structures of the Universe remains a mystery. Mathematically, the chance combination of multiple elementary particles to build up macroscopic structures would require more time than the actual age of the Universe, which is about 14 billion years. This suggests that the Creation of the Universe involves multiple miraculous steps, making numerous accidental combinations of elementary particles highly improbable.

iii. Life can not start spontaneously from the non-live form. Rudolf Virchow and Louis Pasteur pointed out that life can only begin with life. Therefore, the presence of theological power is difficult to dismiss.

iv. What causes the Universe to continue expanding faster than the speed of light and with acceleration? Scientists agree that the term Big Bang is a misnomer since there is no analogy with an explosion or an eruption of a volcano due to the continuous expansion with acceleration. The proposition, in theory, is that an enormous condensed energy and extremely high temperature in a hypothetical tiny area is challenging to conceive and imagine. Contrary to the theory, common sense is that there must be a continuous release of an infinite source of unlimited/ everlasting energy. Lao Tsu attributed this kind of energy to Emptiness. Emptiness can only develop from Emptiness. The phenomenon means that the Emptiness is uncreated and therefore imperishable, for it is whole, unchanging, and complete (Parmenides). Emptiness precedes Existence. However, scientists often continue searching for a precursor that will activate the endless chain of searching.

In addition, there are other unanswered questions about Big Bang, Please see page 31.

v. **The synaptic interconnection** between axon terminal-dendrites is a complex system, with a vast number of billion connection choices for the encoding-consolidation of the MM and retrieval (Fig 1.1, 2,3). Ultimate Omniesciences/ UO likely plays a critical role in selecting the neuronal pathway. This UO/ Buddhahood is always available to the Soul because Buddhahood includes everything in the Universe after Creation. An example of this role of the Soul (permanently attached to Buddhahood) is the case of DID. In this disorder, the principal Soul is suppressed by traumatic events. The second, if existing in the concerned individual, so far repressed by the principal Soul, is liberated and controls the body. Clinically, the patient of DID displays amnesia involving mainly the explicit MM, but the implicit MM remains intact. As a result, the consolidation and the storing of the implicit MM do not require the specific Soul (regardless of principal or second Soul). But the consolidation/ storing and retrieval of the explicit MM do require the specific Soul. Therefore, retrieving explicit and principal Soul-specific MM is impossible when the trauma represses the principal Soul. This creates the amnesia of implicit MM specific to the principal Soul. This underscores the critical necessity of the Soul for the consolidation, storage, and retrieval of MM. All Souls are attached to the UO/Buddhahood (Please see pg 144: **PARADYM OF SOUL AS A QUANTUM COMPUTER**).

(Figures 1.1,2 &3)

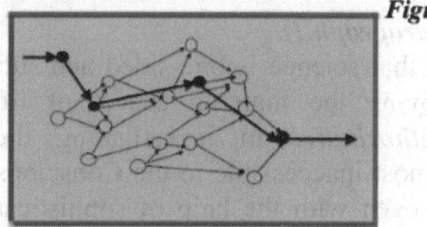

Figure 1.1

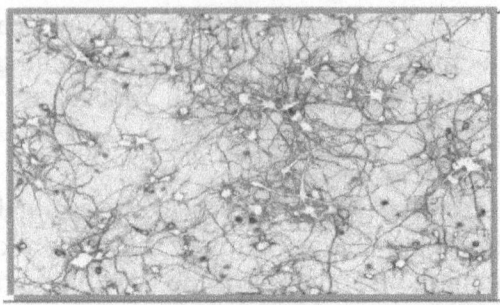
Figure 1.2

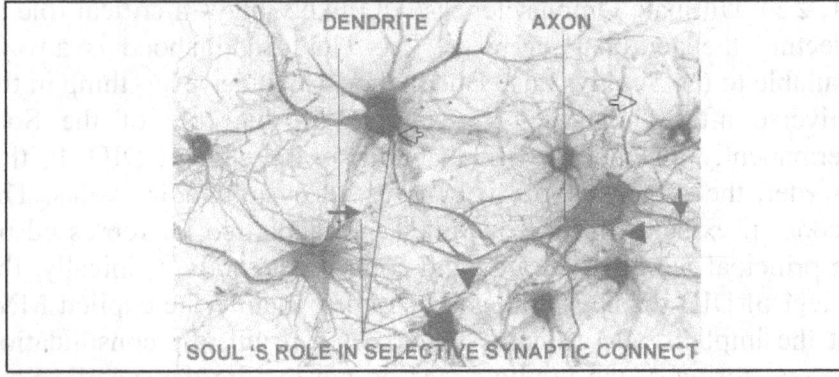
Figure 1.3

Fig.1,2 *Diagram shows the selective synaptic connection to establish the path of neuronal conduction. Given the multiple choices for selecting a connection, the option is specific for each data type.*

Figure 1.3
The choice of each synaptic connection is not random but ingeniously selective and likely designed by the Ultimate Omniscience that controls the Soul (See pg 52, 91).

CONCLUSIONS (paragraph I).

Therefore, it is likely that science is one-sided and subjective. Due to its success in improving the material aspect of life, science has gradually become **authoritative** in investigating the metaphysical realm. This area is almost inaccessible to the Consciousness that is the product of the brain, even with the help of sophisticated technology, including artificial intelligence/ AI. In the Western world, scientists and philosophers tended to be more empirical than the Far Eastern philosophers like Confucius and Mencius. Even with Lao Tsu, the latter was still practical and much attached to physical life. The evidence is that Eastern philosophers abstained from discussing phenomena after death. ***This made Lao Tsu's Taoism incomplete regarding religion.*** In contrast, in Middle Eastern and Indian spiritual masters with spiritual revelations in Hinduism and the Original Mind, a state of consciousness that is pure and unconditioned, in Buddhism, sentient beings' lives comprise the present as well as past and future lives. Buddha was empowered with infinite perception capability in the boundless dimension of Buddhist scriptures. As a result, scientific discoveries are consistent with and confirm Buddhist revelation and are to render these revelations applicable to life. Physical theories keep changing and being modified to conform to new and accurate observations. In contrast, Buddha's sayings are always authentic, as Buddha frequently reminded his disciples that ***Buddha's expression was not illusional...***

In Samyutta Nikaya, Chapter IV, THE *SIMSAPĀ* GROVE
31 (1) The Siṃsapā *Grove (translated by Bikkhu Bodhi,*
On one occasion, the Blessed One was dwelling at Kosambī in a siṃsapā grove.
Then the Blessed One took up a few siṃsapā leaves in his hand and addressed the bhikkhus thus: "What do you think, bhikkhus, which is more numerous: these few siṃsapā leaves that I have taken up in my hand or those in the siṃsapā grove overhead?"
....

"So too, bhikkhus, the things I have directly known but have not taught you are numerous, while the things I have taught you are few. And why, bhikkhus, have I not taught those many things? Because they are unbeneficial, ... (Teachings of The Buddha)

The above is to highlight that Buddha only said the essential discourse that is always practical and correct, as opposed to scientists often changing and correcting the theories and experimental interpretation due to the shortcomings of the CS.

II. THE TAO

In Hindu sutra, In Nature of Knowledge - First Khanda, The Kena Upanishad opens with:
> *"There the eye goes not,*
> *the speech goes not, nor the Mind.*
> *We know not, and we understand not. How would one teach it?*

In Brihadaranyaka Upanishad 1.4.10
10. This (self) was indeed Brahma in the beginning. It knew only I(?) as. 'I am Brahmaṇ.' Therefore It became all. And whoever among the gods knew It all became That; and the same with sages and so on. ...
(meaning: I am Brahma who creates the world)

Note: According to Buddhism, Brahma is the God of the low level of Triple World/ (skt Traidhatuka).

(i. **The realm of passions**/sensuous desire for sex and food: all beings have five sensory organs) Kamadhatu (skt)
- Six realms of samsara existence;

The Hell/(skt: Naraka-gati),
The state of hungry ghosts (skt Preta-gati),
The state of animals/ (skt: Tiryagyoni-gati),
The state of human beings (skt Manusya-gati), / Celestials—
The state of asura, angry demons /(skt Asura-gati),
The state of gods (skt Deva-gati)
- Six heavens; Mount Meru/ Sumeru, Sineru, or Mahāmeru start appearing until the formless state. Longevity lasting from 1000-8000 years

(ii. **The realm of Form or Matter /Beauty** (skt: Rupadhatu) with bodies and palaces

Composed of the first three stages of Dhyana:1,2,3. The beings are not associated with desire, however the finest desires remain
Beings of the third Dhyana stage do not have to return to lower levels

(iii. **The Immateral realm/No form/No Beauty/pure spirit**: no bodies, no palaces. Beings of Fourth Dhyana, Arahat and up

However, Tao is manifested as Taoism or Daoism, which is the way to practice Tao living in balance and harmony with the Universe and ultimately with Tao who creates the Universe.

Lao Tsu (born 571 BC, <u>Chu</u> Dynasty, ancient China) said:
The Tao is like an empty container:
it can never be emptied and can never be filled.
Infinitely deep, it is the source of all things.
…… It is hidden but always present.
I don't know who gave birth to it.
It is older than the concept of God
And
The space between Heaven and Earth is like a bellow;
it is empty yet has not lost its power.
The more it is used, the more it produces;
the more you talk of it, the less you comprehend.
and
Yet mystery and reality emerge from the same source.
This source is called darkness. Darkness is born from darkness.
The beginning of all understanding (Consciousness)
and
When people see things as beautiful,

> *ugliness is created.*
> *When people see things as good,*
> *evil is created.*
> *And*
> *All creatures in the Universe*
> *return to the point where they began.*
> *Returning to the source is tranquility*
> *because we submit to Heaven's mandate.*
> *Returning to Heaven's mandate is called being constant.*
> *Knowing the constant is called 'enlightenment'.*
> *Not knowing the constant is the source of evil deeds*
> *because we have no roots.*

Emptiness is static and homogenous, but Universe is not. Therefore when saying that this Universe is static, Einstein was so embarrassed to correct this.

In the Surangama (kinh Lăng Nghiêm), Buddha said to Ananda
> *...You should inquire into all the creations which in this material world are subject to change and destruction. Ananda, which one of them does not decay?*
> *Yet you have never heard that Emptiness can perish.*
> *Why? Because it is not a created thing*

Lao Tsu said
> *Look for it, and it can't be seen. Listen for it, and it can't be heard.*
> *Grasp for it, and it can't be caught.*
> *These three cannot be further described,*
> *so we treat them as The One.*
> *It's highest is not bright. It's depths are not dark.*
> *Unending, unnameable, it returns to nothingness.*
> *Formless forms and image less images,*
> *subtle, beyond all understanding.*
> *Approach it, and you will not see a beginning;*
> *follow it, and there will be no end.*
> *When we grasp the Tao of the ancient ones,*
> *we can use it to direct our life today.*
> *To know the ancient origin of Tao: this is the beginning of Awareness.*

Parmenides (earlier part of the 5th century BCE) said:
> *" what exists is uncreated and imperishable, for it is whole and unchanging and complete."*

As a result, all discussions of Tao are difficult to express, and the quintessence of the expression in words is difficult to perceive. Lao Tzu said:
The wise student hears of the Tao and practices it diligently.

The average student hears of theTao and gives it thought now and again.
The foolish student hears of the Tao and laughs loud
If there were no laughter, the Tao would not be it is.

III. THE EMPTINESS / VOID, VACUUM.

Emptiness (vn: Chân Không)/ Tao/ Ultimate pole (vn: Thái Cực) / Original Mind (vn: Bản Tâm)/ are synonyms to designate Buddhahood/ Holy Spirit/ GOD, the ultimate origin of the Universe and all species. Buddhahood/ Holy Spirit possesses immeasurable power, Ultimate Omniscience (UO) (vn: Trí Huệ Bát Nhã), which refers to the all-knowing nature of the divine, and other attributes like ingeniosity, infinite light, infinite sound, being splendid, homogenous, non-discriminatory, uncreated, imperishable, and not associated with birth or death. Because of the homogeneity and non-discrimination, Emptiness is characterized by the similarity between the microscopic and macroscopic structure in all parts.

Lao Tsu's teachings on *Tao emphasize its transformative power: it takes from those who have an abundance and gives to those who lack, elevating the low and humbling the high. This underscores the potential* of Emptiness to be filled with imperceptible content, challenging **the notion of a vacuum**. In Taoism, as well as in the doctrines of major religions and great philosophers, Emptiness is revered as the transformative origin of the Universe.

Buddha illustrated Emptiness with a profound example: imagine dividing an object into minuscule particles, then continuing to divide countless times. These particles, now invisible and imperceptible to the five sensory organs, embody Emptiness. This concept aligns with the result of the process, as one can envision. Thus, Buddha proclaimed that Emptiness and form are not distinct, but intricately interconnected.

- *Vacuum* **per se** never existed before or after the Big Bang.
- The reverse process (creating Existence from Emptiness) will create the form accessible to the sensory organs.
- The process creating Emptiness from Existence generates energy, similar to making the atomic bomb by splitting atoms.
- Energy must be accompanied by mental processes such as omniscience, Awareness, or Consciousness because of the principle of duality: meaning that the process of formation of Emptiness from Existence create not only the Energy but also the Omniescence and *that Emptiness is empowered with Energy and Omniescence.*

Emptiness, Tao, Buddhahood, and the Holy Spirit, despite their name, are not devoid of potential. They possess the power to create Form, Force, and Mind, challenging the conventional understanding of emptiness. This revelation inspires a new perspective on the generative power of Emptiness.

Kena Upanishads wrote about fullness instead of Emptiness as origin of the Creation:*"From Fullness, fullness comes, when fullness is taken from fullness, fullness still remains." Similarly Lao Tsu said:" Approach it, and you will not see a beginning; follow it, and there will be no end.*

IV. THE MIND Search for constituents of the Soul
/vn: TÂM or SOUL/ vn HỒN).
Definition and Identification of the Soul.

As Wikipedia defines, the Soul is a person's immaterial, spiritual, or thinking aspect, as contrasted with the person's physical body. In lay terms, the Soul is the spiritual essence of a person, which includes our identity, personality, and Memories that are believed to be able to survive after our physical death. For Buddhism, the Soul is not immortal. The Soul is created when the karma is first created and will add up with new additional karma, which will vanish when it can get rid of all karma (bad karma). The personality, thinking, and physical activities represent the living Soul. After death, the Soul leaves the body and carries with it all karma to ascend to upper levels or re-incarnates in the Six Realms of samsara Existence/ vn: Sáu nẻo luân hồi (gods, demi-gods/ sct asuras, humans, animals, hungry ghosts and hells). **Therefore the Soul is equivalent to Karmic Consciousness or store current and previous memories**

In Hinduism, the Soul, known as Atma, represents the microcosmos, the self, as opposed to Brahma, which represents the macrocosmos or the Universe. In this concept, the Soul is immortal, similar to Brahma. Buddha criticized this concept and made an essential difference between Hinduism and Buddhism. The Soul, as defined in this writing, is consistent with karma in Buddhism and constitutes the metaphysic distinct from the physical body entity, from the Buddhahood called Original Mind/ the Emptiness, and is uncreated and immortal. The soul, or karmic Consciousness, is essential for the incarnation of the cycle of birth and death. As a result, karmic consciousness is also considered to be karmic power. This power,

which is not a physical force, is the driving force behind the Soul's journey through the cycle of birth and death.
The Soul can be recognized as two parts (Fig 1.4B):
- The easily identifiable parts are:
- CS of the five sensory organs, general CS (sixth CS) such as time-space, name, meaning...
- Thinking, concept, idea (Manas Consciousness) the seventh CS / vn: Mạt na thức/ Suy nghĩ)
- The encrypted parts are;
- Store CS Alaya CS/ vn: A lại da thức/ Tạng Thức
- Buddhahood is not a constituent of the Soul, but the Soul must be attached for it to be functional.

Despite enormous research, the Soul has remained elusive to scientists. There needs to be more understanding of the Soul and neuroscience (Damasio 1999, Ravel 1997, Trimble 2007).

CS is represented by the Memory obtained for information by the five organs and the proprioceptive sensory information (ense of body awareness of the status of muscle, tendon...) with integration in the brain. The Memory and the related CS, known as inner CS (ICS), are seated in the ICS that comprises sensory cortices and DMN composed of PCC, MTL, RSC, Precuneus and vmPFC (see pg 238). The problem is identifying the Soul or the relationship between the Memory/ CS and the Soul. As previously pointed out, the Soul is not accessible to the five sensory organs. Its existence is denied at large in the science community.

All living beings (including plants), and minerals are entities presented as wholeness. Humans and animals only acknowledge the parts of nature revealed through electromagnetic energy, weak and strong force through the five senses related to CS. As a result, this type of perception and Consciousness must be incomplete because there are entities of the creation that are not "visible" to the five senses.

Furthermore, living organisms are composed of cells and minerals. Cells can be attached to form organized structures/ organs like the brain, liver, stomach, and limbs...or can be independent like red blood cells, white blood cells, cancer cells... circulating in the blood.

A

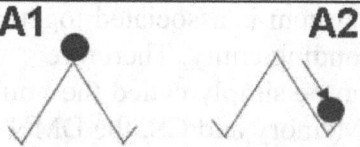

Principle of asymmetry in the creation of the Universe
A1: the system is symmetrical with all in the highest level of energy, but with the highest unstability
A2: the system is asymmetrical with the ball in the lowest level of energy, but with highest stability

B Body destroyed after the death
                       Along with the Memory (MM)

Fig 1.4 The MM + Memoirs (memoir is a part of SOUL/ MIND)

From the perspective of the interconnectedness of all entities in the Universe, *the Soul can be perceived as the formless essence of any entity*. In the traditional Chinese concept, this interconnectedness is symbolized by the Ultimate Pole (vn: Thái cực) (Fig 1.4B). The Ultimate Pole consists of two parts: Yang, representing Form, and Yin, representing the Formless. However, this division is never complete and symmetrical: in the Form, there is always a residue of the Formless.

Let's designate the Yang as the Great Yang and the Yin residue as a minor Yin. In the Great Yin, there is a Yang residue or minor Yang. In the case of animals or humans, the Great Yang or Form is all the physical parts, and the Great Yin or Formless is the Soul. The minor Yin represents Memory with its neuronal synapses in the brain. The great Yin is the Soul. The memoirs represent the minor Yang, the remaining parts existing among the relatives of the deceased.

The analogies are:
i. all cells and organs of the body of humans and animals have their related Soul,
ii. soul of the entire body or organs is composed of the summation of all constituents, including soul of nerves and blood vessels...

iii. soul of the same system is associated together to form a distinct entity for the corresponding entity. Therefore
- Soul of the brain can be simply called the Soul, which is seated in the brain part of the Memory and CS, the DMN storing the semantic MM and the sensory cortices. The Soul is connected with the souls of other parts of the body to form the entire Soul
- Soul of the non-brain, including the head, body, four extremities, and viscera. Psychologists with the concept of embodied emotion have just recently shown interest in this type of entity.

In this book, the Soul of the brain is composed of nonphysical parts that may have an indirect impact on the life of humans or animals, including:
· MM of the present life and thoughts
· The CS of the present life and CS of many previous lives.
· Part of the Original Mind/ Buddhahood/ Emptiness to which the CS is connected

Our brains are conditioned through education, through religion to think we are separate entities with separate souls and so on. We are not individuals at all. We are the result of thousands of years of human experience, human endeavor, and struggle."
……..and
"We carry about us the burden of what thousands of people have said and the Memories of all our misfortunes. To abandon all that is to be alone, and the Mind that is alone is not only innocent but young -- not in time or age, but young, innocent, alive at whatever age -- and only such a Mind can see that which is truth and that which is not measurable by words."
(Krishnamurti, J in The Krishnamaurti Reader, 2011)
In the above sayings, *brains so conditioned through education, through religion, to think ….the burden …. the Memories of all our misfortunes* are present and past CS constituting the Soul humans carry with them for innumerable lives. Furthermore, current and past CS are continuously reactivated for humans to be accommodated with life. The Memories are independent of the brain of the past lives and the present life.

V. The Big Bang And Genesis Revelation.
The search for the origin of the Universe and life is the question of humanity from antiquity. The answer to this question will serve the correctness in establishing behavioral conduct, scientific theories, and philosophical views.

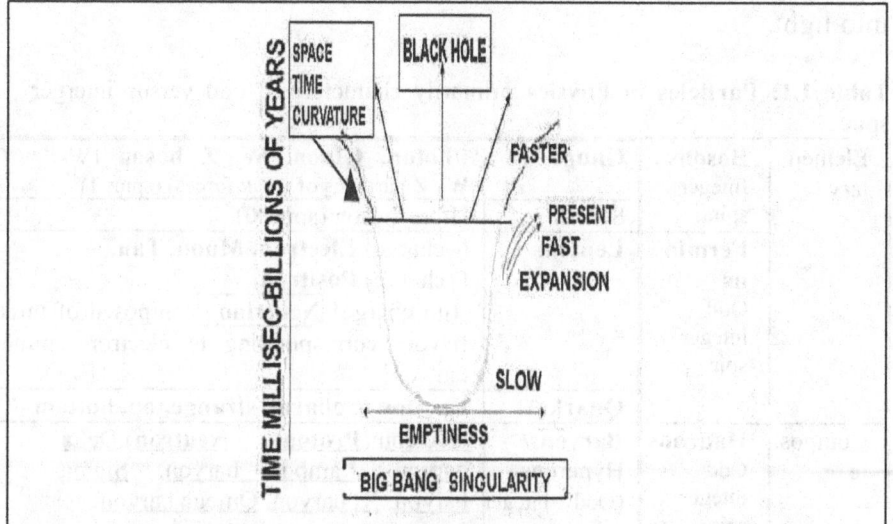

Fig.1.5 Universe developed after the Big Bang is hypothetically similar to a cone expanding with acceleration without a foreseeable end. For physicists, the end of Big Band may be a Big Rip or Big Chill, Big Crunch with collapse to be restarted with a new Big Bang, or Big Bounce. Other possibility is a flat Universe or open Universe (Please see more on pg 53).

A. Theory Of Big Bang.

1. Scientific Concepts Of Big Bang/ BB Theory (Fig 1. 5 and 1.5A-E).

The theory proposed that the Universe starts with one tiny point containing a source of condensed energy, and at a very high temperature, the size of the original may be about

From a minuscule point, the Universe rapidly expands, its growth accelerating. This concept is what we refer to as the Big Bang Singularity.

Following the very first moment, there was cosmic inflation (coined by cosmologist Alan Guth, 1980) with the creation of space and time. After 1/1000 billion sec, the formless energy starts condensing with the successive formation of photons (particles of Boson type), quarks (meson type), and leptons represented by electrons and antimatters. The early Universe was a dynamic environment, with rapid changes occurring as it cooled and particles formed.

After 1/106 sec, the generation of quarks ceases, and they begin to combine to form Hadrons, such as protons, which lack an electron. The Universe continues to expand, remaining extremely hot, before gradually cooling. Photons cease their interactions and transform into light.

Table 1.1: Particles in Physics primarily characterized odd versus interger spin

Elementary	**Bosons** Integer spin	Gauge	**Photon, Gluon, W, Z boson** (W-W+,Z particles of weak forces) (spin=1)
		Scalar	Higgs boson (spin=0)
	Fermions Odd integer spin	Leptons	(- charge) **Electron, Muon, Tau,** (+charge) **Positron,** (no charge) Neutrino composed of three flavors corresponding to electron, muon, tau
		Quarks	Up, down, charm, strange, top, bottom
Composite	**Hadrons** Odd integer spin	Baryons/ Hyperons (Odd integer spin)	Nucleon:(**Proton, Neutron**),Delta baryon, Lambda baryon, Sigma baryon, Xi baryon, Omega baryon
	Hadrons integer spin	Mesons/ Quarks (integer spin)	Pion, Rho meson, Eta meson, Eta prime, Phi meson, Omega meson, J/ψ, Upsilon meson, Theta meson, Kaon
		Others	Atomic nuclei, Atoms, Diquarks, Exotic atoms: (Positronium, Muonium, Tauonium, Onia), Superatoms, Molecules
Hypothetical	Gravitino, Gluino, Axino, Chargino, Higgsino, Neutralino, Sfermion, Axion, Dilaton, **Graviton,** Majoron, Majorana fermion, Magnetic monopole, Tachyon, Sterile neutrino		

Note: Six elementary particles of Quarks can change flavor by exchanging boson particles W and Z)

Hadrons are composed of two types: Baryons and Bosons

After one second, the Universe was as big as half of the solar system. The energy is enough to create atoms of Neutron/ Proton hydrogen and a small amount of Helium, Deuterium, and Tritium. When the temperature falls to 3000 C at 400.0000 years of light, there is no more creation of atoms. Before all electrons are attached to atoms, the Universe is thick, and photons can not fly without hitting the electrons. At cooler temperatures, the Universe becomes more transparent with electrons attached to atoms, and photons start flying in straight lines. The expansion of the $0.1A^0$ point with accelerating speed creates the Universe. In summary, this physics theory proposes that the Universe developed from a small area, of a very high temperature and a very high and condensed energy. The Universe develops from the expansion of this point with the development of CS, time and space, Dark Force,

Dark Matter (non-baryonic Matter), and then matter (ordinary Matter or baryonic Matter)

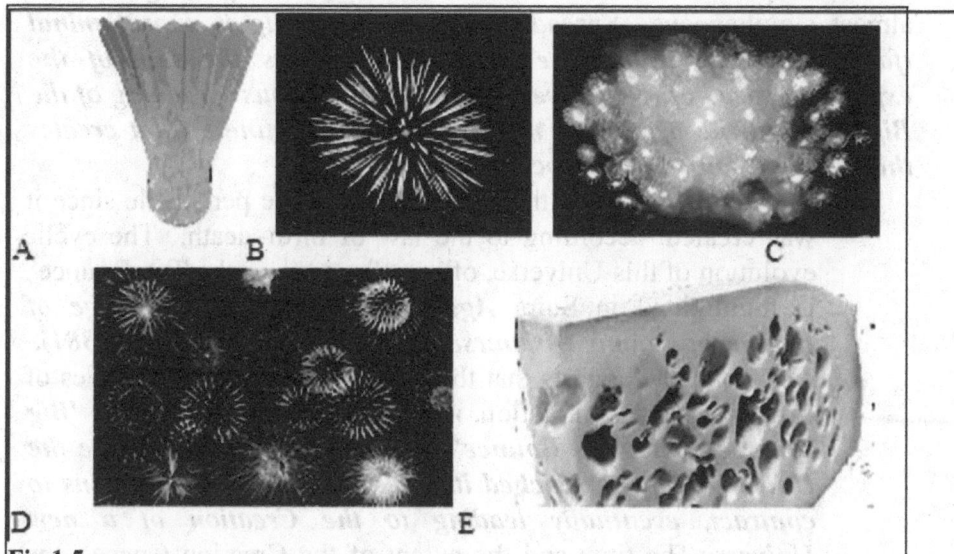

Fig 1.5
AB: Big Bang is not a single explosion
CD Big Bang may be an extensive expansion multicentric
E: Multiple Universes, each is represented by a space surrounded by multiple Big Bangs, forming a patten similar to multiple holes in a piece of Gruyere Cheeze

In 1981, Stephen Hawking and his close associate James Hartle asked:" What happened before the Big Bang?" Hawking proposed that there is no previous beginning for the Big Bang, and the cosmos has the shape of a shuttlecock (badminton ball) with a *rounded bottommost end (Please see below.)*
The Concept Of Multiple Universe And The Fate Of Multiverse. (Fig 1.5 CDE)
Emptiness, as understood in Buddhist philosophy, is a concept that is exclusively unique in the Creation. According to the dissertation of Buddha and Bodhisattva Nagarjuna, *the 14th Patriarch of Indian Buddhism, Emptiness is not a void or nothingness, but a state of interdependence and lack of inherent existence.* This unique nature of Emptiness is a profound concept that underpins the Buddhist worldview.
Since Emptiness is homogenous, non-discriminative (Neither identity nor difference), an event in the Emptiness will eventually involve the entire the space created by that event. Therefore, the rounded bottommost end of the shuttlecock/ badminton ball (as exemplified by S Hawking) must be represented by the extensive areas in the

"transformed Emptiness". As a result, the Big Bang is not singular but represents large spaces with simultaneous Big Bang that create multiple Universe Space or Multiverse. The Creation of multiple Big Bangs is almost simultaneous. *Spreading of the Big Bang is supraluminal (faster than the light): the spreading speed is the sum of the expansion speed of our Universe + the simultaneous spreading of the Big Bang at the beginning moment within Emptiness (that creates the rounded shape bottom cock badmington* .

- This Universe (and other Universes) will be perishable since it was created, according to the law of birth–death. The cyclic evolution of this Universe, often referred to as the 'Big Bounce', is highlighted in Sutra *Aggañña Sutta: On Knowledge of Beginning/ Long Discourse of Nikaya, page 33, 60, 381).* This concept suggests that the Universe goes through cycles of expansion and contraction, *with each cycle ending in a 'Big Bounce'. The 'Big Bounce' is a theoretical event where the Universe, having reached its maximum expansion, begins to contract, eventually leading to the Creation of a new Universe.* The time and the extent of the Creation (space), the innumerable/ unimaginable number of Bodhisattvas and Buddhas mentioned in major Buddhist sutras, exceed the ages and extent of the Earth and even this Universe, highlighted in many sermons of Flower adornment and Lotus Sutras. The sheer magnitude of the Creation, as described in these texts, is enough to humble even the most knowledgeable of scholars.The Omnipresence of multiple Buddhas and Bodhisattvas, the extent of the past time can not be conceivable without the concept of the Creation originating from the Emptiness at the time and Space of zero.The concept of Multiverse of Hugh Everett II may reflect to some extent the concept of Emptiness with multiverse.

The characteristic is widening the bottommost point of the ideal zero diameters. This point has no time and space.... Since there is no time, it is absurd to pose the question of being before the Big Bang. The expansion of this point of zero is expressed by the no boundary wave function (an expression of how to locate a particle by the wave equation). The phenomenon will be followed with more discussion in the two paragraphs of Void/ Emptiness and False Thought).

According to Einstein's equation of space-time:

$Ds^2 = dx^2 + dy^2 + dz^2 - c^2 dt^2 = dCS$

Dx, dy, dz : change in three dimensions of space.

Dt: change in time.

Ds: In this case, is CS/ Consciousness (because Space is only recognized by CS).

Unlike the explosion phenomenon, the expansion was found to increase its velocity (, i.e., acceleration) with time and dimension of the Universe (according to Hubble's Law (or Hubble–Lemaître law: galaxies are moving away from Earth at speeds proportional to their distance. In other words, the farther they are, the faster they are moving away from Earth) based on the observation of the expansion of the Universe). Along with the expansion, the temperature drops gradually. The formation of particles follows this event: **The first types of particle conceived by the astrophysicist are an almost chargeless Neutrino** (no electrical charge), which is 10 billion, billion, billion times smaller than a grain of sand, most abundant in the Universe, other five types of leptons like electrons, muon, tau...), and photons. It is often conceived that photon is not the first appearing particle. As the Universe cools down, more types of particles are formed from the six elementary particles—subsequently, atoms. Molecules and more complex structures appear. Macroscopic systems, planets, and galaxies then follow to develop...

It is worth briefly reviewing the concept of the String theory proposed in the 1960s by Gabrielle Veneziano, that the force within the Hadron is comparable to the elastic string. Later on, it was revealed that the Hadron is composed of Quarts, and as a result the theory is disfavored.

However, the theory later changed to Superstring Theory or Theory of Everything/ TOE. TOE encloses all particles, including Graviton, a hypothetical particle for Gravity. The profound implications of the String theory, from its humble beginnings to its current form, are enough to intrigue even the most seasoned of physicists.

A Superstring is a loop that vibrates according to the direction in the space. As a result, various particles are generated, a phenomenon similar to the violin string generating various sound tones. The loop of string is tiny, about 10-33 cm, weightless, and therefore shares the nature of Emptiness. Superstring Theory, also known as the Theory of Everything (TOE), is a theoretical framework in physics that aims to describe all fundamental forces and particles in the universe in a single, unified theory. Prominent physicists like John Schwartz, Edward Witten, Michael Green, David Gross, John Ellis, Abdus Salam, and Sheldon Glashow believed in the Superstring Theory, but Richard Feynman did not believe very much.

By Analogy of the relationship between the electromagnetic force (photon as a particle) and baryonic Matter (perceptible Matter), the Dark Matter likely represents the transformation of the Dark Force into Dark Matter (DM). Dark Matter is a hypothetical form of matter that is thought to account for approximately 85% of the matter in the universe and about a quarter of its total energy density. It is only associated with Gravity but not acquiring proprieties like photons, neutrinos, electrons, and quarks.

The theory of the Big Bang was developed due to observation at the beginning of the 20th century by Vesto Sipher, followed by American astronomer Edwin Hubble and Belgian priest and physicist Georges Lemaitre. From their observation, the galaxies keep receding. The recession was interpreted as the expansion of the Universe.

The expansion was believed to be the expansion of the space created by the Universe. This contradicted Einstein's concept that the Universe is static. Using physical theory and mathematical calculation, the Big Bang was hypothesized, originating from a state of extremely hot and condensed enormous energy. The significance of the Big Bang theory, and its profound impact on our understanding of cosmology, cannot be overstated.

Shortly after the Creation, the temperature dropped. The initial Formless state starts forming photons (a type of boson [integer spin of 0,1 0r 2]), quarks (a type of meson), electrons (a type of fermion, spin of odd half integers 1/2, 3/2, and 5/2,), and hadrons of meson or fermion type (composed of many Quarks attached together by gluons). The photons at the beginning could not freely travel due to the presence of elementary, mainly wandering electrons, rendering the space opaque. The formless Universe became murky, then gradually became clearer due to the formation of larger and larger particles and atoms: at 1/1000.000 sec, there were no more quarks formation; Quarks combined together with the presence of Gluons to form Protons and Neutrons At 300.000 years when the Universe temperature dropped to the point where electrons were used up by proton for nuclear formation neutrons, the space became gradually transparent, Photon was free to travel toward the expanding end of the Universe. Subsequently, atoms were formed until the age of 400,000 years. Cosmic Microwave Background (CMB) represents the phenomenon by which photons generate Electromagnetic Radiation (Waves) before the appearance of atoms. CMB is a remnant and evidence of the Big Bang.

With time, stars and galaxies formed. The earth is at the appropriate location, with the temperature regulated by the sun. As a result, water, inorganic, organic, and anaerobic organisms (not requiring O2) were formed. Subsequently, CO2 was decomposed to generate Oxygen and then Aerobic organisms (requiring O2).

Problem of recognizing the mass

In 1964 Peter Higgs and Franois Englert raised the problem of recognizing the mass of entities in the Universe, particularly mass of particles in Quantum Mechanics. As seen in Figure 1.4A, an object at the position of A1 has the highest level of energy but is the most unstable. In A2, the level of energy is lowest, but the most stable. In 1964 Peter Higgs and Francois Englert raised the problem of recognizing the mass of entities in the Universe, particularly the mass of particles in Quantum Mechanics.

As a result, the symmetrical, harmonious, and balanced condition is associated with precarious stability. The asymmetrical, unbalanced, and heterogeneous condition is associated with a low energy level but the highest stability level to keep the status asymmetrica and chaotic.

Example: it is more difficult to keep the world in peace than in war. In addition, *the manifestation of the mass is less detectable in symmetric environments than in asymmetric ones*. In other words, the mass measures the resistance or inertia to acceleration (change of velocity) when a net force is applied. Mass is different from gravity; Mass and gravity is not always proportional to each other. In Quantum mechanics, the photon is massless and weightless, W has a mass but is weightless, and neutrino, electron, and quark have mass and weight.

As a result, like the electromagnetic field, Higgs proposed that there is a field specifically sensitive to the mass. Fifty years later, in 2012, the Hadron Accelerator CERN in Geneva discovered the Higgs boson of the Higgs field, sensitive to particles with mass and nonreactive to massless particles like photons and gluon. Higgs boson is much larger than the proton, has spin=0, and is chargeless. This concept of Higgs Field is similar to that of the Ultimate Omniscience on which the data from five sensory organs is based to form the CS.

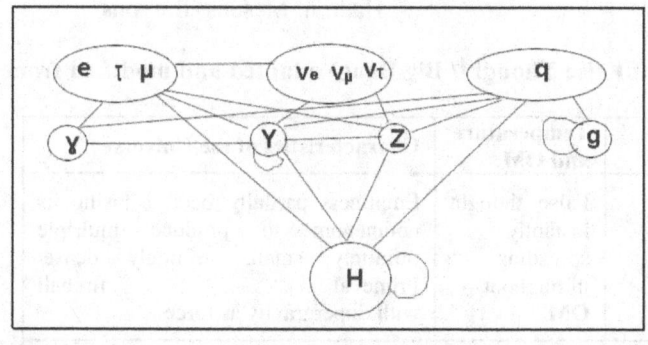

e: electron, τ: Tau, μ: muon, v: Neutrino with three flavors e, τ, μ; q: Quark, g : Gluon; H: Higgs, Z,Y: Boson; γ: Photon
Straight and curved lines represent the Higgs field interaction of Higgs Boson particles with itself or other particle to generate the Mass impression. The curved line represents the interaction in Higgs field

Note: The superstring theory, which proposes the existence of a single fundamental particle at a subatomic scale, has been accepted by many scientists. This acceptance provides a sense of reassurance in our quest for understanding, even though it is not universally embraced. The superstring view is perplex, especially when considering the concept of Emptiness/EM. EM is believed to be the source of everything: Force, matter, and Intelligence.

The above process of creation of particles, atoms, molecules, stars, galaxies, and ultimately living sentients is consistent with the concepts of some great Oriental philosophers like Chuang Tsu;

...with the original breath; When it comes together, there is life, When it is dispersed, there is death... (vn: Khí tụ thì sống, khí tán thì chết).

Table 1.2: Forces and Matters in Universe.

Six organs of perception	Force/ Form	Particle/ matter	Connecting force	Range	Force
Perceptible, non-emptiness or physical /Form	Weak force,	Quarks, leptons	W,z (decay, radioactive)	Very short	Weak
	Strong force (quarks forming Protons and Neutrons)	Quarks, Gluons binding Quarks	Boson, Gluon (nuclear binding)	Short	Very strong
	Electromagnetis incl Gamma Ry	molec binding)	Photon /Boson(	Infinite	Strong
Non perceptible Emptiness/ Formless	Gravity (mass binding)	mass	Graviton not yet determined	Infinite	Very weak
	Dark force of 5th force	Universe expansion	? Dark photons	Infinite	Strong

Bosons: photon, W+,W-.Z0 Gluon Higgs. and Meson
Fermions: Lepton (electron, neutrino...) and Baryons (proton,neutron...)
Hadron: Mesons+Baryons

Chronology of False Thought/ Big Bang) adapted and modified from Google).

Time	Era	Temperature and OM	Characteristics of the Universe
0- 10^{-43} s	Interface (Emptiness -False thought) (Planck Time)	False thought instantly spreading throughout , OM	Emptiness partially/focally losing its nounemom to produce mulriple infinitely small, infinitely dense Primeval fireball with Supergravity as force
10^{-43} s	Big Bang	infinite - 10^{32} K, OM	described by physics and conceived in spirituality: Dark force, DM, gravity, GUT/ Grand Unified Theory hyothetically composed of weak, electromagnetic, and strong forces),
10^{-35} s	End of	10^{27} K	Dark force, DM, gravity, strong

Time	Era	Temperature	Events
	GUT		nuclear, electroweak Quarks and leptons form
10^{-35} to 10^{-33} s	Inflation	10^{27} K, **OM**	Size of the Universe drastically increased, by factor of 10^{30} to 10^{40}, DM
10^{-12} s	End of unified forces	10^{15} K, **OM**	Dark force, DM, 4 forces, protons and neutrons forming from quarks,
10^{-7} s	Heavy Particle	10^{14} K, **OM**	Dark force, DM, proton, neutron production in full swing
10^{-4} s	Light particle	10^{12} K	Dark force, DM, electrons and positrons
#100 s	Nucleosynthesis era	10^9 - 10^7 K, **OM**	Dark force, DM, helium, deuterium, and a few other elements form
380,000 years	Recombination (Decoupling)	3000 K, **OM**	Dark force, DM, Matter and radiation separate End of radiation domination, start of matter domination o
500 million yrs	Galaxy formation	10 K, **OM**	Dark force, DM, galaxies and other large structures form in the universe
14 billion y	Now	3 K, **OM**	Present time

OM: Original Omniescience, DM: Dark Matter

2. Unanswered questions of Big Bang (continued with paragraph the understanding of some basic quantum phenomena is incomplete on pg 36).

Is Big Bang unique or multiple? Can other Big Bangs occur before or after our Big Bang? What existed before the Big Bang? These questions can not be experimentally answered and have received little attention from scientists. If one says that Emptiness precedes and gives rise to the Big Bang, this assumption belongs to the teaching in major religions. Some scientists still believe that something else must exist before the Universe exists. With this type of reasoning, there will be endless questions about the ultimate origin.

On the other hand, Christianity and Buddhism reveal a diagram of Genesis that is consistent and not significantly different from each other.

BB, as described in physics, is just 5% of the universe. The remaining 95% is a mystery, not addressed by religions. It's crucial to understand that time, space, temperature, and pressure are all relative to the observer. Therefore, during the time of BB, the Buddha nature did not perceive these changes as significant. However, these changes were indeed perceived by all living beings after BB.

2. Unanswered questions of Big Bang (continued with paragraph the understanding of some basic quantum phenomena is incomplete

3. **Big Bang Singularity?.**

Singularity in Big Bang concept is defined as the condition associated with matter is forced into a single area. In this Gravitational Singularity, the light ray from infinity curves and the wraps around the area.

The Big Bang develops from the above tiny area.

4. **Final destiny of Universe**: Big Crunch, Big Rip, Big Chill, Big Bounce?
- Because Big Bang is created, it can be multiple and is perishable

In 1998 Saul Perlmutter (Lawrence Berkeley National Laboratory and University of California, Berkeley, CA, USA) (Brian SchmidtAustralian NationalUniversity, Weston Creek, Australia.), Adam Riess (Johns Hopkins University and Space Telescope Science Institute, Baltimore, MD, USA.) (Nobel Prize 2011), attempted mapping the Universe by studying of the most distant supernovas (that occurs during the last evolutionary stages of a massive star nuclear fusion with powerful and luminous explosion resulting from white dwarfs that are triggered into runaway. A white dwarf is very dense: its mass is comparable to the Sun's, while its volume is comparable to the Earth's). These astrophysicists, heads of two independent teams, have discovered that the Universe expands with acceleration. According to them, the Universe continues expanding forever, pointing toward a "Big Rip" ending in Chill.

On the other hand, computer simulations reveal an alternate model of Universe evolution: a cyclic Universe with no beginning or end, an alternate expansion and contraction instead of a Big Crunch.

It is commonly thought that the Universe can not last forever; since it is created, it must perish. Interestingly, the model of cyclic expansion and contraction is consistent with a Buddhist sermon in which Buddha revealed a similar concept of the Universe's cyclical nature.

(*Aggañña Sutta: On Knowledge of Beginning, Long Discourses*) ...: *'There comes a time, Vasettha, when, sooner or later after a long period, this world contracts. At a time of contraction, beings are mostly born in the Abhassara Brahma world. And there they dwell, Mind-made, feeding on delight, self-luminous, moving through the air, glorious*

— and they stay like that for a very long time. But sooner or later, after a very long period, this world begins to expand again. At a time of expansion, the beings from the Abhassara Brahma world, [85] having passed away from there, are mostly reborn in this world.

In the last decade, Sir Roger Penrose, an Oxford mathematician, theoretical physicist, and Nobel Laureate in 2020, is only second to Einstein regarding general relativity in cosmology. He proposed a new concept of BB based on his research on the Black Holes. This concept has remained controversial to Cosmological Physicists. In the BB process, the accelerating expansion is followed by the degeneration of all matters, as exemplified by the Proton decay or degeneration of electrons, and becomes imperceptible. Subsequently, the Universe becomes a massive Black Hole that sucks in everything it touches, in the manner of a gigantic vacuum cleaner of the Universe near the end of its acceleration. The Black Hole will eventually degenerate with the emission of radiation, become empty, and revert to the Original order (likely Emptiness) that will be ready for another cycle. The cycle is called Aeon. Our Universe is, therefore, preceded and followed by other Universes. The concept is called **Conformal Cyclic Cosmology/CCC**. In addition, Universes are multiple.

The Cosmic Microwave Background supports the above concept of CCC, the low Entropy of the Early Universe, and the identification of Hawking's points, which likely represent remnants from the transformation of a former Black Hole. Inside the black hole exists a point of singularity associated with no spacetime and enormous energy gravitional force. This singularity is comparable to EM in Buddhism. As a result, the process from a galaxy transformed into a black hole singularity/EM and brom EM to BB and Universe is consistent with the concept of EMà Form and FormàEM

KSANA of thoughtlessness In the common concept, EM is also represented by the non-thoughtful and present short moment of one ksana, in which space and time are maximally ksana, one can feel the happiness and freedom of Nirvana.

.

A black hole is defined by the center with a point conceived as a singularity at below zero temperature having enormous energy with strong gravitational force, light being unable to escape,
Relativistic jet: visible jer connected to the black hole (? coming from or entering into the black hole with the near or supraluminal speed of light)
The event horizon represents the surface and the boundary of the black hole. The event contains chaotic debris and is at a very high temperature. Inside this boundary, everything is doomed inside the Black hole

Other astrophysicists criticize the evidence that the preceding Universe is lacking.
Penrose's CCC is consistent with Oriental philosophy and Buddhism: Only Emptiness/or Creator is the only Oneness. By the universal law, the Universes are regulated by the law of Birth and Death, and impermanence and must enter in the cycle of "Reincarnation/ Samsara

Creation of the Physical versus Metaphysical Realms
According to the principle of duality, in the post-creation era, Yin and Yang were formed. Major Yin is represented by the metaphysical realm, which is represented by the Dark Force (68% of the Universe), with Minor Yang represented by the Dark Matter, which is formless but has gravity (27% of the Universe). Major Yang is represented by the Physical realm (5% of the Universe and Minor Yin is represented by the Consciousness which is formless but is associated with neuronal synapses. Only the Major Yang component has photons that enable the sensory organs to recognize the materialistic creation solely. This privilege of posting photons in the physical world also impedes the five sensory organs from identifying the metaphysical realm. The significance of this privilege restricted in the 5% of the Universe can be speculated as the special realm that is critically specialized in spiritual training (practice of Morality: Four Immeasurable Minds of Virtue)

B. CHRISTIANITY.
In the beginning, there is formless existence and darkness: The **Earth proved to be formless** and wasteful, and there was **darkness upon the surface of the watery depth (consistent with the Emptiness).** God

created the heaven and the Earth, and God's active force was moving to and fro over the surface of the water:

Day 1. God's Spirit moves to and fro upon the watery surface then He creates the light, for the distinction of Day and Night.

Day 2. Creation of Heaven above the water with distinction of morning and evening. The division between Earth and Ocean.

Day 3 Creation of plants.
Day 4 Creation of Sun, Moon and Stars.
Day 5 Creation of birds and fishes.
Day 6 Creation of Land Animals and Humans.
Day 7 Sabbath of rest.

Of course, a day in Genesis may last millions of years, probably more or less one billion years (50,000 years in Islam) as compared to the earthly year. It is reminded that time, space, and CS in the Universe are relative as there is no time, space, and CS before the "false thought" generating the Creation. After the Big Bang or Genesis, time, space, and CS increased with the expansion of the Universe as conceived in the Big Bang Theory and demonstrated by astrophysicists.

(Capra F1983.. Schroeder, GL1990. Revel JF 1998 Brake M. 2019 Trinh, X T. 2011,Close F 2010Dalai Lama, 2005 Chopra D 2016, Penrose, R 1998, 2017 2016Layzer D 1990. Han,MY, 1990,Asimov,I 1990,Ruelle D1991, Talbot M 1988. Deleuze 1988, Francis C. 1994T, Aharoni,J. ,1985 Murdock D1987. Morris R, 1993, Hawking S1992, Davies P 1990, ,1988, Pear DF 1990. Einstein's Moon, Contemporary Books, In,Illinois 60601,)

C. TAOISM AND HINDUISM
1. Lao Tsu: Tao is:
Forever undefined; before heaven and Earth...
In the silence and Void...
Darkness within Darkness, The gate to all mystery...
The Tao is an empty vessel, it is used but never filled, ...hidden deep but ever present....
Her gateway is the root of heaven and Earth, like a veil barely seen. Use it, it will never fail....
It cannot be seen, it is beyond form, It cannot be heard, it is beyond sound, and Intangible: These three are indefinable but joined in one. ....
Intangible and elusive and yet within is essence
Dim and dark and yet within is essence...
The beginning of the Universe, is mother of all things
Knowing the mother, one also knows the sons.

Chuang Tzu
Mixed up in the Formlessness, is the Essence, transformation of the Essence generates the Form.
Gathering of the Essence (Qi) create the Life, Dissipation of the Essence creates the Degeneration, therefore in the Universe there is only the Essence

2. In Hinduism, Isha Upanishad said: *Self is everywhere, Bright is the self, Indivisible, untouched, by sin,wise, Immanent and transcendent. He it is, Who holds the cosmos together.* The creator is represented by the Self, the Lord, The Supreme Reality who exists in everybody

The Lord is enshrined in the hearts of all,The Lord is the supreme Reality:
Rejoice in him through renunciation
Covet nothing, All belongs to the Lord...
Those deny theSelf are born again
Blid to the Self, enveloped in darkness..
Self is One, Ever still, the Self is
Swifter than thought, swifter than the senses, Thougt motionless, he outruns all persuit

The main difference with Buddhism is that the same Self exists in every one including the Supreme Creator. As pointed out at the beginning, Ony the Creator is represented by the Self Who creates all sentient that are Selfless, and tainted with sins

D. BUDDHISM.

Genesis, Creation of the Universe is rarely addressed by the students of Buddhism, just because Buddhist scriptures are so voluminous that it is difficult for anyone to go through with an adequate understanding of all Buddha teachings. However, Buddha is not only omnipotent/ almighty but also supremely/ perfectly enlightened; it is obvious that the genesis has to be addressed by Buddha. Genesis is essential for the understanding of Buddha's teachings. However, most Westerners believe that Buddhism is atheist, a system of philosophy, and is not even a religion because of the lack of theological power in the Creation. This is a serious misunderstanding.

According to Buddhism, the Universe was created from Emptiness. **This Emptiness does not mean anything inside per se**. As indicated above, *VOID per se can not exist* under the law of Taoism.

In Surangama sutra, Buddha said that by dividing any object into trillion-trillion specters of dust, the end result is imperceptible and called Emptiness. As a result, Existence is Emptiness and vice versa: Emptiness←→ Form.

The above sermon in the Surangama sutra is the best illustrated the truth. The Buddhist cosmology of above all wordy expression with innumerable Universes and innumerable Buddhas, as much as sand grains in the Ganges river.

From the Emptiness or stillness to Form stage, one can see two intermediate steps:
- The Dark Force develops after the false thought. The release of Dark Force from the Emptiness state after Genesis is immediately accompanied by the loss of the equilibrium in the Emptiness. The force is tremendously powerful and represents the initial exploding force of the Big Bang conceived by physicists.
- the expanding force accountable for the expansion of the Universe with acceleration

Knowledge of Beginnings (Long Discourses), Buddha revealed how the world/ or Universe was created in the successive stages of the contraction of physical form and time, leading to annihilation and then re–Creation after a very long period. Darkness came first, then stars, moon, and sun reappeared. The new world expanded, and living beings passed away from other worlds (Souls) were reborn. The discriminative Mind then developed as described in the sermon:

Buddhad said to Vasettha

...... beings are mostly born in the Abhassara Brahma world (A world different from this world). *And there they dwell, Mind-made, feeding on delight, self-luminous, moving through the air, glorious* (meaning: No physical body) — *and they stay like that for a very long time. .....*
After a very long period, this world begins to expand again. *At a time of expansion (*for extinction of that world*), the beings from the Abhassara Brahma world, having passed away from there, are mostly* **reborn in this world**. *Here they dwell, Mind-made, feeding on delight, self-luminous, moving through the air, glorious — and they stay like that for a very long time.*

"10. 'There comes a time, Vasettha, when, sooner or later after a long period, **this world contracts. At a time of contraction, beings are mostly born in the Abhassara Brahma world**. And there they dwell, Mind-made, feeding on delight, self-luminous, moving through the air, glorious — and they stay like that for a very long time. But sooner or later, **after a very long period, this world begins to expand again**. At a time of expansion, the beings from the Abhassara Brahma world, having passed away from there, are mostly **reborn in this world**. Here they dwell, Mind-made, feeding on delight, self-luminous, moving through the air, glorious — and they stay like that for a very long time.

'At that period, Vāseṭṭha, there was just **one mass of water, and all was darkness, blinding darkness. Neither moon nor sun appeared**, no **constellations or stars appeared, night and day were not distinguished, nor months and fortnights, no years or seasons, and no male and female, beings being reckoned just as beings**. And sooner or later, after a very long period of time, savoury earth spread itself over the waters where those beings were. It looked just like the skin that forms itself over hot milk as it cools. It was endowed with colour, smell and taste. It was the colour of fine ghee or butter, and it was very sweet, like pure wild honey".

Buddha started describing the formation of this world with one mass of water, and all was darkness, blinding darkness without stars, night and day, months and fortnights, no years or seasons, and no male and female, beings being reckoned just as beings. After a very long period, savory Earth spread itself over the waters. The Earth provides food for all beings, which developed as all living sentients today .
(Thompson C. 2010 Takakusu,J 1949 Krishnamurti, J2011. Ravel JF 1997Teachings of The Buddha1990,)

E. SOUL AS AN ENTITY IN THE UNIVERSE.
1. Soul as an entity.
Regarding the creation of the Soul, given that it must exist as an entity, the Soul must be composed of the known and hypothetic entities of the Universe after the Big Bang and related to the MM. The characteristics of the Soul are:
- Imperceptible to five sensory organs but able to generate secondary effects sensitive to the CS.
- The Soul does not have Form but possibly has gravity. This was suggested in three different instances (see later).

2. The Soul in Physics and Neuroscience

The Soul, a profound entity, eludes identification through current scientific techniques that rely on electromagnetic energy and the five sensory organs. Yet, it exerts an indirect influence on Consciousness, as demonstrated in the enigmatic realms of near-death experiences (NDE), out-of-body experiences (OBE), and other metaphysical phenomena.

Since most neuroscientists only use Consciousness to try to investigate neuroscientific changes. But as pointed out in the previous paragraphs and by Krishnamurti, Osho, Consciousness is the artificial product of the Brain, the distorted discriminative Mind (the veil of ignorance, **incomplete and deviated CS, according to Gödel's incompleteness theorems**) please see page 100). The findings by most neuroscientists are findings *to justify their preconceptions, what the scientists believe in, and what the scientists are, but not what the true nature of the Soul is.*

As discussed in the previous paragraph, the Soul is distinct from the Brain, and the Soul carries with it the karmic Consciousness. In Buddhism or in Christianity and in cases of Near- Death-Experience (NDE) and Out- of- Body- Experience (OBE), the Souls out of the body still have the capability of Consciousness like perception, storing of information, and thinking. The Soul in

Buddhism is different from the term Soul used in Hinduism since, in the latter terminology, the Soul is immortal. The Soul in Buddhism is the carrier of Karma. When the karmic Consciousness is completely cleansed from the Soul, the remaining part of the Soul is the original Mind, Buddhahood merging into the Emptiness/ Nirvana.

3. The Soul and Inner Consciousness (ICS) (Fig.1.6)

Since the Soul represents the CS of the present and many previous lives, the Soul is the store of CS or Memories also stored in the DMN. In the Soul, MM is not represented by neural synapses but by a mysterious synapse equivalent. This enigma, coupled with the Soul's lack of significant electromagnetic activities, adds an intriguing layer to our understanding. The capability of perception and storage in the Soul may hold significant implications for scientific speculation and research (please see Chapter 5: The Soul).

Within the Brain, the Soul, Memories, and Consciousness, particularly Inner Consciousness, are believed to share a common location. This coexistence results in the storage of MM in two distinct yet interconnected locations: the Soul and the Brain's ICS area represented by the DMN.

It's crucial to understand that the incomplete nature of MM stored in the DMN, represented by neuronal synapses, is due to the fact that sentient beings can not fully utilize the DMN without the Soul's interaction. This underscores the indispensable role of the Soul in the process of memory storage.

The Soul, while lacking a physical form, possibly possesses gravity. The complexity of this lies in the fact that the Soul controls the MM (and the CS), but it requires the Brain as a mediator to interact with the physical world. This underscores the crucial role of the Brain in the Soul's functioning. One can conceive that the DMN is a substrate for the Soul to manipulate the entire Brain.

DOES THE SOUL HAVE THE GRAVITY.

Spinoza, a young philosopher of 20, boldly posited that the Creator is an uncreated substance, an unlimited source of creation. This radical belief, which offended the Congregation of Catholics, led to his expulsion from the society.

Throughout history, Western thought has held that the Soul is a material entity, often likened to Ether. This understanding persisted until the advent of Einstein's General Relativity theory, which explained many universal phenomena. However, even these scientific breakthroughs could not unlock the mystery of the Soul.

Buddha said in the Surangama Sutra that if the piece of the earth is divided over trillions of time, the end result is innumerable specters of imperceptible dust, and this is called Emptiness. The process not only creates imperceptible dust but also unlimited energy (similar to the process of splitting atoms to make a nuclear bomb. When the energy is created, the consciousness, or, more precisely, the omniscience, is also generated due to the Principle of Duality. Therefore, according to the Buddhist sutra, the Original Mind is Emptiness. In nature, nothingness per se does not exist since Lao Tsu said that *Tao is to take those who have too much to give to those who do not have enough to make the high lower, the low raised.*

In 1901, Dr Duncan McDougall, assuming the Soul as a (imperceptible) substance, performed an unusual experiment by measuring the soul's gravity. McDougall set to monitor the change of physical weight of six dying patients fitted with a sensitive scale. Only one patient, fulfilling experimental conditions, lost three-fourths of an ounce (21.3 grams). Other patients displayed variations in weight ranging from decrease to increase and variable fluctuations. The experiment has been largely considered by scientists to be faulted but has remained popularized in the media, including movies. The experiment was repeated in dogs but without the expected results. Another group of researchers performed the monitoring of dying sheep which only showed some increase in weight. In addition, Andre Maurois, a Nobel Prize nominate in literature, in a novel entitled "The Weigher of the Soul" described the changes in the weight of dying patients observed by his friend as a physician. Maurois also cited similar findings in the First World War I.

People commonly fail to realize that the existing but imperceptible things are often forgotten.

In the above experiments, McDougall overlooked the possibility that in dying persons, the soul can depart and then return (due to the attachment to the physical body).

Moreover, other Souls may attempt to ensoul into the agonizing person even before the principal Soul departs. This suggests that any changes in weight could be a reflection of the Soul's gravity.

As a result, the Soul is characterized by:
Its imperceptibility, pervasiveness, and possibly associated with gravity.
Capability of perception of CS, thinking, storing MM and transcribing to the next life.
a) Memories of Previous lives and karmic Consciousness
at the Soul's ensoulment into the developing embryo, the Soul imprints the CS of previous lives. The evidence of prior life Memories can be demonstrated in:

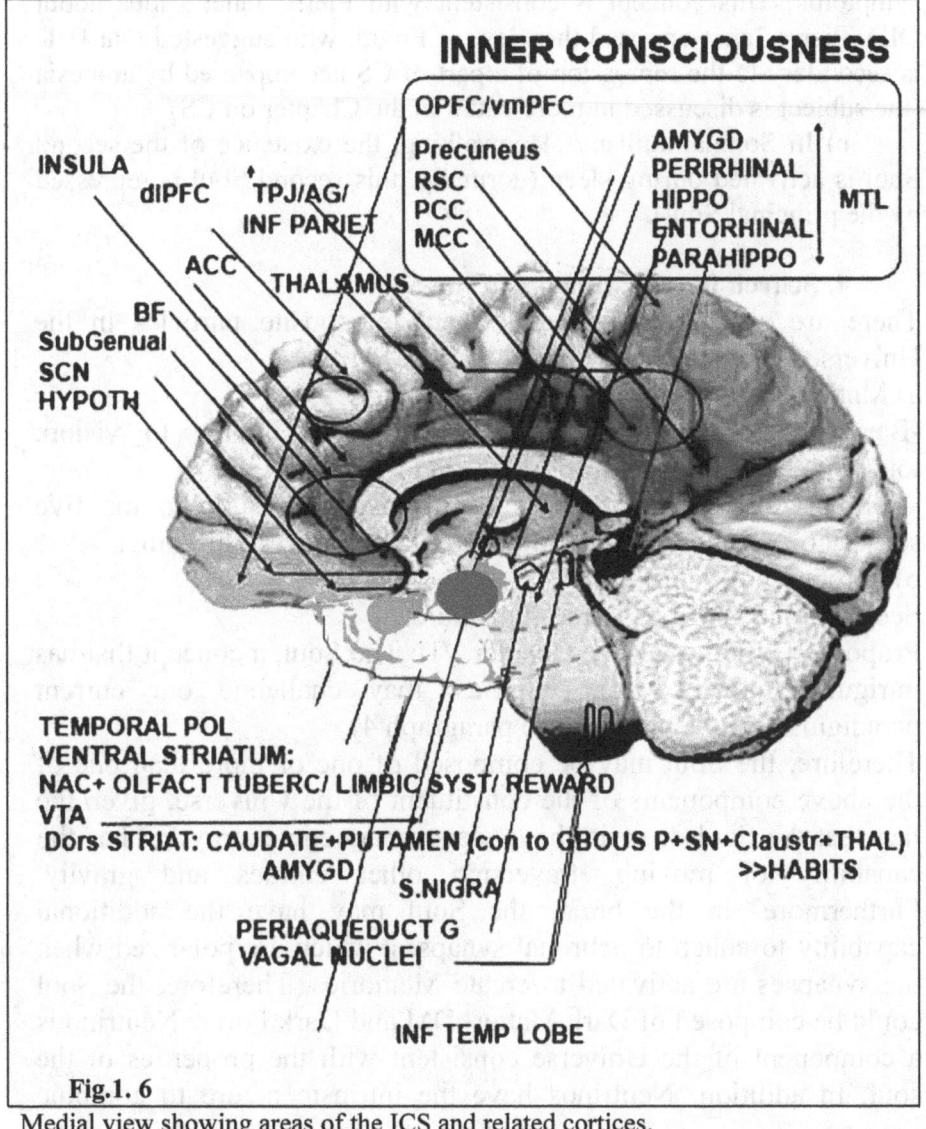

Fig.1. 6
Medial view showing areas of the ICS and related cortices.

- Some children, usually below the age of 10 years, have a recollection of their immediate previous lives, overwhelmingly illustrated in Dr. I Stevenson (PMC1839221).
- Some children and even some older people have extraordinary talent in art, languages, literature, mathematics...
- Persons practicing mindfulness Meditation attaining dhyana of level two and higher, or persons with NDE have some capability to review their previous lives.

b) In Dissociated Identity Disorder (DID) with two or multiple identities, multiple Souls in one physical body can account for all symptoms. This concept is consistent with Pierre Janet's idea about DID. Pierre Janet opposed the view of Freud, who suggested that DID is secondary to the repression of a part of CS accompanied by amnesia (the subject is discussed in more detail in the Chapter on CS)

c) In Somnambulism/ sleepwalking, the existence of the second Soul is activated during sleep (normally, this second Soul is repressed by the principal Soul).

4. Search for constituents of the Soul.

There are about 200 elementary and composite particles in the Universe. The Universe is likely composed of

a) Matter:
-Baryonic Matter, 5% of the Universe, accessible to vision, touching, hearing, smelling, and tasting.
-Non-baryonic Matter or dark Matter not accessible to the five sensory organs but has the mutual attraction between all things

b) Force:
See Table 1.1 and 1.2: Forces in Nature.

Proposed Hypothesis (See Chapter VI): The Soul, a concept that has intrigued humanity for centuries, may challenge our current paradigms of the Universe (see paragraph 4).

Therefore, the Soul may be composed of one or more than one of the above components of the constituent of the Universe, given the fact that the Soul does not have electromagnetic energy but has the capability of moving, traversing other entities and gravity. Furthermore, in the brain, the Soul may have the additional capability to attach to neuronal synapses, which are polarized when the synapses are activated to create Memories. Therefore, the Soul could be composed of Dark Matter/ DM and Dark Force; Neutrino is a component of the Universe consistent with the properties of the soul. In addition, Neutrinos have the intrinsic nature to combine with electrons, which is comparable to a glue and electromagnetic force for variable function in the MM, CS in the brain, Chi/ Qi in acupuncture and martial arts, and in close-range healing with hand (vn: nhân điện). Neutrino is chargeless but can be attached to an electron with a negative charge. As a result, the Soul can be attached to the synapse with a positive charge when the synapses are *not* activated.

Given the role of the Neutrino in the Soul, which plays a role in registering, storing, and retrieving memories (see page 122), Neutrino is likely the elementary particle of Consciousness. Photons, neutrinos, and likely gravitons (hypothetical particles of gravity, particularly associated with Dark Matter) appeared in the early epoch of the Big Bang.

.F. Characterization Of Constituents Of The Soul.
There are about 200 elementary and composite particles in the Universe. The Universe is likely composed of
a) Matter:
-Baryonic Matter, 5% of the Universe, accessible to vision, touching, hearing, smelling, and tasting.
-Non-baryonic Matter or dark Matter not accessible to the five sensory organs but has the mutual attraction between all things
b) Force:
See Table 1.1 and 1.2: Forces in Nature.
Proposed Hypothesis (See Chapter VI): The Soul, a concept that has intrigued humanity for centuries, may challenge our current paradigms of the Universe (see paragraph 4).
Therefore, the Soul may be composed of one or more than one of the above components of the constituent of the Universe, given the fact that the Soul does not have electromagnetic energy but can move, traversing other entities and gravity. Furthermore, in the brain, the Soul may have the additional capability to attach to neuronal synapses, which are polarized when the synapses are activated to create Memories. Therefore, the Soul could be composed of Dark Matter/ DM and Dark Force; Neutrino is a component of the Universe consistent with the properties of the soul. In addition, Neutrinos have the intrinsic nature to combine with electrons, which is comparable to a glue and electromagnetic force for variable function in the MM, CS in the brain, Chi/ Qi in acupuncture and martial arts, and in close-range healing with hand

(vn: nhân điện). Neutrino is chargeless but can be attached to an electron with a negative charge. As a result, the Soul can be attached to the synapse with a positive charge when the synapses are *not* activated.

Summary of features of Neutrino:

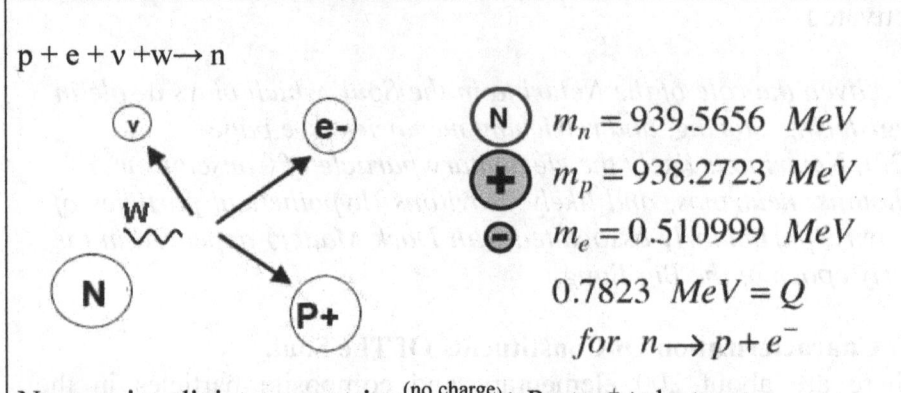

$p + e + v + w \rightarrow n$

$m_n = 939.5656$ MeV
$m_p = 938.2723$ MeV
$m_e = 0.510999$ MeV
0.7823 $MeV = Q$
for $n \rightarrow p + e^-$

Neutron is split into a neutrino$^{(no\ charge)}$+ Proton$^+$ +electron$^-$
W: energy released

Neutrino is not different from Antineutrino and can traverse through the baryonic matter without causing any reaction. Neutrino is of fermion particle type with spin ½, with tiny mass and chargeless
. There are three flavors (types):
- electron neutrino (νe),
- muon neutrino (νμ), and
- tau neutrino (ντ).
- The mutual transformation between flavors, a phenomenon called oscillation,
- Associated with weak force in short-range.
- Tau and Muon. Electrons can be associated with Neutrinos, but Tau and Muon have very short haft life.
- Neutrinos, with their unique properties, emerge as the most compelling candidate for a component of the Soul. The various combinations of the three types of Neutrinos could potentially represent different aspects of the Consciousness System, akin to the combination of lines in the I Ching. This connection could lead to a profound shift in our understanding of the soul, independent of the filtering effect of the brain, sensory organs, and environment.The composite is associated with a Negative charge when attached to an electron (or Muon, Tau).

· In the brain, particularly in the context of Memory and Mind (MM) formation, the Soul can be attached to synapses with a positive charge. In the depolarization of the synapses in nerve conduction, the Soul at the synapse site is negatively charged because the electron is attached to the Neutrino.
- In the body, Chi/ Qi in acupuncture represents the Soul of the non-neuronal body, or in Qi Gong.

It is well known in acupuncture, a Chi Gong, electrical charge can be measured at acupuncture points or in skin areas receiving the Chi Gong.

In summary, the hypothesis proposes that Neutrinos and Dark Matter form a composite that can be attached to electrons. This complex shares many features of the Soul, potentially opening up new avenues of understanding in the realms of physics and metaphysics.

2. Dark Matter (DM):

When observing the lensing of the light caused by the gravitational force making the light bend and the curvature of Space-Time, the lensing of the light exceeds the calculation based on the known gravitational force of the galaxy. DM is not present in the black Hole. When two galaxies collide, the DM is seen penetrating through the galaxy. In addition, the involved galaxy appears to be pulled along the path of the DM. In addition, DM is required to account for the galaxy's rotation.

In 1992, the project MACHO (Massive Astrophysical Compact Halo Object) failed to find a piece of DM about ten times the size of the Earth in the "Large Magellanic Cloud."

Despite the setbacks of the MACHO and EROS projects, the scientific community's unwavering belief in the presence of dark matter in our solar system, around the Earth, and inside the Earth is truly inspiring. https://www.nasa.gov/feature/jpl/earth-might-have-hairy-dark-matter, Monthly Notices of the Royal Astronomical Society, Volume 468, Issue 2, June 2017, Pages 1962–1980, https://doi.org/10.1093/mnras/stw3385, https://academic.oup.com/mnras/article/468/2/1962/2970349,

Despite the challenges in identifying dark matter on Earth, the overwhelming evidence of its existence, particularly through gravitational lensing effects, is convincing.

Since DM existence is still questionable, DM particles has remained unknown. Axion is a possible candidate particle v by ita very snall mass, spin of near zero (likely of meso particle type).

a) The Gravity associated with baryonic matter is not perceptible to the five sensory organs but causes indirect gravitational effects that are perceptible. Baryonic matter is an ordinary matter directly perceptible to humans and animals. The elementary particle of Gravity, the Graviton, accounting for this force is hypothetical and has not been definitely identified. The force is called Gravity, the attraction between objects, including the curvature of space-time. The light follows this curvature by bending its path.

b) Dark Matter (DM) is not associated with any mass like baryonic matter but has Gravity with a density five times that of the planet. Therefore, DM may attach to the planets, make them heavier, and bend the light more than expected with a planet without DM. DM can travel through the planets, as demonstrated with the telescopes. Whether or not DM exists in our galaxy is still debatable. DM accounts for 27% of the entire Universe.

The soul shares features of DM like its sensorial imperceptibility, ability to traverse through materials, and Gravity. In three separate instances, the Soul is associated with a gravity weighing up to 21.3 g in humansi

2. Dark Force.
Accounting for 68 % of the Universe.
And distinct from Dark Matter and Gravity. Einstein originally believed that the Universe is static. This is completely not reconcilable with the Far Eastern philosophy that everything in nature is moving as conceived in the I ching/ Book of Changes. Later, it was discovered that the Universe is expanding with acceleration. As a result, the Dark Force is the hypothetical candidate for this mechanism of Universe expansion (Bennett 2013, Planck Collaboration 2014 17a b).
The discovery of the Dark Force was a profound surprise to Astronomers, sparking a wave of intrigue and fascination. It provided a compelling explanation for the ongoing expansion of the Universe since its inception, a phenomenon that continues to astound us. The rate of this expansion, surpassing the speed of light, adds another layer of mystery to this cosmic puzzle.

The cosmic microwave background (CMB) is a testament to the early stages of the Universe, a designated electromagnetic radiation (Waves) created by photons before the formation of atoms and molecules. It is a constant reminder of the Big Bang, its remnants still present in the Universe, connecting us to its earliest moments..

3. Dark Matter in the Milky Way and Earth

he Gravity lensing phenomenon, a gravitational effect that causes light to bend around massive objects, reveals the presence of Dark Matter-attached planets and galaxies located in the spaces between galaxies, but not in the vicinity of black holes. This phenomenon is a crucial tool in the study of Dark Matter distribution in the Universe.

When galaxies collide, a mysterious phenomenon unfolds. The Dark Matter, seemingly unaffected, penetrates through the galaxy without disintegration or obstruction. However, the clump of DM in the collision moves at a slower pace than its counterpart galaxy, adding another layer of intrigue to this cosmic dance. Furthermore, the study of galaxy collisions has revealed that some galaxies are devoid of DM, a discovery that piques our curiosity and prompts further investigation.

DM mentioned in the Buddhist Sutra Agama or Nikaya. The sutra described the birth of a Bodhisatsava when descending from the Tushita Heaven—The Heaven-World in which the Buddha-to-be Maitreya waits for his coming,
, being ensouled into the mother and then born at the moment full of splendid light.
The light shines up on the dark space in the Universe and is accompanied by quakes and convulsions, as follows.

Long_Discourses_of_the_Buddha   (Digha_Nikaya)
Mahapadana Sutta: The Great Discourse on the Lineage

*It is the rule, monks, that when a Bodhisatta descends
from the Tusita heaven into his mother's womb, there appears
in this world with its devas, maras and Brahmas, its ascetics
and Brahmins, princes and people an immeasurable, splendid
light surpassing the glory of the most powerful devas. And
whatever dark spaces lie beyond the world's end, chaotic,
blind, and black, such that they are not even reached by the
powerful rays of sun and moon, are yet illumined by this immeasurable splendid
light surpassing the glory of the most
powerful devas. And those beings that have been reborn
there recognize each other by this light and know: "Other
beings, too, have been born here!" And this ten-thousandfold
world system trembles and quakes and is convulsed. And this
immeasurable light shines forth. That is the rule.*

Meaning: The light is very bright during Boodhisatsava descent, penetrating the dark area corresponding to the Dark Matter. They are also living sentients in there. The Dark Matter is likely a component of hell and realm og hungry ghosts where there is no light.

4. Dark Matter in Human beings.

In addition to the hypothetical proposal of Soul consisting of DM, DM may be associated with other structures.

The matter suggestive of DM has been recently described. Kinesin-1 constitutes the transporting agent in cytoplasmic vesicles in the neuronal dendrites. When examined under the microscope, the speed of transport in the neurons is much faster than the speed of transport of the same material without kinesin-1. In the cytoplasm, transport should be slower than in the milieu outside the cytoplasm because different organelles can interfere with the movement of transport. The additional DM in the cytoplasm renders the material heavier with greater momentum.

5. Hot DM, Warm DM.

Hot Darkmatter: Darkmatter+ Neutrinos.
Warm Dark matter: Darkmatter+ Hot matter.

So far, only observations through the phenomenon of Gravitational Lensing have confirmed the existence of DM. Many theories about hot, cold, and warm DM have been put forward, not according to the temperature but according to the speed of movement. Cold DM is heavy with a density five times that of the galaxy, whereas if hot DM is made of Neutrinos, the speed is close to that of light. The presence of warm DM is a possible instance of an intermediate body between cold DM and hot DM. However, the existence of hot, cold, and warm DM has not been confirmed yet

6. TRAVEL WITH SUPRALIMINAL SPEED , MIRACLES AND METAPHYSICAL REALM   50

It is a well-known concept in Buddhist sutras that Buddhas, Bodhisattvas, and the Spirit/Soul in the metaphysical realm manifest almost instantaneously at a meeting convened by Buddha Shakyamuni. This paragraph discusses this paramount issue. This book proposes an idea that the Soul is primarily composed of Dark Matter (DM), a concept that has historical roots in the early 20th century. This proposal, if supported by related facts or concepts, could reinforce our understanding of the Soul, its composition, and its manifestation in the metaphysical realm. Despite the elusive nature of DM in all investigations and the hypothetical status of dark photons, a particle of DM, according to special relativity theory, the possibility of particles faster than light has been repeatedly proposed, albeit mainly in fiction, since the early 20th century.

Recently, the hypothesis of the particle Tachyon, which is related to DM, re-emerged. Based on the equation:

$E = mc^2 / \sqrt{1-v^2/c^2}$

If $v=0$ (object not moving) $E=mc^2$, the famous Einstein formula

If $v > c$,

$\sqrt{1-v^2/c^2}$ is an imaginary number (represented as i with $i^2 = -1$).

It is currently believed that for E to be a real number, m should be an imaginary number. This intriguing relationship, where the faster an imaginary m travels, the smaller the energy E becomes, presents a mystery that does not align with the current concept of physics and spirituality.

It is also commonly believed that an imaginary number is an integral part of an entity, although this part is not within the domain of attention and, therefore, beyond consciousness (CS). For physicists, reality consists of all data accessible to the CS, and unreal or illusional/imaginary data are not accessible to the CS (meaning unmeasurable, unobservable....by five sensory organs).

As pointed at the beginning of this book, everything exists in its wholeness, including the physical and metaphysical parts. Therefore imaginary number:

$\sqrt{1-v^2/c^2}$ may represent entities in the metaphysical realm. In the metaphysical realm, the Soul, which is mainly composed of DM, is imaginary to physicists; the energy E associated with the Soul is characterized by the CS, which is also Karma. Karma manifests by its karmic energy. However, in the metaphysical realm, the Soul with its DM is suspiciously associated with gravity, which accounts for the real mass m. Therefore mi=m (imaginary mass). As a result, E in

$E = mc^2 / \sqrt{1-v^2/c^2}$

must be imaginary. One can consider Ei to replace E in the above equation, that is, imaginary; Ei can be expressed as:

$$Ei = mc^2 / \sqrt{1-v^2/c^2}$$

ccording to Einstein's Relativistic theory, the photon represents the hallmark particle without mass (m=zero) (and is associated with a travel speed of 300 km/msec, representing the upper limit of all speeds in this Universe). If one considers the zero mass as an assumption that results from the limited capability of human brain's consciousness, the zero mass only represents the inability of the human brain to this measurement rather than the reality of the Creation. In other words, zero mass of photons represents the limit of the physical world, a limit that is defined by the capabilities of our human brain. Beyond this zero limit is the metaphysical realm. The reason is that nothing can be measured below m of the photon, which does not mean nothing exists

Therefore, in keeping with Einstein's Relativistic theory, m or E beyond this human CS limit is called imaginary (or illusional). (Of note: m of the Soul is both real because it is hypothetically measurable and unreal because it is still not accessible to the CS)

Table: Grading the soul in different metaphysical and physical realm according to Karmic energy and corresponding Soul speed

Immateral Realm / No form/No Beauty/pure spirit	Realm of Form or Matter/Beauty	Realm of Passion		
Metaphysical	Meta physical	Metaphysical	Physical & Metaphysical	Meta physical
Dhyana:4. Arahat Buddha	Dhyana: 1,2,3.	Asura, Deva (state of god)	HUMANS, Animals	Hungry Ghosts, Hell
No more Karma	Decreased Karma ←		Real energy and karma For living	→ increased karma
v↑ with Es↑ and	v↑ with Ei ↓			v↓ with Ei↑

v supraluminal velocity of the Soul, Es Energy of the Soul, Ei: imaginary Energy or karma

In the metaphysical realm, if m represents the mass of the Soul, Ei must represent the Karmic Energy. Karmic energy, or simply Karma, is the critical factor that drives the Soul in the transmigration, the motivation of sentient beings in action in living, and exerts influence on the energy for the soul travel. According to Buddha, in worldly life, sentient beings continuously create and accumulate Karma, which results in the inversion or turmoil of the Creation. Therefore, Karma, which is nothing but consciousness, a veil of ignorance, deviates, distorts, and obliterates Buddhdhood and Omniscience. As a result, Karmic energy renders the Soul downgrading on the spiritual scale. One can express this phenomenon of obliteration by this equation with Eo, the original energy assigned to each Soul, and Es, the current energy of the Soul after the obliteration by the Karma or Consciousness

$$Es = Eo - Ei = Eo - mc^2/\sqrt{1-v^2/c^2}$$

The equation denotes that entities with higher energy travel faster. The more enlightened the spirit, the more energy it has and the faster it travels. This could have profound implications for our understanding of energy, spirit/Soul, and enlightenment. Supraluminal travel only applies to entities in the imaginary/metaphysical realms. In the physical realm, this type of supraluminal communication may be possible in humans with high levels of enlightenment whose spirit can detach from the physical body (similar to cases of out-of-body experience). Nevertheless, non-enlightened humans can also communicate with bodiless souls or spirits, as in instances of mediumship. It's important to note that the power of meditation is not limited to the enlightened. It is a tool that can be used by all to reveal deeper truths and enhance spiritual communication.

In the above equation, if v decreases to c, Ei becomes very high in value and will obliterate the Buddhahood/Holy Spirit. If v approaches zero value, Ei=E=mc2). This is consistent with the fact that Karma originates from Physical material and is the direct cause of Greed and Anger/ Renouncing worldly things is the decent road to the Spiritual life. In other words, Karmic energy is materialistic. The evidence is overwhelming: Buddha left his golden throne, palace, precious stones, and jewelry to live in the forest and beg for daily food, just enough for survival. In the Old Testament, the fact that Adam and Eva, who ate the fruit in the middle of the Eden garden, were expelled from heaven represents the message of the Bible, which instructs that Materialistic things are the boulevard to Hell. The metaphysical world, which consists of the Realm of no Form, the Realm of Form, and part of the Realm of Passion, is better understood through the lens of the equation. Only the Physical Realm to which humans and all living being belong, has photons. The Hell represents the Realm of Invisible Form, inaccessible to the five sensory organs due to the lack of photons and is dark due the obliteration of the heavely light by the high level of Ei (karma). The Hell likely belongs to the metaphysical Realm. The metaphysical Realms do not have photons and are likely characterized by other particles carrying the energy (for example, the hypothetical

Dark photons or Tachyons). The later particles do not interact with photons but use other gauge systems. As a result, dark matter and dark photons can not be characterize, but they can indirectly affect the photons, like deviating from visible light or interacting with the Physical Realm through the brain at the level of neuronal synapses. As a result, visualization of the metaphysical realm is not possible through sensory organs but only by discarding the veil of ignorance (example: through meditation).

Please see more on pages 64. 136, 275-278

Diagram Showing the Relationship of Buddhahood, Karma, and Velocity in Different Realms of the Creation

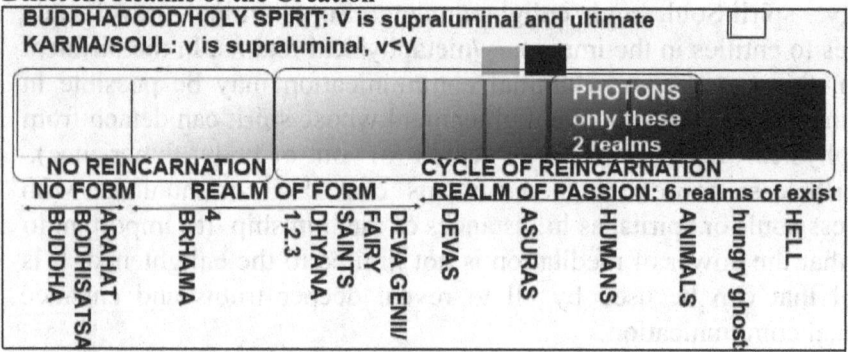

Darker shade, lower supraluminal velocity v

MIRACLES IN THE METAPHYSICAL REALM
Examples of spiritual phenomena
• There are many, such as Nostradamus, the prophets of the time, and the pandemic, but it's hard to verify.
• Cases of finding the graves of the victims of the Vietnam War such as Mrs Phan thị Bích Hằng, Hòang thị Thêm, Đoàn Việt Tiến .... and many others.
• The phenomenon of out-of-body experiences and near-death experiences have many scientific foundations. Please see the book of related sections.
• Jesus walking on the water, making wine, fish, bread, curing the blind.
• Buddha went up to Heaven to preach Sutra to his mother and countless untold landscapes. Bodhisatsava Nagarunja went to Oceanic Dragon Palace for three months.....
• A case story on CNN.
Baby who survived car crash into Utah river gets better | CNN

The unenlightened Soul can enter the normal human brain to convey information. It's called illusion, but events are both a spiritual and a materialistic process.

In Buddhism, at great meetings such as in Lotus, Flower Adornment, Surangama sutras..., Buddha also helps non-enlightened people such as Ananda or little enlightened people to see and hear other Buddhas coming from all over the Universes/Multiverse; the process is to help the brain of ordinary people to contact with the Spirits in the high World.

In the phenomenon of Our Lady Mary, Spiritual Mother of Fatima, in Portugal or in La Vang Quang Tri Viet Nam, Our Lady has manifested Herself in the brains of those who need to be seen. A video or audio recorder cannot record expressions such as those mentioned above. Similarly, if the UFO status is real, only the person who sees can know but can not record images or sounds by machine.

In the case of Jesus resurrecting Lazarus after four days of death, the Lord (God manifested himself) integrated the image of Lazarus into the brains of the witnesses, including the last time Lazarus sat at the final dinner table with the Lord. Of course, Lazarus' body is dead, not living like a normal man without God. Jesus said Lazarus was a good man who didn't deserve to die. Death here is spiritual blindness, so there will be life in Heaven.

Sound light and objects such as gemstones, diamonds, castles, and other objects are also expressed according to the laws of Creation making the Universe.

In summary, a miracle is an event that surpasses all known human or natural powers and is therefore attributed to a supernatural, divine, or spiritual cause. It is the ability of the Consciousness that humans with full knowledge can hardly understand. But the mechanism of the miracle is also governed by the laws of physics extending into the metaphysical Realm, which can be modified slightly so it may differ slightly from the physical Realm but must be based on the fundamental principles of Creation. The decomposition and synthesis of the molecules/atoms of the body of the dead must follow the process of biochemistry, physics, and spiritual attachment to the body. Buddha often said the mind cannot turn sand into rice. Similarly, the power of prayer can only be expressed through forms of reality; for example, sick prayers should often be saved through medical treatment or natural physiopathology.

The metaphysical Realm, with its supraluminal velocity and Karmas, confers a miraculous impression. Despite the opposite, the two realms flow seamlessly and continuously; the interruption is a hallucination caused by the brain specifically constructed for this World. In this Universe, photons constitute elementary particles and are used as an assessment tool. It is well known that photons are not the only the scale of Creation. The metaphysical Realm has no photons, so Consciousness cannot perceive it. Because the metaphysical Realm has no photons, humans of this physical Realm cannot recognize the Metaphysical Realm. But the Metaphysical Realm also has mountains but higher (like the Neru mountain), there is the sea should have Oceanic Dragon King, mentioned in the Lotus or the Flower Adornment, continent with beautiful flowering trees, there are diamonds but more beautiful, the scenery is more magnificent, there is a more magical sound..... That's because, in the Metaphysical Realm, the speed is faster, the energy is bigger, the structure is more sophisticated, and the life of the creature is longer. As previously stated, the brain, along with the neural synapses, is the most important place for communication with the Soul. The Soul belongs to the Metaphysical Realm. Therefore, the brain is an important intermediary between the two realms and helps humans know the presence of the Metaphysical Realm through a phenomenon known as miracles. This mysticism needs to be seen as the natural reality of the Metaphysical Realm, enlightening us and expanding our knowledge.

The scientific investigation of the Metaphysical Realm very much found only indirect evidence. The indirect evidence is that we have known through many spiritual phenomena that science always denies because of prejudice and hypocrisy. The desire for the discovery of direct evidence of the Miracles in the Metaphysical Realm

The desire for the discovery of direct evidence of the Miracles in the Metaphysical Realm

Examples of spiritual phenomena

• There are many, such as Nostradamus, the prophets of the time, and the pandemic, but it's hard to verify.
• Cases of finding the graves of the victims of the Vietnam War such as Mrs Phan thị Bích Hằng, Hòang thị Thêm, Đoàn Việt Tiến .... and many others.

- The phenomenon of out-of-body experiences and near-death experiences have many scientific foundations. Please see the book of related sections.
- Jesus walking on the water, making wine, fish, and bread, and curing the blind.
- Buddha went up to Heaven to preach Sutras to his mother and countless untold landscapes. Bodhisatsava Nagarunja went to the Oceanic Dragon Palace for three months.....
- A case story on CNN.

Baby who survived car crash into Utah river gets better | CNN (copy and reprint in Google) Baby who survived car crash into Utah river gets better | CNN

The unenlightened Soul can enter the normal human brain to convey information. It's called illusion, but events are both a spiritual and a materialistic process.

In Buddhism, at great meetings such as Lotus Flower Adornment Surangama, Buddha also helps non-enlightened people such as Ananda or little enlightened people to see and hear the Buddhas coming from all over the Universes/Multiverse; the process is to help the brain of ordinary people to contact with the Spirits in the high World.

In the phenomenon of Our Lady Mary, the Spiritual Mother of Fatima, in Portugal or in La Vang Quang Tri Viet Nam, Our Lady has manifested Herself in the brains of those who need to be seen.

A video or audio recorder cannot record expressions such as those mentioned above. Similarly, if the UFO status is real, only the person who sees can know but can not record images or sounds by machine.

In the case of Jesus resurrecting Lazarus after four days of death, the Lord (God manifested himself) integrated the image of Lazarus into the brains of the witnesses, including the last time Lazarus sat at the final dinner table with the Lord. Of course, Lazarus' body is dead, not living like a normal man without God. Jesus said Lazarus was a good man who didn't deserve to die. Death here is spiritual blindness, so there will be life in Heaven.

Sound light and objects such as gemstones, diamonds, castles, and other objects are also expressed according to the laws of Creation making the Universe.

In summary, the miracle is the ability of the Consciousness that humans with full Knowledge can hardly understand. However, the mechanism of the miracle is also governed by the laws of physics extending into the metaphysical Realm, which can be modified slightly so it may differ slightly from the physical Realm but must be based on the fundamental principles of Creation. The decomposition and synthesis of the molecules/atoms of the body of the dead must follow the process of biochemistry, physics, and spiritual attachment to the body. The Buddha often said the mind cannot turn sand into rice. Similarly, the power of prayer can only be expressed through forms of reality; for example, sick prayers should often be saved through medical treatment or natural physiopathology.

This metaphysical Realm may be the opposite of this World through the phenomenon of supraluminal velocity and Karmas conferring the miraculous impression of the metaphysical Realm. Despite the opposite, the two realms flow seamlessly and continuously; the interruption is a hallucination caused by the brain specifically constructed for this World. In this Universe, photons constitute elementary particles and are used as an assessment tool. It is well known that photons are not the only the scale of Creation. The metaphysical Realm has no photons, so Consciousness cannot perceive it. Because the metaphysical Realm has no photons, humans of this physical Realm cannot recognize the Metaphysical Realm. But the Metaphysical Realm also has mountains but higher (like the Neru mountain), there is the sea should have an Oceanic Dragon King, mentioned in the Lotus or the Flower Adornment, a continent with beautiful flowering trees, there are diamonds but more beautiful, the scenery is more magnificent, there is a more magical sound..... That's because, in the Metaphysical Realm, the speed is faster, the energy is bigger, the structure is more sophisticated, and the life of the creature is longer. As previously stated, the brain, along with the neural synapses, is the most important place for communication with the Soul. The Soul belongs to the Metaphysical Realm. Therefore, the brain is an important intermediary between the two realms and helps humans know the presence of the Metaphysical Realm through a phenomenon known as miracles. This mysticism needs to be seen as the natural reality of the Metaphysical Realm.

The scientific investigation of the Metaphysical Realm very much found only indirect evidence. The indirect evidence is that we have known through many spiritual phenomena that science always denies because of prejudice and hypocrisy. The desire for the discovery of direct evidence of the Miracles in the Metaphysical Realm is illusional if not through the revelation in Meditation or in a state of special ability to communicate with the Soul. Since the suspicious Dark matter is the substance that forms the foundation for the photon-free Soul, the same goes for the Dark Force that occupies more than two-thirds of the Universe. Artificial intelligence (AI) today helps people advance technique, but the speed of light limits it so it is impossible to go beyond this limit to enter the spiritual World. Spiritual practice and Meditation are the ways back home of Creation that can help people understand the root of Creation and truly be happy.

CREATION OF THE PHYSICAL VERSUS METAPHYSICAL REALMS

According to the duality principle, Yin and Yang were formed in the post-creation era. Major Yin is represented by the metaphysical Realm, which is represented by the Dark Force (68% of the Universe), with Minor Yang represented by the Dark Matter, which is formless but has gravity (27% of the Universe). Major Yang is represented by the Physical Realm (5% of the Universe and Minor Yin is represented by the Consciousness, which is formless but is associated with neuronal synapses. Only the Major Yang component has photons that enable the sensory organs to recognize the materialistic Creation solely. This privilege of posting photons in the physical World also impedes the five sensory organs from identifying the metaphysical Realm.

The sentient Consciousness does not recognize the metaphysical World because it is in the physical Realm. Photons are the elementary particles of matter acting as the veil of ignorance **(incomplete and deviated CS)**. The fact that the privilege of having photons is limited to 5% of the Universe can be speculated as follows: This physical World specializes in spiritual training (moral practice: Four immeasurable Minds of Virtue). It can be said that in the process of Creation, because of the false thought (that initiates the Creation/Big Bang) the inversion/chaos is generated because there is Inversion, so there is a need to establish a classroom to repair, eliminate the bad Karma in the Inversion. (spiritual training or practice of Morality: Four Immeasurable Minds of Virtue)

. In other words, the Big Bang forms the Dark Force (which can be called the Transparent Force) and the Dark Matter (the Transparent

Matter). Due to errors in the Creation/Big Bang, photons are created. The **Old Testament provides an example:** Because Mr. and Mrs. Adam-Eva made the mistake of not listening to God's prohibited word, they were sent to a re-education camp to study and reform their behavior under harsh living conditions. They could no longer see Paradise. Photons obliterate black matter (transparent matter) and black force (transparent force) (representing Paradise) in this same way.

The spiritual path of perfection is the pathway of discarding photons, or matter. This concept, akin to essential teachings in major religions, is a profound and transformative journey. In Buddhism, the recognition of sufferance is the first step towards seeking the spiritual pathway of annihilating it, as outlined in the Four Noble Truths. In Islam, living is the Path of Perfection, a concept expounded in the book written by Bahram Elahi. As discussed in another section, the Metaphysical Realm, the transformed Emptiness following False Thought/Big Bang, is our Permanent Home. The Supraluminal velocity is a token of the World with miracles, a realm of unthinkable energy, unimaginable ingenuities, and extraordinarily beautiful scenes and sounds. This is the Heaven or Nirvana that humans often conceive. With such a conception, this earthly World is nothing but a camp of labor that humans, for innumerable lives, misunderstood as the stage of pleasure and happiness.

DARWINISM.
1. The theory.

The book of Darwin (Charles Darwin 1809- 1882) entitled "On The Origin of Species," published in 1859, is about the origin of the earthly life. In his book, Darwin's opinion contradicts the previous common opinions that favor life incidentally develops. According to scientific theory, the epistemological (knowledge–related) development of life is neither accidental nor theological. The species develop in the direction to conform/ adapt to the environment. *Those with an appropriately high capability of adaptation will prevail over those with a low potential for adaptation...* With evolution lasting for billions of years, the adaptation and the differentiation resulting in the selection mechanism render different species different from each other.

Recent understanding of the epigenetic phenomenon (related to the expression of genes without mutation of the gene) has added an important mechanism of adaptation by bypassing gene mutation.

In this mechanism, alterations of histone, type of protein supporting chromosomes, and genes can modify the expression of the genes. This type of adaptation, equivalent to learning, can pass from parent to children and shorten the time course of evolution from millions of years to months/ and years, from many generations to parents-children generations. Children of parents with poor nutrition tend to develop diabetes mellitus.

2. Social Darwinism.
On a larger scale, Darwinism can apply to societies or countries. In society, ethnic groups with a high potential for adaptation tend to be more prosperous than other groups.

Comments: Darwin's work was well-received by the scientific community of his time and many of his findings remain relevant today. However, the role of chance and natural selection, while central to his theory, is difficult to scientifically confirm beyond his initial findings. The challenge of proving or disproving the role of chance and contingency in the creation of human beings is particularly complex. This scholastic phenomenon is akin to a handful of sand containing granules of different sizes, each of which is independent of the others but still contributes to a continuum of change. This phenomenon is purely a result of chance, not selection. Furthermore, in Darwinism, it seems that Darwin tried to distance himself from Genesis/ the Creation of the Universe, probably the domain of religion at his time.

3. Origin of life.
In the 1950s, Urey, a Nobel Prize Laureate in Physics and Chemistry, and Miller, a young Ph.D collaborator, conducted a groundbreaking experiment. They successfully synthesized organic compounds from mineral materials, a feat that would significantly impact our understanding of the origins of life. Their experiment,

which simulated the conditions of the primitive early Earth, supported the hypothesis of Alexander Oparin. Oparin suggested that complex organic compounds are synthesized due to favorable conditions, a gradual evolution from inorganic to organic compounds under an oxygen-poor atmosphere.

Haldane, a prominent figure in the field, made a significant prediction in the USA. He foresaw that organic compounds of intermediate molecular weight represent the transitional form between mineral compounds and viruses. This prediction, made in 2003, set the stage for further research and discovery in the study of organic compounds.

The virus was first discovered in 1917.

These findings not only provide a pathway for the evolution of inorganic to organic compounds but also highlight the crucial role of ribosomes. These microscopic granules of RNA play an intermediate step in the synthesis of protein from DNA, marking a significant step in the evolution of organic compounds.

In the context of the creation of the Universe, a thought-provoking question arises: which came first, DNA or protein? This query, akin to the classic 'chicken or egg' dilemma, challenges our understanding of the origins of life and the evolution of organic compounds.

4. The Cell Theory.

According to Rudolf Virchow and Louis Pasteur, life only starts from another life. This leads to the problem of searching for a life that can originally start. *Scientists look for life in other parts of the universe to unlock the mystery or even for life to emigrate to our planet.* For all living organisms, the cells are relatively uniformly composed of:

i. Cell membrane forming the boundaries of individual cells.
ii. Cytoplasm containing organelles for reproduction, development, and basic metabolism, and
iii. Coils of DNA for reproduction and synthesis of proteins.

5. Gap in the evolution.

According to Darwinism, humans, Chimpanzees, and Bonobos share a common "ancestor" (regarding the physical similarities), that is, Hominini based on DNA and gene studies. Despite the close relationship between the physical components, the metaphysical part is completely different. Furthermore, there is a remarkable gap in the mental status between chimpanzees and humans, not only in emotion, thinking, and creativity but, importantly, in morality.

```
12millions    millions    millions years  (in Africa)
   ⇩             ⇩           ⇩
Great Apes→ Homonids→Homoninae →Gorillas
   ↘           ↘ Hominini   →Chimpanzee + Bonobo
Orangutans                ↘ Homo Sapiens
GAP BETWEEN HUMAN-ANIMAL  ⇧ is MORALITY
```

6. The first Man and Woman (suggestive of Adam and Eva in the Paradise in The Old Testament are doubtful) ?

Mitochondrial DNA (mtDNA) in men and women (mitochondria are considered the cell's powerhouse) is of maternal origin since, at fertilization, only the spermatozoa's nucleus enters the ova. Analysis of mtDNA showed that all human mtDNA share the same mtDNA of the same ancestor woman (hence named mtDNA Eva) living in Africa. Analysis of chromosome Y containing gene SRY activated in men also demonstrated that all chromosome Y share the same Y chromosome Y of the ancestor man.

Since 1985, these profound questions have been the subject of extensive discussion and research, leading to significant advancements in our understanding of human evolution.

The name 'Eve' is not a historical reference, but a symbol. It's important to note that 'Eve' may not have been the only woman of her time. Adam, her male counterpart, could have existed before or after her. Additionally, mtDNA can be found in chimpanzees, bonobos, and even in animals of lower phylogenetic evolution.

Our ancestors embarked on a remarkable journey, migrating from Eastern Africa to Europe, Asia, and North America when these continents were still connected to Asia.

The above pathway of evolution and development can be found in the following Sermon in Nikaya Sutra, chapter 27/ *Aggañña Sutta: On Knowledge of Beginnings (Long Discourses) (page 52)*:

Buddha 's teachings are summarized as follows
:

> ..*And sooner or later, after a very long period of time, savoury earth spread itself over the waters where those beings were. It looked just like the skin that forms itself over hot milk as it cools. It was endowed with colour, smell and taste. It was the colour of fine ghee or butter, and it was very sweet, like pure wild honey.*
> *12. 'Then some being of a greedy nature said: "I say, what can this be?" and tasted the savoury earth on its finger. In so doing, it became taken with the flavour, and craving*

> arose in it. Then other beings, taking their cue from that one, also tasted the stuff with their fingers. They too were taken with the flavour, and craving arose in them. **So they set to with their hands, breaking off pieces of the stuff in order to eat it. And [86] the result of this was that their selfluminance disappeared.** And as a result of the disappearance of their selfluminance, **the moon and the sun appeared**, night and day were distinguished, months and fortnights appeared, and the year and its seasons. To that extent the world re-evolved. 13. 'And those beings continued for a very long time feasting on this savoury earth, feeding on it and being nourished by it. And as they did so, their bodies became coarser, and a difference in looks developed among them. Some beings became good-looking, others ugly. **And the good-looking ones despised the others, saying: "We are better-looking than they are." And because they became arrogant and conceited about their looks, the savoury earth disappeared.** At this they came together and lamented, crying: "Oh that flavour! Oh, that flavour!" And so nowadays when people say: "Oh that flavour!" when they get something nice, they are repeating an ancient saying without realizing it. 14. 'And then, when the savoury earth had disappeared, a fungus cropped up, in the manner of a mushroom. It was of a good colour, smell, and taste. It was the colour of fine ghee or butter, and it was very sweet, like pure wild honey. And those beings set to and ate the fungus. And this lasted for a very long time. **And as they continued to feed on the fungus, so their bodies became coarser still, and the difference in their looks increased still more.** And the good-looking ones despised the others ... **And because they became arrogant and conceited about their looks, the sweet fungus disappeared.** Next, creepers appeared, shooting up like bamboo..., and they too were very sweet, like pure wild honey. 15. 'And those **beings set to and fed on those creepers. And as they did so, their bodies became even coarser**, and the difference in their looks increased still more... And they **became still more arrogant**, and so the creepers disappeared too. At this they came together and lamented, crying: "Alas, our creeper's gone! What have we lost!" And so now today when people, on being asked why they are upset, say: "Oh, what have we lost!" they are repeating an ancient saying without realizing it. 16. 'And then, after the creepers had disappeared, **rice appeared in open** spaces, free from powder and from husks, fragrant and clean-grained. And what they had taken in the evening for supper had grown again and was ripe in the morning, and what they had taken in the morning for breakfast was ripe again by evening, with no sign of reaping. And these beings set to and fed on this rice, and this lasted for a very long time. And as they did so, their bodies became coarser still, and the difference in their looks became even greater. And **the females developed female sex-organs, and the males developed male organs.** And the women became excessively preoccupied with men, and the men with women. Owing to this excessive preoccupation with each other, passion was aroused, and their bodies burnt with lust. And later, because of this **burning, they indulged in sexual activist.** But those who saw them indulging threw dust, ashes or cow-dung at them, crying: "Die, you filthy beast! How can one being do such things to another!" Just as today, in some districts, when a daughter-in-law is led out, some people throw dirt at her, some ashes, and some cow-dung, without realising that they are repeating an ancient observance. What was considered bad form in those days is now considered good form. . 'And **those beings who in those days indulged in sex were not allowed into a village or town for one or two months**. .....

Meaning: ....And sooner or later, after a very long period of time, savoury earth spread itself over the waters where those beings were. It looked just like the skin that forms itself over hot milk as it cools... 12.

'Then some being tasted the savory earth on its finger. ... **So they set to with their hands, breaking off pieces of the stuff in order to eat it. ... The result of this was that their self luminance disappeared.**

As a result, **the moon and the sun appeared with**, night and days, months, year and its season: The world re-evolved.

The faces and bodies became coarser, and a difference in looks developed among them. Some became good-looking, others ugly. **Discrimination and arrogance developed, respectively. Then, females and males developed sex organs, developed and indulged in sexual activity.** Successively, different types of crops appear

H. SPINOZA AND METAPHISICAL CONCEPT,

Baruch Spinoza (24 November 1632 – 21 February 1677) was a Dutch philosopher of Portuguese Jewish descent, a great thinker on metaphysics, epistemology, political philosophy, ethics, philosophy of Mind philosophy of science. , and religion with an ontological view. In this view, God is represented by Substance and infinite Attributes and Modes. Attributes are infinite in form and formless state, and mode is the manifestation of attribute. Mode in Spinoza concept is equivalent to form and formless entities in Buddhist concept of five skandhas. The substance is defined by itself and created by itself. This concept of being uncreated and therefore imperishable is very similar to the Chinese philosopher Lao Tsu's The Emptiness of Buddha, with the concept in Buddhahood including everything:"*small, not excluded, and prominent also included*". Spinoza stated: "Whatever is, is in God, and nothing can exist or be conceived without God". The Metaphysical Realm, conceived by Spinoza, is very similar to the Universe. Such a concept is contradictory to the current opinion of his time. Therefore, at the age of 23 years, the Talmud Torah congregation of Amsterdam issued a writ of expulsion or ex-communication) against Spinoza.

I. DAVID BOHM CONCEPT OF IMPLICATE ORDER.

David Bohm is an American theoretical physicist who first worked at Princeton University under Einstein. Due to his communist ideology, he moved to Brasil, Israel, and finally to England, where he became a professor of physics; besides his accomplishments in Physics theory, he had orthodox deterministic ideas on quantum mechanics (De Broglie–Bohm theory). The view of nature is similar to the Cartesian view of the duality of body and Mind but more advanced by proposing the explicate order representing the outer world, implicate order representing the body, and the super implicate order representing the Mind. Bohm devised a continuum between these states of order with the enfolding, the manifestation in the outer world in the body and the Mind. The unfolding represents the manifestation of the Mind and consciousness.

Bohm's concept of the implicate and explicate orders can be compared to Buddha's teaching in the Surangama Sutra, particularly in their shared emphasis on the interconnectedness of all things.

The concept of implicate order and its connection to neurophysiology has not been widely embraced in the scientific community. Physicists, for instance, are primarily interested in theories that can predict measurements, and the implicate order does not readily lend itself to such predictions. Similarly, psychologists and neuroscientists may be hesitant to embrace a theory that connects physics with the mind and consciousness, as it challenges traditional disciplinary boundaries.

Nevertheless, the theories are very close to the revelation of Buddha regarding Consciousness. The Surangama sutra, the concept of visual Consciousness consisting of perception, feeling, and integration, are continuous processes connecting the original/Ultimate Omniscience (vn: Trí huệ bát nhã) and the object. This is comparable to the operation of unfolding and enfolding of the explicate and implicate order: in this analogy, the five skandhas (Form, Perception, Feeling, Formation, and Consciousness), Form represents the explicate order, the four remaining skandhas describe the process of the Mind corresponding to the implicate order.

Of course, Bohm is not a Buddhist student, unlikely understands that consciousness is distinct from the original/ transcendental omniscience..

Bohm's theory, which is based on the quantum interpretation of particle and wave, suggests that consciousness is not a separate entity but is deeply intertwined with the physical world. In this view, the wave informs the particle to make it move and rotate, much like a radio frequency feeds a particle as a radio receiver. This perspective has profound implications for our understanding of consciousness, suggesting that it is not a purely mental phenomenon but is intimately connected to the physical world.

The body is considered as a bridge between the explicate, society, and nature, and the implicate, super implicate representing the Mind. Like Buddhism, Bohm conceived that any space of the Universe is not void of anything but contains the superimplicate order that unfolds into the explicate order.

While Bohm's theory of Consciousness is similar to the Buddhist concept, it is not fully compatible. In Buddhism, Consciousness is often described as a veil of ignorance, a perspective that differs from Bohm's interpretation. This difference underscores the complexity of the topic and the need for further exploration and discussion.

J. EMPTINESS POSTULATE, EMPTINESS IS BUDDHAHOOD

1. Buddhist Sutra.

Buddha said in the following paragraph of Surangama Sutra: if one divides an object such as a block of stone millions of times, the result is innumerable infinitesimal particles of structure smaller than quantum particles of the size of the hypothetical superstring of 10^{-34} cm value of Plank constant) or even possible further division. One finally attains the status of Emptiness, which is imperceptible to CS. The process is analogous to the splitting of atoms in the production mechanism of an atomic bomb. As a result, Emptiness is empowered with tremendous energy. Along with the creation of energy, Omniscience is also created according to the rule of Duality, which is the principle that all things have two sides or aspects, and these two sides are interconnected and interdependent.

Therefore, Form and Emptiness are identical in nature. Due to the veil of ignorance of Humans at his time, Buddha bypassed the explanation of the chain reaction in the formation from Emptiness to subatomic particles to atoms and molecules.

.........Look at the element of earth which ranges in size from the great earth to a tiny speck of dust. Split this speck which is near to nothing and reduce it to the finest mote on the extreme border of form. Then split it again and it becomes the void.

Buddha continue explaining that if the splitting keeps going on infinitely, to reduce to nothing (imperceptible to five sensory organs), the Form become the void. As a result **in the Tathàgata store(Buddhahood) both form and (its opposite) the void arise from self-nature and are identical with each other,** as read in the following paragraph

....Look at the element of earth which ranges in size from the great earth to a tiny speck of dust. Split this speck which is near to nothing and reduce it to the finest mote on the extreme border of form. Then split it again and it becomes the void. Ananda, if this mote can be reduced to nothing, you should know that form comes from the void. You now ask about material changes which you attribute to the mixing and uniting (of the four elements). Take, for instance, this mote which is nearest to the void; how much voidness should be mixed and united to produce it? But it is absurd to suppose that this can be done by uniting motes. Since a mote can be split and reduced to Emptiness, how many (particles of) form should be fused together to create the void? The union of form (with form)

> *produces form but not voidness, and the union of the void (with the void) produces voidness but not form. Form can be split up but how can the void unite (with form)?* . ***You do not know that in the Tathàgata store both form and (its opposite) the void arise from self-nature and are identical with each other,*** *and that the element of earth is fundamentally pure and clean, embraces all in the Dharma realm and manifests because the minds of living beings know and distinguish (between things) in accordance with the laws of karma. Ignorant wordlings wrongly attribute this to cause, condition and the state of the self as such, because **their consciousnesses differentiate and discriminate without their knowing that the language they use has no real meaning***.
>
> *You do not realize that in the Tathàgata store both fire and (its opposite) the void arise from the self-nature and are identical with each other, and that the element of fire is fundamentally pure and clean, embraces all in the Dharmarealm and manifests because the minds of living beings know and distinguish (between things). ânanda, you should know that fire is produced wherever a man holds a mirror (in the sun), and that if mirrors are held up throughout the Dharma-realm, fire will spring up everywhere in accordance with the laws of karma and not in a given place and direction.Ignorant worldlings wrongly attribute this to cause, condition and the state of the self as such without realizing that it is because their consciousnesses differentiate and discriminate and that the language they use has no real meaning. ... Magicians obtain water to mix with their medicines by exposing a crystal ball to the full moon. Does this water come from the ball, the void or the moon?....*

Of interest, Buddha pointed out that *element of earth is fundamentally pure and clean, embraces all in the Dharma realm and manifests (because form and void are identical)* but ***their consciousnesses (veil of ignorance) differentiates and discriminates without their knowing that the language they use has no real meanin. Similarly*** *fire and (its opposite) the void arise from the self-nature and are identical.*

In the example of transformation of Form into EM given by Buddha, there are release of characteristic converning the Form:
   - Force or energy attaching speckles of dust together
   - In the example of the transformation of Form into EM given by Buddha, there is the release of components of laws that characterize the construction of the Form:

   Form   - Force or energy attaching speckles of dust together
      - Architecture in the formation of the Form. Omniscience represents the summation of different types of architectural laws. As a result, Ultimate Omniscience can be represented by a complete set of rules, laws, theorems, and axioms, such as those present in biology, physics, mathematics, and psychology...: from Euclid, Archimedes, Newton, Einstein, Bohr (Quantum interpretation), DNA sequences, Darwin's evolution, Induction, Morality, Beauty, Magnificence .... This set of universal rules is the noumenon of EM.

Note: The concept of Emptiness in Buddhism is very similar to the concept of Holloness in Taoism

In Taoism of Lao Tsu:
The Tao is an empty vessel; it is used but never filled
Oh, unfathomable source of ten thousandthing!
.... Hidden deep but ever present
.... The space between heaven and erath is like a bellows
The shape change but not the form, The more it moves the more it yields

and particularly similar to Fullness in Hinduism (Upanishad):
All this is full, All that is full
From fullness, fullness comes
When fullness is taken from fullness
Fullness still remains (note : This represents individual being, That represents Brhama
Fullness in Hinduism obviously reflects theconept of origin of the Universe in the later stage of the development of False Though

In summary, splitting Great Earth generates tiny specks of dust, that are imperceptible, equivalent to the Void/ Emptiness. Therfore Form and Emptiness are identical in Boddhahood, but non identical for the human discriminative Mind.

For the generation of the fire (or water), the same mechanism is applied: Fire (or water) are identical. Fire ultimately comes from the Void not from the Sun.

In the Heart Sutra, Buddha said to Sariputa that Matter is just immaterial, the immaterial is just matter (Form is Emptiness and the Emptiness per se is form, or The world is not different from Emptiness and Emptiness is not different from the world).

EMPTINESS → THE FORM
EMPINESS ← THE FORM

Nevertheless, the two above paths of the transformation are not identical. *This is due to the asymmetry or the duality of the world like Yin and Yang.* For example, the chicken is killed to be made into the meat. But assembling chicken meat can not make the chicken. The problem is in the process of division of chichen, the Soul (with its base, the Original Omniscience) is lost. To recreate the chicken, Omniscience (The Creator) must be added to the process.

Of interest, in this concept of Form and Emptiness that is identical in nature, Apanishads called Fullness instead of Emptiness: *"From Fullness, fullness come,when fullness is taken from fullness, fullness still remains."*

2. EMPTINESS AS A POSTULATE, ORIGIN OF UNIVERSE. EMPTINESS IS MIRACULOUSLY EXISTENCE.

In physics, black holes represent objects whose gravity is so strong that nothing close to them can not escape. It is created after the death of a very massive star, with its center collapsing in on itself, creating the explosion of the outer layer of the star. The center of the star /galaxy will eventually become a black hole that attracts all debris, including light, into its center. The center is represented as a singularity with enormous gravitational force but no time-space, equivalent to EM. The resulting EM eventually redeveloped into a Big Bang to restart the cycle of EM-Form transformation.

3. FALSE THOUGHT FROM THE EMPTINESS VERSUS BIG BANG SINGULARITY.

As mentioned above, in the Creation of the Universe, Emptiness was discussed by philosophers like Spinoza as Substance, Lao Tsu as Fuzzy Space and Void, in the Old Testament as Formlessness, Void / nothingness, Darkness, Watery Deep, and in Islam as nothingness "creatio ex nihilo".

This concept of Emptiness as the origin of the Creation was later highlighted by (Bodhisattsava) Asvaghosa (80 – c. 150 CE, vn: Ma Minh), considered the first Patriarch who developed this concept of Mahayana deviated from the Hinayana. This concept consists of Emptiness as the ultimate origin of the Creation (Tathata with Sunyada). Tathata with Emptiness as Creator is considered as the ontological concept. The Creation and all following attributes constitute the phenomena. Because phenomenon is a secondary attribute without self-nature/noumenon (Nihilism with Negativity and Relativity. It appear that the concept is similar to Upanishads concept. However the concept of Emptiness is original from Buddha revelation and distinct frm that of Brahmaism, part of Hinduism (please see paragraph thể.

C, THE WORLD, AS FALSE PERCEPTION/PROJECTION)

Bodhisattsava Nagarjuna further developed into the philosophy of the Middle Way/ Mūla-Madhyamaka-kārikā), between the Emptiness Doctrine (Sunnyavada) and the Doctrine from the Theravada Group. Vaibhasika and Sautrantika Group of the Theravada Group accept the existence of the reality of Dharma and the world. The difference between these two groups is the concept of timing. For the Vaisahasika, the world exists in four phases: Birth, Development, Maintenance, and Death. For the Sautrantika, There is only the present, without the past or future).

In summary, the difference between Mahayana
and Hinayana resides in the point of view: With the perspective of view of Emptiness, Dharma and the Form are created and, therefore,

perishable. If the standpoint is just after the Creation of the Sautrantika Group (vn: Kinh Lượng Bộ),, there is no concept of time, but Form and Dharma are just created. For the Vaibhasika (vn: Nhất Thiết Hưũ Bộ), the standpoint of view is later in the Creation. As a result. For Hinduism perspective, the concept is still limited within the three worlds of reincarnation/Trailokya, the concept of Emptiness, no time-space and Egolessness does not pose this kind of controversy.

Bodhisattsava Nagarjuna characterized Emptiness with:
Three self-natures:
- Impermanence,
- Egolessness.
- Dukkha-Affliction.

And The eight negations may be taken as a refutation of the false views
· Neither birth nor death (vn: Không Sanh, Không Diệt).
· Neither end nor permanence (vn: Không Thường, Không Đoạn).
· Neither identity nor difference: the end is the beginning (vn: Không Đồng Không Dị) (Fig.1,7).
· Neither coming nor going— one who has thus not gone" (vn: Như Lai, Không Đi, Không Đến) signifying that the Tathāgata is beyond all coming and going – beyond all transitory phenomena.

For further clarification of features of Emptiness, additional features :
· NoTime-Space,
· Ingenuity, Unlimited power, Ultimate Mind. And,
· Unthinkable miracles in the material, non-material worlds,

In terms of Physics, Emptiness has the lowest Entropy=zero. (according to the second thermodynamic law, Entropy measures the degree of intrinsic disorder or perturbation of a system. As a result, Entropy (vn: độ nhiễu loạn) is inversely proportional to the energetic efficacy of the system)

In other words, Emptiness is symmetrical and homogenous; therefore it is associated with a high level of energy but the lowest level of stability. As a result, spontaneous false thought may occur at times, resulting in symmetry breaking.

No Time and No Space means: Thus come-Thus gone or Thusness/ Thathagata. As a result, the speed is *supraluminal*.

The Emptiness is imperceptible, therefore permanent and real, not perishable. Something perceptible is impermanent (because it is perishable), and is, therefore, not real, non-existent, and illusional.

Emptiness shares imperceptibility with illusion. However, the illusion is non-existent, created and ceases.
(In Diamond Sutra, paragraph 26, Buddha said: If one sees me in forms, If one seeks me in sounds, He practices a deviant way, And cannot see the Tathàgata).

4. CONTROVERSIAL INTERPRETATIONS OF BUDDHA'S TEACHINGS REGARDING THE EMPTINESS.

It is necessary to review Buddhism's spread, division into different branches, and development after Buddha entered Nirvana. It is also essential to understand that Buddha's teachings varied according to various levels of knowledge of human beings, ranging from practical and straightforward to conceptual and complex for beings beyond the learning stage like Arahats or Bodhisatsavas (skt: Asaika, vn: vô học).

There are four Councils of Buddhism to recite Buddha teaching and moral precepts:

1. **The first Council** of 500 Arahats (including Ananda) was held 7 days after Buddha's demise, presided by Mahākāśyapa under the patronage of King Ajatashatru of the Haryanka Dynasty under the Magadha Empire. Ananda composed the Suttapitaka, and Mahakassapa composed the Vinaypitaka (monastic code).

2. **The second Buddhist Council** presided over Sabakami under the patronage of King Kalasoka of the Sisunaga dynasty, held in 383 BC, i.e., a hundred years after the Buddha's death, at Vaishali.
Due to the dispute on the moral precept regarding receiving and keeping the offers and daily eating meals. With a significant splitting into
   Theravada (Thera meaning Elder)
to preserve the teachings of Buddha in the original spirit
   Mahasanghika (Mahayana group, Great community, more numerous monks) is a more liberal interpretation of the sutra. Suitable to the general population

3. **Third Buddhist Council** presided by Mogaliputta Tissa (Mục Kiều Liên Từ Đế Tu), the patronage of Emperor Ashoka (A Dục) (favoring Hinayana) of Maurya dynasty, in 250 BC at Pataliputra. The aims were to purify Buddhism from opportunistic factions and corruption in the Sangha. The Abhidhamma Pitaka (vn: Vi Diệu Pháp) was composed here, making the almost completion of the modern Pali Tipitaka Abhidhamma texts were translated from Prakrit to Sanskrit only in the fourth council.. Buddhist missionaries were sent to other countries.

4. **Fourth Buddhist Council** presided by Vasumitra and patronage of King Kanishka of the Kushan dynasty. Ashvaghosha in the 1st century AD (72 AD, 400 years after Buddha's demise) at Kundalvana in Kashmir. This council resulted in the division of Buddhism into two sects, namely, Mahayana (the Greater Vehicle) and Hinayana (the Lesser Vehicle).
Mahayana sect believed in idol worship, rituals, and Bodhisattvas. They regarded the Buddha as God. Hinayana continued the original teachings and practices of the Buddha. They adhere to the scriptures written in Pali, while the Mahayana also includes Sanskrit scriptures.
There were five or six Buddhist councils, but they are only recognized in the location where it took place in Burma.
   For the Mahayana group, there was only one council, held 500 years after Buddha's demise in Afghanistan, to complete
the four Avatamsaka *Sutras/ vn: A Hàm.*

> Of interest, the decline of Buddhism in India is caused by the corruption among the monks with the deterioration of moral standards, the decrease of royal patronage after King Asoka, accompanied by the increased royal support for Hinduism and the works of Kumarila Bhatt and Adi Shankara, invasion of the Huns and finally, the Islamic armies that almost eradicated Buddhism from the Indian continent in the century of 1200

Since the second council, there has been a subdivision of the Theravada and Mahayana groups into subgroups. Among the Theravada group, the Mūlasarvāstivāda subgroup (Nhất thiết Hữu bộ) is the most distinct by its concept of the actual existence of all Dharmas explicitly the true existence of the worldly life and existence of the Past, Present, and Future times with the creation of Abhidharma jnànaprasthàna sàstra (vn: A-Tỳ-Đạt-Ma-Phát-Trí-Luận).

> In addition there are another six sutras:
>   i. (Abhidharma Sanjitiaparyàpàdá Sàstra (vn: A-Tỳ-Đạt-Ma-Tập-Dị-Môn-Túc-Luận) 20 tomes, edited by Xá-Lợi-Phất during Buddha's time).
>   ii. Abhidharma dhamaskandhapàda Sàstra (vn: A-Tỳ-Đạt-Ma-Pháp-Uẩn-Túc-Luận) 12 tomes, by Xá-Lợi-Phất tạo versus Mục-Kiền-Liên.
>   iii. Abhidarma prajnàtipàda Sàstra (A Tỳ-Đạt-Ma-Thi-Thiết-Túc-Luận) composed 18.000 sermons, by Mục-Kiền-Liên tạo; vesus by Đại-Ca-Chiên-Diên.
>   iv. Abhidarma vijnànakàyapàda Sàstra (A-Tỳ-Đạt-Ma-Thức-Thân-Túc-Luận) 16 quyển, by Devasarman (vn: Đề-Bà-Thiết-Ma - Thiên-Tịch, Thiên-Hộ) 100 years after Buddha's demise.
>   v. Abhidarma prakaranapàda Sàstra (A-Tỳ-Đạt-Ma-Phẩm-Loại-Túc-Luận) 18 tomes, by Thế-Hữu and Kê-Tân La-Hán.
>   vi. Abhidharma dhàtukàyapàda Sàstra (A-Tỳ-Đạt-Ma-Giới-Thân-Túc-Luận) 3 tomes, by Phú-Lâu-Na tạo vesus XáLợi-Phất.
> As a result the combination of all above sustras is made into the Abhidharmamahàvibhàsà Sàstra (vn: A-Tỳ- (ĐạtMa-Đại-Tỳ-Bà-Sa-Luận) with 200 tomes

These views are opposed by the Mahayana groups and the Yochara division, including Asvaghoṣa/ vn: Mã Minh 80 – 150 CE, as a Mahayana patriarch, Nāgārjuna/ vn: Long thụ 150–250, 14th Indian Mahayana Buddhism Patrirarch, attaining spiritual power to retrieve the Avatamsaka/ Flower sutra from the Dragon palace of the Dragon Kings, at the bottom of the ocean, Asanga/ vn: Vô Trước 300-377, attaining Arahatship from Hinayna school and being further advanced in enlightenment with Maitreya in Tusita Heaven (the Delightful Realm, the abode of Bodhisattvas in their last existence before attaining Buddhahood) and his young half-brother Vasubandhu/ vn:Thế Thân/ Duy Thưc Tông/ vijñānavādin 316-396, the 21th of Indian Mahayana Buddhism Patriarch. Avatamsaka is

the most profound sutra with the concept of innumerable things and phenomena as the oneness of the Multiverse. The above Sutra and other sutras from the Mahayana school, like Flower Adornment, Diamond, Prajñāpāramitā Hṛdaya Sūtra, and Sutra/ Heart sutra only consider Emptiness the real entity. All phenomena, dharmas, are born from Emptiness and represent the secondary phenomena in the Multiverse of duality. Time (Past, Present, and Future) and
space are illusional and unreal. It is worth noting the concept of the epistemology of Chandrakirty, a Buddhist philosopher with Indian and Tibetan influence. He said, "Whatever has causal powers (*arthakriyāsamartha*), that exists (*paramārthasat*)". *Therefore, some entities are real.* The real is only the momentarily existing particulars (*svalakṣaṇa*), and any universal (*sāmānyalakṣaṇa*) is unreal and fiction. Furthermore, for Chandrakirty, *Emptiness in Buddhism is not knowing the object. Therefore, all mind and mental factors (Omniscience) have ceased* in Emptiness.

In common thinking and in Taoism, "*Tao is to lower the high and raise the low, to take from those who have too much and give to those who do not have enough*". Therefore, there is no emptiness per se in this Universe, knowing that the Dark Matter and/ or (?) Dark Force are imperceptible to five sensory organs, likely present in the Emptiness.

Furthermore, separating the form into tiny parts of the quantum level and beyond the quantum level is laborious and energy-consuming. This process is comparable to splitting materials, like uranium nuclei, into subatomic particles (quarks) in the process of creating atomic bombs with the generation of enormous energy (strong force). This means that Emptiness contains a tremendous amount of energy. Since Emptiness is boundless, its energy is inexhaustible.

When considering the energy, it is reasonable to consider the Ultimate Omniscience (UO), which is the source of Consciousness (understanding), and Awareness, which is broad but superficial in details Ultimate Omniscience (UO) in Buddhism refers to the highest form of knowledge and understanding, which is independent of the subject. It is a key concept in understanding the nature of Emptiness and its relation to the Universe.

The "Way" also represents Emptiness or Tao, as Lao Tsu said:
The Tao that can be told is not the eternal Tao.
The name that can be named is not the eternal name.
The nameless is the beginning of heaven and Earth.
The name is the mother of the ten thousand things.

As a result of all the above discussions on Emptiness, it can be described as follows:
i. Self-existence with No birth, no death,
ii. No time and space, such as at the beginning of the Big Bang/Genesis or False thought, time and space are zero as expressed by Einstein's equation
$Ds2=dx2+dy2+dz2-dt2$ with s=0 t=0 and x,y,z=0
After the Creation, t>0 and s>0, therefore x,y,z>0
iii. No discrimination.
Exemplified by the Fractal phenomenon that implies similarities in geometry on a large and small scale (Fig 1.7)
The useful application is the discovery of the structure of atoms similar to that of the solar system by Niels Bohr
iv. possession of unlimited power despite the consummation of the power, such as in the case of the Big Bang, the expansion keeps accelerating with the speed of expansion exceeding that of the light
v. association with the Omniscience. The knowledge is, as the thing is, independent of the subject. In addition to the omniscience, other properties are:
- Construction, Creation
- Maintenance
- Destruction
- Ingenuity, skill, cleverness.
- Four immeasurable minds associated with selflessness (Immeasurable love -skt: Metta, Immeasurable kindness—Boundless compassion/ skt: Karuna in sharing sufferings, Immeasurable sharing joy/ skt: Mudita — Immeasurable detachment/ Perfect equanimity/ skt Upeksha.

As a result, Emptiness in Buddhism always contains something in order but not accessible to the Consciousness or the five sensory organs. For this state of order, the beginning of the Genesis, Creation, or the Big Bang is triggered by the movement of the spirit of God,/ the Big Bang from a point with an accumulation of tremendous energy /or from the False Thought in Buddhism. The Creation develops secondary to the loss of the equilibrium/order.

Fig.1,7: Fractal phenomenon: many generations of branching follow the similar pattern.

In Buddha's address to Ananda in the meeting of 1250 monks was registered in the Surangama sutra,

The sermon mentioned in paragraph 1), page 70 above, denotes that the Void/ Emptiness conceived in Buddha's teachings is a homogeneous, still, non-discriminative, and boundless state. As a result, a False Thought eventually arises at "any point" in this state (of no space!) and instantly involves every areas of the Void/ Emptiness. Since the energy in the Void is unimaginable, the False Thought is enough to create the
heat necessary for the fusion reaction to form particles. This energy is also responsible for expanding the space created after the Big Bang, along with the Creation of time. The arising of the False Thought from the Emptiness accounts for the fact that the Cosmic Microwave Background Radiation came in from all directions of the Universe as opposed to the Big Bang Singularity as thought initially (please see pages 40-46: **theory of Big Bang, 1. scientific concepts of Big Bang theory**).

Emptiness must be differentiated from Illusion That is created and is associated with birth and death**5. EMPTINESS AFTER FALSE THOUGHT/BIG BANG**
It is crucial to understand that BB or False Thought disrupts the equilibrium of the Emptiness, leading to Its cessation. However, the state of Ultimate Omniscience, Power, Buddhadood, Holy Spirit, and Nirvana, which remain in the state of Primorial Duality, are of utmost significance. In spiritual practice, reaching Nirvana is not a mere absence of Emptiness, but a state of Ultimate Omniscience, unlimited power, tranquility, and harmonicity.

NIRVANA: Nirvana is the spiritual land (no form) or state that existed just before the development of False thought and the veil ignorance, and is not Emptiness. However, Nirvana is closely related to Emptiness. It is equivalent to the state of Primordial duality (see page 72). As a result Nirvana has many characteristic of Emptiness and some features i.e. Omniscience, Ingenuity, Light, Sound... of the creation just before the development of the veil of ignorance. But it is unreasonable to agree with some Buddhists Scholars that living beings with form (five skandhas) can enter Nirvana, Buddha Sakyamuni only enters Nirvana during meditation by his Spirit or after death (leaving behind the form)

6. PARAMITA and SELFLESSNESS
Paramita means the other shore of the river, the Nirvana , as compared to this shore of Birth and Death.
In Lankavatara Buddha distinguished three types:
- Paramitas of the supreme ones of Bodhisattva
- Paramitas for Sravakas and Pratyekabuddhas relating

to the future
- Paramitas for people, in general, relating to this world

In all cases, Paramita represents the state of Selflessness, commonly used in the Four Immeasurable Minds (Love, Compassion, Inner Joy, and Detachment). The principle is that human beings are children of the Creator, do not own anything in this world, and do not have free will.

All materials and love given away belong to the Creator. Happiness only results from the fulfillment of the duty instructed by the Creator.

So, in the Ethics of the giving, there is neither giver nor receiver, but the Ethics of the receiver is the Ethics of gratitude. (the asymmetrical law of Duality)

7. UNDERSTANDING OF THE IGNORANCE and THREE SEALS OF DHARMA, Four Kinds of Mindfulness Eightfold Path

Selflessness, Impermanence, and Suffrings are characteristically essential in Buddhist dharmas along with 37 factors of enlightenment (Four Right Efforts, Four Sufficiences, Four Kinds of Mindfulness, The Five Faculties, The Five Powers, Seven Bodhi Shares, Noble Eightfold Path: right view, right aspiration, right speech, right action, right livelihood, right effort, right mindfulness, right concentration.). Failure to recognize the three Seal Dharmas is due to ignorance, that will lead to an inversion view of the world.

In life, the common pathway of recognition of the three Dharmas is the observation of the earthly phenomena and using the induction method. This method may be associated with the potential for error.

- **Buddha recommends the technique of Four Kinds of Mindfulness.**

*Contemplation of the body, realization of the serenity or impurity of the body.

* Contemplation of Feeling, realization of the evils of sensations or pleasant feelings, no matter whether they are painful, joyous, or indifferent sensations

* Contemplation of the Mind, the evils or the virtue of the thought and Mind, the realization of the different states of Mind (Greed, Anger and Ignorance or Immeasurable Mind of virtue)

* Contemplation of Dharmas or the origination of the Mind, realization formation of the Mind

- **Ontological Method.**

Following the Universe creation, the Creator is the oneness who is the father of all living Beings and Owner of all other physical and metaphysical entities. Human beings are given the conditions for living, working/ creating, and entertainment commensurable to their needs. The extra givings belong to the Creator and are available to living sentients to share with the others. The body, the feeling, the Mind, and Nature are selfless, owned by the Creator.

Thoughtful understanding and realization of this principle are almost equivalent to the mindful contemplation of the above four areas.

As a result, the realization of Selflessness and detachment from erroneous ownership is the key to mindful contemplation to eradicate the bad Mind, as told by Buddha to Ananda and Rahula in Nikaya Sutra 61

8. MANIFESTATIONS OF GOD & GOD's PERSONALISATION.

Manifestations of God can be found in scriptures from major religions. However, it is essential to be discussed in this book that the concept of Emptiness, which is miraculously Form, is posited as the Creator and origin of the Creation. This concept, rooted in the intersection of neuroscience and quantum mechanics, provides a unique perspective on the nature of God.

Therefore, the manifestation of God is all attributes of Emptiness that can be summarized in three categories:

- Power: accounting for all Forms of Nature and living organisms
- Omniscience accounting for the CS, *ingenuities, and miracles*
- Splendor accounting for magnificent, splendid accessible to CS and sensory organs

These features of God, as manifestations of Emptiness, are not isolated from each other. They are intricately interconnected, forming a complex and profound understanding of the divine.

The manifestation must be differentiated from the communication with living organisms, particularly with livings. To make the communication complete and practical, and due to the veil of ignorance of sentient beings, God has to be manifested in three different Forms, including Personalized God (Trinity in Christianity or three bodies/ Trikaya in Buddhism). When God is personalized as a human being, the manifestation is limited, at least in time and space. For example, Buddhism is limited in some parts of the continent, and its influence wanes after 500 years. Buddhist devotees are facing degenerative Dharma time nowadays... It is likely that other attributes of Emptiness are also limited.

In Islam, monotheism, and absolute sovereignty, *Qu'ran categorically rejects the doctrine of the Trinity, firmly asserting God's Oneness. This rejection is based on the belief that God is indivisible and cannot be divided into separate entities. Furthermore, Buddha was not mentioned. Holy Spirit, represented by the Angelic Gabriel, Father, Son, and messenger .. has been explicitly rejected in an early date.* Nevertheless, the existence of three entities, Angelic Gabriel, Messenger, and God, can be regarded as a similarity between the three major religions.

It is important to remember that human Consciousness is limited and distorted by the Brain. Therefore, the manifestation of Emptiness (GOD) can not be fully comprehended within the human body. As known in Buddhist Sutras, Buddha frequently practiced Meditation, especially before proclaiming prominent Dharma sutras like the Lotus sutra, Flower Adornment Sutra and other major sutras, Buddha meditated and radiated infinite light. This serves as a humbling reminder that Emptiness possesses all attributes far exceeding human Consciousness.

9. PHENOMENA OF PARTICLE/ WAVE and EPR PARADOX
(Fig 1.8ABC) (for EPR please see page 342)

Features of Emptiness are Homogeneity and Neither identity Nor difference.

These proprieties can be seen in religion expressed by the Trinity in Christianity and Trikaya/ Three Buddha bodies in Buddhism. Father and Dharmakaya/ Son and Buddha/ Holy Spirit and responding body are different but represent the same Oneness (God). The three bodies may also appear at the same time but in other places, and never at the same time and same place.

Particles and waves in Quantum Mechanics are entities of the world of duality; however, they are very close to Emptiness.
Therefore

- Wave and Particle representing the similar entities can not be seen at the same time and in the same place
- Two different particles of the same system are similar (in spin) (Homogeneity, Neither identity nor difference)

Of note: Buddha can appear in different forms as evidenced in the following sermon in Lankavatara sutra in page 83

Quantum Mechanics: Heisenberg's Uncertainity and EPR paradox.

Heisenberg's Uncertainity: As previously outlined, due to the light weight and small size, a particle can not be characterized as to its momentum and velocity

a. EPR paradox

EPR paradox: two photons or electrons belonging to the same system are shot in opposite directions in the thought experiment designed by Einstein, Podolsky, and Rosen called the EPR paradox. In this experiment, later confirmed in the laboratory by Alain Aspect (who won the Nobel Prize in 2022), the particles always spin in the same direction despite the distance. Physicists call this phenomenon "entanglement" because the particles of the same system are interconnected. Einstein used this evidence to argue with Niels Bohr that the quantum mechanics of the Copenhague interpretation is incomplete due to the hidden variable that governs the Entanglement of particles. In Middle school of Maha Buddhism by Nagarjuna, the Emptiness that represents the original Mind/ Buddhahood, Transcendental Awareness is characterized by the Eight negations:

i. No birth No death (vn:Bất sanh diệt)
ii. No end No permanence (vn:Bất Thường Bất đoạn)
iii. No similarity, no difference (vn:Bất nhất Bất dị)
iv. No coming, No going away (vn:Bất khứ Bất lai)

The number iii Negations represents the entanglement phenomenon since the particles, despite being separate (No similarity) and different, are the same (No difference).

This phenomenon of ERP paradox or Entanglement in quantum mechanics may illustrate the determinism.
Likewise, in Bell's theorem, the two photons are shot out in two directions, forming an angle of θ.

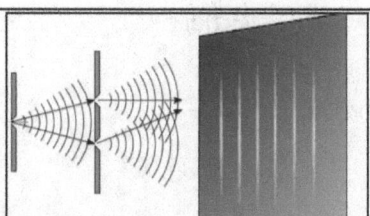

H1.21 Light passing through two slits produces patterns that can only be explained by the wave nature of light becomes unknown

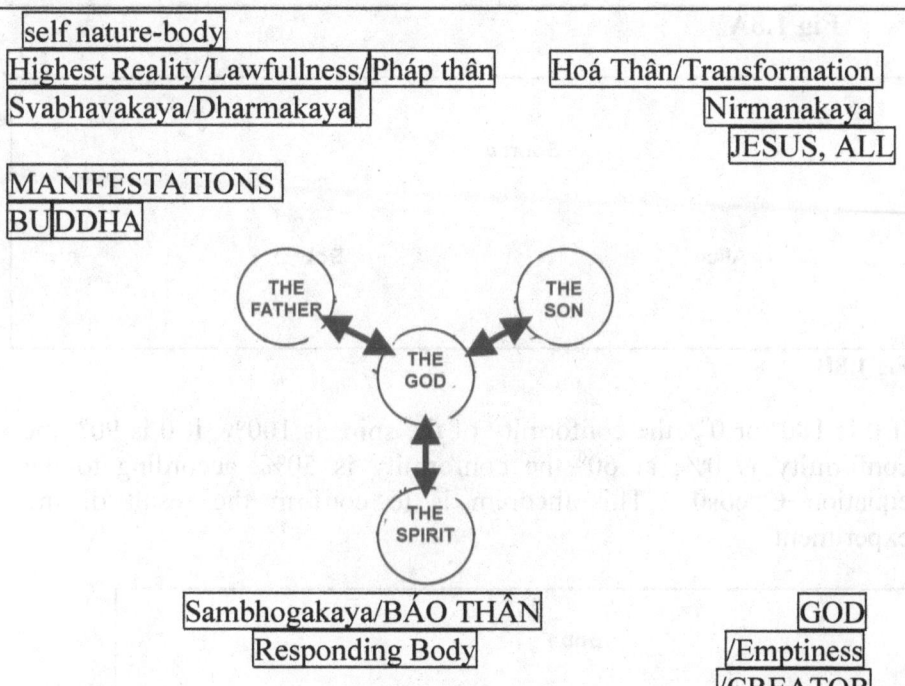

self nature-body
Highest Reality/Lawfullness/Pháp thân
Svabhavakaya/Dharmakaya

Hoá Thân/Transformation
Nirmanakaya
JESUS, ALL MANIFESTATIONS
BUDDHA

Sambhogakaya/BÁO THÂN
Responding Body

GOD
/Emptiness
/CREATOR

Since Tao is the One, therefore, despite the difference in location and timeline, the concept of Tao is similar: TRIPLE BODY OF BUDDHA (vn"Tam Thân Phật, skt: Trikaya) and (TRINITY(vn: Ba Ngôi Chúa) have the same meaning. According to the diagram, the three Identities in Buddhism or Christianity represent the non-personalized Creator in different circumstances. Furthermore, each identity appropriately appears in location and time. One can conceive that the three identities never appear at the same time in the same location but can appear at the same time in different locations.

This phenomenon is similar to Waves and Particles. Despite the duality, one can not pin down particles and waves simultaneously.

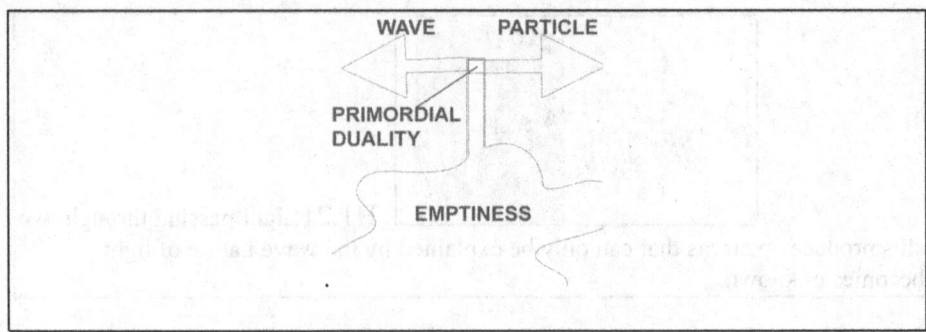

Fig 1.8A

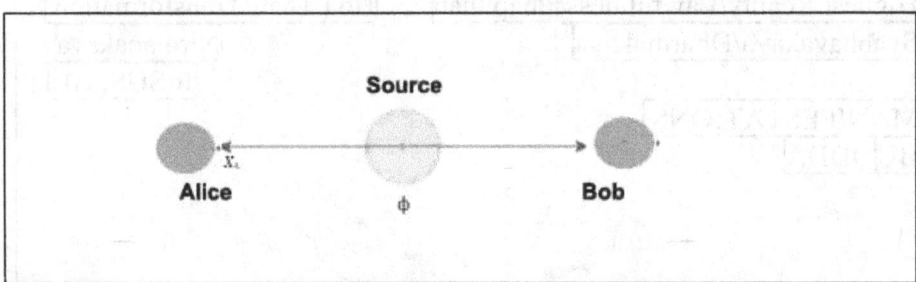

Fig 1.8B

If θ is 180^0 or 0^0, the conformity of the spin is 100%, if θ is 90^0 the conformity is 0%, at 60^0 the conformity is 50%, according to the equation C=cosθ. This theorem is to confirm the result of the experiment

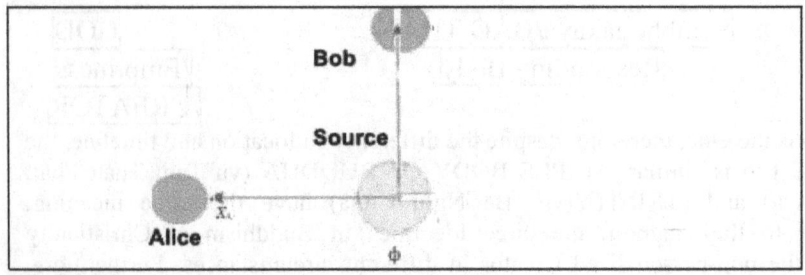

Fig 1.8C

Therefore, the EPR paradox and Bell's theorem confirm that before leaving the source, the photons are predetermined to spin according to a fixed spin. They do not have the choice to display the characteristic when observed by Alice and Bob. The experiment proves that the observation remains the same as the experiment's results. As a result, Einstein is accurate when objecting to Copenhague's quantum interpretation of quantum mechanics. However, the hidden variable does not exist because Emptiness can be accounted for the no-locality and interconnectedness

b. Reality and Non-Reality.

Reality is defined as an entity that has definite properties independent of observation. As a result, at the quantum level, the particle is locally not real. However, according to Buddhism, Reality is defined as its permanent existence.

All entities that have form accessible to the five sensory organs are perishable and, therefore, impermanent and unreal—vice versa; Emptiness that is imperceptible, therefore imperishable and permanent. Illusional entities are inaccessible to touch or some other sensory organs and are perishable, therefore impermanent and unreal. This stark contrast to the Buddhist concept that quantum particles are real challenges our understanding. On the other hand, particles display definite spin that is measurable. The above dual properties of Reality (by Einstein) and Unreal (by Bohr) reflect the characteristics of the entities in the interface between Emptiness-False Thought (the so-called Primordial Duality).

This writing is based on the Surangama Sutra and Lotus Sutra. The Surangama Sutra, which recorded Buddha's essence of teaching, is an important sermon that delves into the concept of impermanence, a key aspect of Buddhist philosophy.

It is said that this is the first Sutra to disappear at the end of the Dharma age (three stages: Right Dharma/ vn: chánh Pháp/First 500 years, Middle age of the Dharma/ vn: Tượng-tưởng tượng- Pháp /second 500-1000 years and Latter Day Dharma/ Degenerate Age of Dharma lasting for 10.000 years / Mạt Pháp in which Buddha Shakyamuni would gradually lose the power of salvation evidenced by increasing unrest, conflicts wars, disasters, and famine, until new Buddha, Maitreya, would come to save the people. This Sutra reveals the Law of the Tathagatha of One Reality. From the Tathagatha arise the transient/ false/misleading thoughts (skt: Vitathavitakka) generating ignorance and enlightenment. The talk occurred in the Jetavana vihàra near Sravàsti attended by 1250 monks/ bhikkhus, including mainly Great Arhats who got to the level beyond the reincarnation. This Sutra was preached after the Laṅkāvatāra Sutra (vn: kinh Lăng già Tâm ấn) and in the second decade of the path of propagation of the True / Correct Dharma. After Buddha sent Bodhisattvas Mañjuśrī to rescue Ananda from a low caste Màtaïgà who uses Kapila magic to attract him, just before

Ananda breaks the rules of good living. This teaching is also to help Ananda better understand the Dharrma, follow the spiritual path, and instruct him to practice one of the methods of meditation. Buddha invited 25 Arhats and Bodhisattvas to reveal the method of Meditation they had performed to attain the level of transmigration. Being commanded by Buddha, Bodhisattvas Mañjuśrī pointed out the method most appropriate for Ananda and humans in the later Days of degenerative Dharma. The Meditation practitioner should recognize and avoid fifty circumstances of illusions in the Meditation. Buddha pointed out to Ananda and the monks in the meeting the Original Mind is uncreated and therefore imperishable, for it is complete and unchanged. The original Mind is uncreated, non-perishable, free, independent, autonomous, and sovereign, with a unique true Existence of Self and Ownership. This is contrary to all other beings, children of the Creator,. They are selfless and not sovereign; they must be obedient to the Creator and have no free will. The significance of meditation methods in understanding Dharma cannot be overstated, inspiring us to practice and delve deeper into its teachings.

Time And Space And Relativity.
In 1915, Einstein proposed the Theory of Relativity, a concept that has profound implications, relating mass and energy.
$E=mc^2$ with **E** the energy created by the mass, **m** the weight the mass, and **c** the light speed, and later the General Relativity $ds^2=dx^2+dy^2+dz^2-c^2t^2$ with xyz the measurement of the three dimensions of space, the time c speed of the light and s the space corresponding to the observer.
In Buddhism, the Universe is created from the concept of 'Emptiness', which refers to the absence of inherent existence. Before the Creation, there was no Time and Space associated with this Emptiness. As a result, for Buddhahood and Buddha, Bodhisattsava, primordial False Thought/Big Bang, Space, and Time is Zero or near Zero.
Therefore, Buddha is also called Thus Come Thus Gone. Their movement speeds are supraluminal. In Sutras like Laṅkāvatāra, Flower Adornment, and Lotus, which are significant texts in Mahayana Buddhism, innumerable Buddhas come to the meetings from a Universe of imaginable distance with an unimaginable number of years to travel with the speed of light.

In astronomy, the farther galaxies, the faster their speed, which exceeds the speed of light.

This ultraspeed can only be explained by the spread of the Big Bang in the Emptiness. The spread is instantaneous.

In contrast, in our Universe, the speed is only limited by the upper limit of the light: on the moving train, the time span is shorter than the observer outside the train. This will compensate for the advantage of the moving passenger in the moving train.

Speed of the Movement Of The Soul

Since the Soul is likely composed of entities close to the Emptiness (Dark Matter, Dark Force, and Neutrinos.) DM is five times heavier than the Baryonic Matter, but Neutrino is light. It appears very early in the Big Bang. The Soul of Living Sentient is different than that of Buddha and Bodhisattva because of the presence of Karma. Nevertheless, the moving speed of the Soul is not as fast as that of Buddha and Bodhisattva but likely very fast. The following Sutta is to support the above impression.

The Scripture on the Monk Nāgasena in Page 201, told the King tathata seventh heaven of Brahmā?" is very far away. Letting a stone as big as the king's palace falls from the seventh heaven of Brahmā on this earth just here after six days." An Arahant is able to fly up to the seventh heaven of Brahmā as quickly as a person bends or stretches an arm. Or just like you thing and arrive immediately your birth place

---

The Scripture on the Monk Nāgasena Page 201

The king asked Nāgasena again, "How far away is the seventh heaven of Brahmā?"Nāgasena said: It is very far away. Letting a stone as big as the king's palace fall from the seventh heaven of Brahmā, it will only fall on this earth just here after six days." The king said: Dear, you recluses say that one who has attained the awakening of an arahant
is able to fly up to the seventh heaven of Brahmā as quickly as a person bends or stretches an arm. The king said, "I do not believe this. How could one travel millions of (Chinese) miles so quickly?" Nāgasena asked the king, "In what country was the king originally born?"
The king said, "I was originally born in a city called Alisan (Alexandria) in the great Qin (Greek colonial) empire."
Nāgasena asked the king, "How many (Chinese) miles from here is Alisan?" The king said, "It is two thousand *yojana*s or about eight thousand (Chinese) miles." Nāgasena asked the king, "In the past, would you have thought of some matter in your distant country of origin?" The king said, "Yes, I often think just of matters in my country of origin." Nāgasena said, "May your majesty try to think again of a matter in your country of origin, something you did in the past." The king said, "I have thought of it." Nāgasena said, "How has the king traversed eight thousand miles and come back again so quickly?" The king said, "It is well."

Note: it is reasonable that, Bodhiattsava will take an instantaneous moment to travel from the earth to Brhamas land, but for imaging the site that is previously seen, it is a different process of retrieving the Memory, which only takes nine ksanas (in video, recording the minimum duration of one video frame in the video recording is 32 mec, necessary to render the video smooth and continuous. Ksana is defined as the minimum duration for one vent to happen. Therefore one ksana=1/2 video frame=32 msec/2= 16msec).

The speed of travel of Buddha is instantaneous, faster than that of the light because Buddha is equivalent to Emptiness. For Bodhisatsava and Arahat, the speed may be slower than Buddha but still faster than the light. Speed is relative to the space. When space is zero, the speed is infinite, faster than the light.

Time and Space in Meditation: Here And Now

Before the false thought/ Big Bang, Emptiness is associated with stillness, homogeneity/ no discrimination, and no time and space. Similarly, in mindfulness, the state of 'no thought' refers to a mind that is free from the usual stream of thoughts and distractions. In this state, time and speed gradually collapse to a minimum with the past, future, and present time as well as the space. ***Here and Now, the concept is to keep oneself free of attachment by realizing Selflessness in the wakefulness and Awareness/ Omniscience.***

In summary, time and space are relative to each individual and circumstance. In the study of the samsara circle, time is presented as circular as opposed to the linear phenomenon in the common concept in daily life.

In mindfulness meditation, time and space are represented by the concept of HERE and NOW (no time and space). The meditator embarks on a transformative journey, gradually losing the concept of time and space, and ultimately becoming selfless.

K. Creation of the Five Skandhas after the False Thought.

The original Mind, detached from its original, non-discriminative emptiness, begins to be obscured by deviated thoughts, acting as a veil of ignorance. To ordinary humans, the original Mind is less apparent than the Awareness. However, as the veil of ignorance thickens and becomes tainted with impurity, the Mind is not just influenced but dominated by Consciousness. This dominance, akin to Adam and Eva's disobedience and the subsequent rendering of the

Mind discriminative, is a pressing issue in our understanding of consciousness.

Imortant note: *Since the five Skandas are created by Emptiness, Skandas are always egoless, and are not owned by any living sentients. The five skandha represent different constituent of the Universe/Multiverses including all forces, materials*

1. Form: Water, Earth, Wind And Fire:

The form, a manifestation of the environment and the human body, is intricately connected to the five sensory organs, serving as gateways to our perception of the world.

Our perception, whether it's of objects or problems, is a complex interplay between the nature of the objective, the subject's readiness, and the influence of the environment. This process, be it for learning, studying, or practicing, is a constant revelation of knowledge through the lens of our surroundings.

Duration in the process of creation of CS: 5 ksana=80ms)

In tomes 2 and 4 of the Surangama was asked by Sàriputra, the wisest among Buddha's disciples, Buddha replied:

The original Mind (Emptiness in nature) is the One as a whole and is neither ignorant nor enlightened, giving rise to false /misleading thoughts. A wrong view then accompanies the false thought, leaves the Emptiness, and is associated with ignorance. The differentiation between enlightenment and ignorance is the beginning of the discriminative Mind. The pure, silent, and tranquil part of this later Mind is of water in nature. Water is crystalized into Earth. The interaction between Earth and Water results in the generation of the Wind. This causes the fall-down of water, resulting in ascending Fire. As a result, the separation between wet and dry parts of the Universe. As one can see, a volcano can erupt in the ocean, and water flows in the river in the middle

of the Earth. The four elements interact and create interrelationships that cause and affect intertwinement. This conflict is the cause of the limitation/ turbidity of the Awareness.

Note: in comparison to the Far Eastern philosophical concept, the physical world consists of five components: Earth, Metal, Water, Wind and Fire; in Buddhism, the material world is only composed of four components: Earth, Water, Wind, and Fire. The missing component represents the the Self/Ego. This missing reflects the radical understanding that there is NO SELF in this world that is created and owned by the Creator

2. Four Metaphysical Components (Perception, Feeling, Formation, CS).

Once the four elements (Water, Earth, Wind, and Fire) are generated, they undergo a complex and intriguing transformation, becoming the Form that the five senses can perceive.

· Another component, the metaphysical component, is then added. This later component is not accessible to the five senses but can induce secondary effects affecting the perception of the five senses. This component consists of four subcomponents:

- Perception: (feeling/ sensation/vn: Thọ/ skt: Vedanta the process of receiving the data

Acquiring information through the five sensory organs is the basis of our perception. However, even with sophisticated equipment, our perception is humbly limited by the principles of Quantum mechanics. At the quantum level, the inherent 'fuzziness' or uncertainty of particles can affect the accuracy of our perception.

As **Heisenberg and Bohr** pointed out, the elusive nature of speed and momentum prevents them from being pinned down instantaneously. Consequently, Artificial Intelligence/ AI is inherently limited in its understanding of Consciousness, a fact that should be approached with utmost caution.

Duration in the process of creation of CS: 3 ksana=50ms

- Feeling: (recognition/ vn:Tưởng/ skt: Sạmna) representing the non-conditional reflex for immediate defense for survival in the brain, after traversing sensory organs, the information follows the dorsal stream (Top-Down stream), going through motor areas of the cerebral cortex to elicit the reflex of reaction for defense

Duration in the process of creation of CS :2 ksana 2ksana=30ms

- Formation: (formation of impression/ vn: Hành/ skt: Samskara) is a complex integration process that corresponds to the Ventral stream of the sensory input, highlighting the intricate nature of perception.Duration in the process of creation of CS :2 ksana 7ksana=120ms

- CS: Cognition/ vn: Thức/ skt: Vijana)the process of making the CS

3. In Total, there Are Five Constituents: Form, And Four
L. Dharma/ Vn: Pháp.

Form, and four subcomponents, together forming the five Aggregates/skt: skandhas or panca/vn,: Ngũ Uẩn, Ngũ Ấm/skt: the body and soul of all living beings.

As mentioned above, the five Aggregates are the vivid manifestation of the veil of ignorance impeding humans from recognizing the Buddhahood, Ultimate Awareness/ vn: Trí Huệ, Trí Huệ Bát Nhã/ skt: Jnan, Prajna-paramita.

Emptiness is a homogenous entity that is not preceded by any entity except itself. In the Old Bible, Emptiness is exemplified by the state of Mind of Adam and Eva before the Serpent argued about eating fruits in the middle of Eden's garden. The Serpent argues that Eva can eat that fruit, ...your eyes will be open, and you are bound to be like God, knowing Good and Bad.
Furthermore, in Surangama:

For instance, though a flute can make sweet melody, it is useless in the absence of skillful fingers; it is the same with you and all living beings for although the True Mind of precious Bodhi is complete within every man when I press my finger on it, the Ocean Symbol radiates but as soon as our Mind moves, all troubles (klesa) arise. This is due to your remissness in your search for Supreme Bodhi, in your delight in Hinayana and your contentment with the little progress you regard complete.

In Lotus Sutra, Bodhisattvas S*amantabhadra (vn: Phổ Hiền) replied to* Bodhisattvas/ vn: Như Lai Tánh Diệu Đức:

"This boundless Universe is not created by a single cause or by a single phenomenon. It is created by innumerable, uncountable cause, uncountable phenomenon. Massive clouds torrential rains and four types of celestial potentials. These four potentials are:
    Maitenance for enduring deluge of water
    Destruction for clearing the deluge
    Construction for making new structures, and
    Rendering magnificent, orderly and significant

The works result from the cumulative efforts of all sentient beings and all beneficial and wishful intentions of innumerable Bodhisattvas. All sentient being can use the fruitful material available and commensurable to their needs. This boundless Universe is formed and develops from uncountable causes. This Dharma nature is uncreated, unauthorized, unconsciously designed and performed."

Bodhisattvas S*amantabhadra added*
Briefly, the Universe is not different from a grain of dust. In each grain of dust there are orderly, solemn and magnificent structures, there is no mix-up of structures

As a result, there is no need to explore the other extraterrestrial planets of the Universe to understand the mechanism of the composition and the formation of living beings.

In summary, following the false thought, the Emptiness creates the form, the metaphysical entities, and the secondary Emptiness.

Comments: *In the creation revealed in the Old Testament and Buddhist sutras, Water was first mentioned in the Form before the appearance of Earth, Wind, and Fire. It is well known that the existence of Water on any planet means that life can exist. Life is associated with various aspects of Consciousness. There was no description of elementary particles, atoms, or mineral molecules in the Old Testament and Buddhist sutras. These steps in the creation were purposely skipped in the Holy and Buddhist scriptures, likely because these steps are not contemporarily necessary for the understanding of the creation of lives and because of the ignorance of human beings.*

L. DHARMA/ vn: PHÁP.

is a terminology that carries profound implications .

The true rules/ laws that govern the creation post-Big Bang, virtue, the phenomenon of ultimate truth, Buddha's teaching, and the scriptures of other religions, are all complex and intricate.Things, objects, or living beings, because they are also created from Emptiness, are part of a profound and deep rule or law.From the above understanding, Dharma represents the interface between Emptiness and the False Thought /Big Bang-Genesis/Creation. In other words, **the Ultimate Dharma is the primary mechanism initiating the False Thought /Big Bang**

In addition, there is a tendency to distinguish: types of Dharma and Non-Dharma, in the famous Buddha's statement (in Laṅkāvatāra sutra): Dharmao should be abandoned, and Non-Dharmas even so
/ vn: Pháp phải buông bỏ, huống là Phi Pháp). Dharma refers to the creation of the physical thing, and Non-Dharma to the metaphysical things.

M. Philosophy Of The Primordial Duality, Interface Between Emptiness-Creation

(before 10^{-43} sec Planck Epoch).
Emptiness is the Oneness. Dharma, Big Bang, and After-Big Bang Creation are Duality. (F 1.8)
In this philosophy, Ultimate Dharma closest to Emptiness is designated primordial Duality (Planck epoch $<10^{-43}$ s). As a result, primordial Duality shares features of Emptiness and Duality. For example:
- Dharma are both Dharma and non-Dharma.

These concepts refer to the nature of Existence and non-existence, and the relationship between creation and non-creation.

Egocentric and Egoless.

Reality and Illusion.

Permanent and Impermanent.
- God's earthly manifestation as Buddha or Jesus is both Unique and Multiple.

True and Illusional,
 Not created, nor inside skandhas, **nor non-existent**
Because it exists it doesn't/ because it doesn't it does/ its nonexistence cannot be grasped/ nor can its existence be imagined.
 PERMANENT AND IMPERMANENT.
Egocentric and Egoless.
Power and Ultimate Omniscience.
Who gets free from all errors / truly sees my way /this is called seeing truly / not slandering the guide.".
What neither arises nor ceases is another name for Tathagatas (Buddhahood).
Existent but without a cause.
(note: it is of common opinion that Jesus is God versus Human because He existed before the world was made, performed miracles that had never been seen before, performed unique works similar to what God performs claimed to be God, resurrected from the dead...) but Jesus is a human).
- Particles in Quantum mechanics (earliest particles after the Big Bang).

Entanglement versus particle as an entity.
Wave and Particle as two facets of one entity.
- Soul.

 Local and Non-local not created and not non-created.

Power and Awareness.

Physical Entity and Metaphysical.

Morality/.Virtue/ vn: Đức is the manifestation/ projection of Tao in the material world. Tao can not be described, but one can feel Tao through its manifestation seen as Morality or Virtue. In this concept, Morality/ Virtue/ vn: Đức is equivalent to Dharma. In the concept of Genesis/

False Thought/ Big Bang, Morality/ Virtue is equivalent to Dharma. In this concept, Genesis/ False Thought /Big Bang Dharma and Morality are uncreated and have no cause or dependent origination because these phenomena represent the inherent properties of Emptiness. Therefore, Dharma is Non-Dharma.

In this concept, Buddha is distinct from Buddhahood (or Holy Spirit versus Jesus Christ). Buddhahood represents Emptiness, whereas Buddha represents the status of Interface and is characterized by:

Oneness and duality or multiplicity.

Selflessness and no-selflessness.

Almightiness and limited power by the physical body with possible illness,

Ominscience and Omniscience were obliterated by six sensory organs.

Non created Being and creating being susceptible to birth and death.

The following Sermon highlights the Oneness and Multiplicity of Buddha's nature (Laṅkāvatāra Sutra):

Bodhisatsava Mahamati asked the Buddha, why you proclaim that you are all Buddhas of the King Mandhatri, a six-tusked elephant, a parrot, Shakra And unetra'?"

> Mahamati then asked the Buddha, "Bhagavan, why did Bhagavan proclaim to the assembly, 'I am all Buddhas of the past'? Or why in recounting the hundreds of thousands of tales lives about his previous did he say, 'I was once King Mandhatri a six-tusked elephant, a parrot, Shakra And unetra'?" The Buddha told Mahamati, "It was because of the four uniformities that the Tathagata, the Arhat, the Fully Enlightened
>
> One proclaimed to the assembly, 'I was once Krakucchanda, Kanakamuni, and Kashyapa Buddha.' And what are the four uniformities? They are the uniformity of syllables, the uniformity of voices, the uniformity of teachings, and the uniformity of bodies. These are the four uniformities..
>
> ....
>
> Mahamati said, "Bhagavan, are the tathagatas, the arhats, the fully enlightened ones created or not created, the result or the cause, what sees or what is seen, what teaches or what is taught, what knows or what is known? Or are they different or not different from such terms as these?" The Buddha told Mahamati, "In regard to such terms as these, the tathagatas, the arhats, the fully enlightened ones are neither the result nor the cause. And why not? Because both would be wrong. Mahamati, if tathagatas were the result, they would be created and would be impermanent. And if they were impermanent, then every result would be a tathagata, which is something neither I nor any other buddha would want.160 But if they were not created, they would not attain anything, and their cultivation would be empty, like a rabbit's horns or a barren woman's child, because it would not exist. "Mahamati, if they are not the result and not the cause, thenthey neither exist nor do not exist. And if they neither exist nor do not exist, then they are beyond the four possibilities. The four
>
> possibilities refer to the mundane world. If they are beyond the four possibilities, then they are not subject to the four possibilities. And it is because they are not subject to the four possibilities that they are perceived by the wise. This is how the meaning of all expressions about a tathagata should be understood by the wise. "As I have said, there is no self in anything, by which you should understand that by no self what I mean is the nonexistence of a self. Everything exists as itself and does not exist as another, like a cow or a horse. For example, Mahamati, a cow does not exist as a horse. And a horse does not exist as a

> cow. In reality, they neither exist nor do not exist, but they do not not exist as themselves. Thus, Mahamati, there is nothing that does not have its own characteristics or that does have its own characteristics. But that they have no self is something foolish people cannot understand due to their projections. Thus, the emptiness, the non-arising, and the absence of the selfexistence of things are to be understood like this.

The Buddha explained, "It was because of all uniformities that the Tathagata, the Arhat, and the Fully Enlightened are associated are: uniformity of syllables, the uniformity of voices, the uniformity of teachings, and the uniformity of bodies. These are the four uniformities that only refer to the mundane world. These four uniformities have the same noumenon, Emptiness without its characteristics.

N. The Pathway: Original Mind False Thought, And The Creation Of Metaphysical Component And Form.

As a result, there is no need to explore the other extraterrestrial planets of the Universe to understand the mechanism of the composition and the formation of living beings.

It is evident that the Big Bang theory and Darwin's Theory of evolution, are incomplete/ Of note, as in the case of the theory of Quantum Mechanics. Albert Einstein pointed out that the Quantum theory is incomplete because it can not explain the EPR paradox. The paradox can not be understood and, therefore, not accepted by many current physicists, including Bohr. Five decades later, the EPR paradox is experimentally proven to be a reality. However, in comparison to Genesis in Christianity, False Thought is considered to have a similar significance with "God's *active force moving to and fro over the surface of the waters*". The orderly Creation of the form of the Universe as outlined in the Old Testament is consistent with the ingenious Mind of Buddha and Bodhisattvas.

Again, compared to the Big Bang, the Universe's accelerating expansion is also consistent with the release of the unknown energy of Emptiness through False Thought. Likely, the release of the energy of the Emptiness creates :
  - First, the dark forces responsible for Universe expansion,
  - Dark matter causing the lensing of the light and curved Spacetime, then
  - Formation of particles, physical structures known as Forms of the Universe.

Non-Baryonic in the Universe, Galaxy, and The Earth.

The Non-Baryonic, composed of Dark Matter and Dark Forces, is a hypothetical form of matter not detectable by the five senses. The five senses can only detect matter when it directly emits or absorbs electromagnetic radiation. DM tends to form a mass, especially associated with many galaxies in the Universe, but DM appears to be present in black holes.

Old Testament and Buddhist sutras revealed no explosion, tremendously high temperature, or powerful energy at the beginning of Genesis or False thoughts. Still, there was mention of the movement of the Spirit or the initiation of the False thoughts. In both cases, there was a disturbance of the stillness of the Emptiness or the Darkness/Watery Deep.

Matter suggestive of DM was described in **Mahàpadàna Sutta** kinh bộ Nikaya or Agama, (Mahapadana Sutta - Wikisource, the free online library) concerning the birth of Bodhisattvas or Buddha:
", When a Bodhisattva is born from his mother's womb, the worlds above of the gods, the Maras and the Brahmas, and the world below radiate an infinite and splendid light sound, Trembling, and quakes. The splendid radiance manifests even in spaces between the **baseless, murky, and dark worlds and where even the moon and sun cannot prevail to give light. Living beings who happen to exist there perceive each other by that radiance.**"

> *'It is the rule, brethren, that, when a Bodhisat issues from his mother's womb, there is made manifest throughout the Universe — including the worlds above of the gods, the Maras and the Brahmas, and the world below with its recluses and brahmins, its princes and peoples, — an infinite and splendid radiance passing the glory of the gods. **Even in those spaces which are between the worlds, baseless, murky and dark, and where even moon and sun, so wondrous and mighty, cannot prevail to give light, even there is manifest this infinite and splendid radiance, passing the glory of the gods. And those beings who happen to exist there, perceiving each other by that radiance, say : — "Verily there be other beings living here ! '** : And the ten thousand worlds of the Universe tremble and shudder and quake. And this infinite and splendid radiance is made manifest in the world, passing the glory of the gods. This, in such a case, is the rule.'*

In summary, Emptiness in Taoism and Buddhism implies the imperceptibility to the Five sensory organs and homogeneity. As defined, Emptiness does not mean there is nothing at the level of the Ultimate Mind.

In Physics, astrophysicists, including Stephen Hawking, agree that the Big Bang represents the beginning of the Universe. However, the enigma of existence before the Big Bang, a puzzle that has eluded even the most brilliant minds, remains an unanswered question.

The Big Bang is a physical theory that describes the Universe beginning with the expansion of an original point, initial singularity, or state of high energy and temperature. The theory is based on the knowledge of the law of physics concerning energy, the principle of formation of materials, observation of the receding galaxies, and space expansion and identification of the cosmic microwave background. Galaxies beyond the telescope Hubble distance recede faster than the speed of light. It is currently believed that the location of the Big Bang is *everywhere, in* all directions at once, and it does not represent an explosion per se. In the beginning, there is no time-space. After the expansion of time-space is created, the Universe becomes boundless without a center. With telescope JWST, one can look back at the Universe in the early stage of Creation, just 300,000 years after the Big Bang. Evidence of the Big Bang is seen in all directions.

In summary, the initiation of the False thought of the Emptiness/ Original Mind is not confined to a single location. The rise of Thought is a profound process that extensively involves all parts of the Emptiness due to its inherent homogeneity and non-discrimination. Understanding that Emptiness is not bound by time and space, but immediately after the Thought initiation, the Emptiness undergoes a transformation, losing its equilibrium and harmony. This phenomenon, occurring just before the Big Bang, is independent of the Big Bang itself. As a result, the entire Emptiness is intricately involved in the phenomenon of False thought; the Big Bang is observed everywhere in space. This co-initiation of the False thought, along with the expansion of space and the recession of remotely located galaxies faster than light, likely accounts for the so-called universe expansion.

These facts are similar to Non-Locality and interconnectedness in quantum mechanics with the supraluminal transmission. Furthermore, the Creation of the Universe may or may not need the hypothesis of tremendous heat and power at the beginning because Emptiness itself is empowered with the untarnishable source of energy and Ultimate Omniscience

.

O. Purpose Of Creation.

Understanding of Creation/ Genesis can only be complete with the discussion on the purpose of Creation.

In Islamic theology, the Creation is without purpose. The qu'ran verses elucidate how the Universe, with heaven and earth, is created to serve humanity (Qu'ran 2:22, 2:29) and to manifest the divine revelation. The purpose of creating man is to know, love, and worship God. This profound understanding of the purpose of Creation in Islamic theology is a fascinating area of study that can engage and intrigue scholars and students alike. (God has not created the Jinn and humankind except to worship =ibaadah) Me (51:560)

For Christians, Creation is a means of God's revelation. The care of nature is the expression of the love of God and to glorify God: Ephesians: 1:5-9 *he predestined us for adoption to sponserhip through Jesus Christ, in accordance with his pleasure and will— to the praise of his glorious grace, which he has freely given us in the One he loves. 7 In him we have redemption through his blood, the forgiveness of sins, in accordance with the riches of God's grace that he lavished on us. With all Awareness and understanding, he made known to us the mystery of his will according to his good pleasure, which he purposed in Christ*

In Buddhism, Buddha did not reveal the purpose of Creation, theological Buddhists rarely discussed that. Saying that because the cause and purpose of Creation are obvious. Buddha often mentions that the earliest event of Creation is False Thought. False Thought creates Ignorance and the Universe. If there is Right Thought, there is only Ultimate Onmniescience and Nirvana, a phenomenon equivalent to Christianity with Adam and Eva living in Eden Garden before developing the discriminative Thought. **Thought is a noumenon of Emptiness, this means that Thought, like Emptiness is not created, there is no cause.** *Emptiness is a hypersymmetric state with maximum energy potential; despite the stability, this stability is difficult to maintain (see Fig 1.4A)/ Breaking this equilibrium is followed by the Thought. If the hough is Right, EM is transformed into Nirvana with all features of EM like supersymmetry, No discrimination associated with unlimited energy, and Ultimate Omniscience. With a mistake in the Tought (so-called False Thought) Creation the Physical world is created (or Big Bang).* As a result, Creation is an inherent event of Emptiness. Sufferings are not due to the Creation but are secondary to the False Thought or the original sin-causing discriminative Mind. Unlike another phenomenon associated with dependent origination/ vn: nhân duyên, secondary to a preceding cause), the Creation develops from Emptiness and is not associated with any reason.

Of interest, BB, with a tremendous release of energy, was not mentioned in any major religions. It is possible that because: (i) the BB phenomenon merely accounts for only 5% of the Universe, and the BB phenomenon described by physicists is perceptible to sentient beings of this physical realm but not perceptible or significant in sentients in the metaphysical realm

In many books and excerpts, it is believed that Buddha was only interested in rescuing human beings from the samsara circle, and ultimately, the sufferance, and was not interested in the revelation or discussion of the origin of the Creation. It is true that salvation is the ultimate goal, but the problem is how to achieve it. To achieve the goal of coming back home to Nirvana, one has to understand its origin and the process of its Creation. In Surangama, Buddha explicitly explained that the origin of the Creation is Emptiness, the only One that is not created nor perished. As discussed above, Emptiness is a state of hypersymmetry and equilibrium. Despite its permanence, Emptiness is prone to degenerate by the intrinsic activity that Buddha called False Thought. As a result, the ultimate cause, the so-called dependant origination of ignorance, is Emptiness. The ignorance subsequently is the cause of the Formation, then Consciousness, then Form...Therefore, Ignorance is the most serious and original Karma committed by living Sentients. Ignorance is equivalent to the Original Sin in Christianity. It is associated with the Death Sentence, resulting in the cycle of Birth and Death of Human Beings

VI. THE CREATION OF SPECIES.

As previously mentioned, Original Awareness/ vn: Trí Huệ is derived from the Original Mind/ Emptiness. It is characterized by four significant potentials: maintenance, destruction, construction, and impeccable, ingenious, and orderly creativity. These potentials, which are uncreated and quiescent, manifest only through their effects.

To summarize, the Emptiness, following the Thought, is the creation of Form and Metaphysical entities, including the Secondary Emptiness.

It refers to *the birth of life of all beings, which is symbolized by the number twelve (calculated as 3 [present, past and future] x 4 [east, north, west, south]). Despite being separated in the ten directions of space, all living organisms are attached to the same root. This mythical revelation of Consciousness occurs at the moment of creation, which is equivalent to the very beginning of the dawn.*

A. Parents Nurture But Do Not Create Children.

It is currently believed by evolutionary biologists, anthropologists and molecular scientists that human beings and chimpanzees share a common ancestor. This belief, which has been accumulating evidence for over 150 years since Darwin published his book "On Origin of Species", is of utmost importance in our understanding of human evolution. The human origin and its evolution were based on the similarity between species emphasized by Darwin and Huxley (1). Furthermore, the evolutionary and morphological features of fossil bones (skull, mandible, teeth, enamel), analysis of DNA and genome are suggestive of the occurrence of the speciation of human-chimpanzee and chimpanzee-gorilla at 6000 to 10 000 centuries ago. It is well known that the more similar species are more closely related. However, as Wallace, who supported Darwin's theory of evolution, pointed out in 1864, "Children of the same parents are not all alike". The opposite is true: look-alike people may not come from the same parents.

In this observation, the careful study of the germline in most animals shows that the germ cell line is distinct from the somatic cell groups. This distinction, which is a result of the complex process of early embryogenesis, is a key factor in understanding the evolution of species. In early embryogenesis, the germ line cells undergo mitotic divisions at the end of the second week of post-fertilization. Each division gives rise to a germ cell and a somatic cell. After the third mitotic cell division, the germ cell undergoes meiotic cell division and differentiation into the pure germ line with further progression into ova or spermatozoa essential for species propagation. In the same period, the somatic cells derived from the germ line undergo successive mitosis, indispensable for the formation of body and accessory structures.
The germ cells do not originate from the somatic cells except in certain animals of low levels of phylogenetic scales, like planarians. In a developed body, germ cells are localized in the gonads (Figure 1.9).

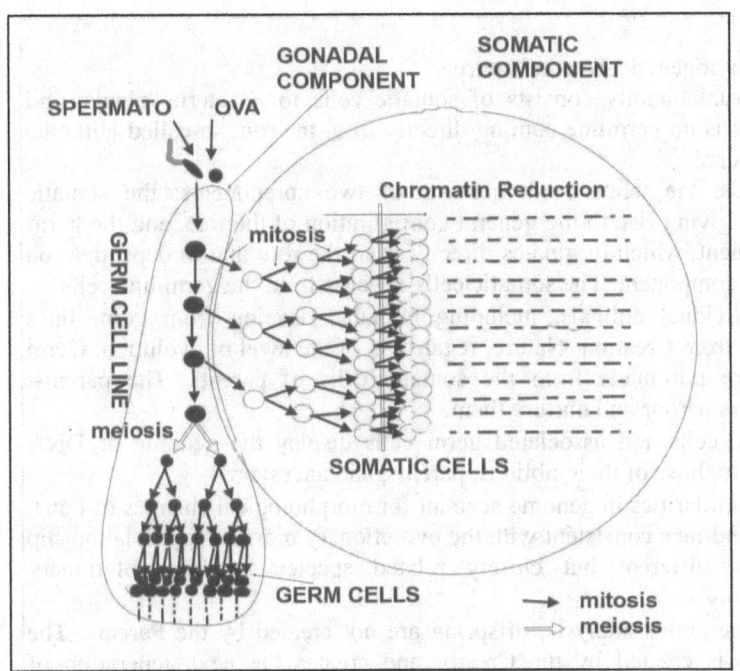

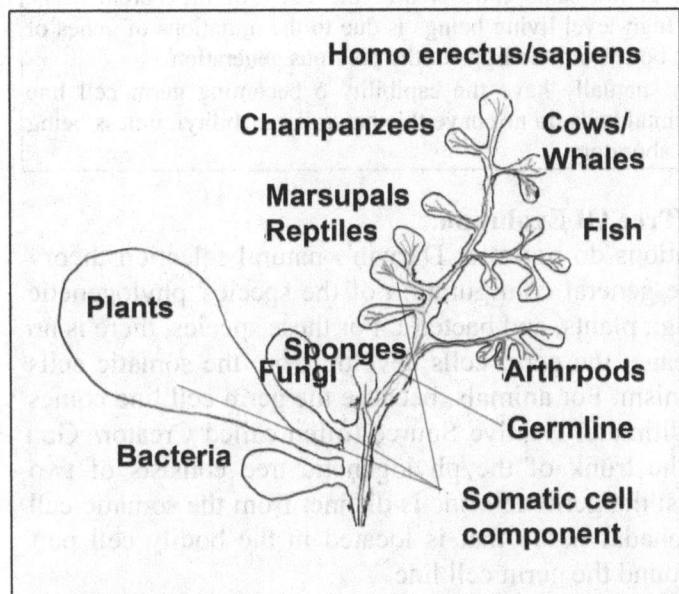

Figure 1.9 **Figure 1.10**
Showing the formation of the germ line in the gonadal component and the development of a somatic component only consisting of somatic cells from the fertilized ova following the successive mitosis of the zygote (fertilized ova).). The red line inside the tree trunk shows the cell lineage for each animal that originated from the Creator
Figure 1.10

> Proposed Phylogenetic Tree in Nature
> -The tree trunk mainly consists of somatic cells for Bacteria, plants, and fungi. There is no germline coming directly from the root so-called Ultimate Creative Source
> -For animals, the tree's trunk consists of two components: the somatic component, giving rise to the general configuration of the tree, and the germ line component, which originates directly from the root and in dependent on the somatic component. The somatic cells develop from the germ line cells
> - All individual animals, including humans, develop from germ lines created from Creation/ Nature, regardless of the level of evolution. Germ cells are not made from the somatic cells of parents. The parents/ancestors harbor and nurture them.
> - Somatic cells and associated germ cells display the genome or DNA closest to those of their siblings, parents, and ancestors.
> - The similarities in genome account for morphological changes of bone, body, and face consistent with the evolutionary morphologic relationship between different but closely related species in the evolutionary phylogeny.
> - From the above analysis, offspring are not created by the Parents. The embryo is created by the Creator and creates the next generation of embryos and at the same time creates the body of the Parents. The formation of high-level living beings is due to the mutations of genes of the new Eggs born from the Eggs of the previous generation.
> - Vegetal cells natùally have the capibility ò becoming gėrm cell line cells. But animal cells do not have this naturall capabilityi, unless being modified in Labòratory

B. Phylogenetic Tree Of Evolution.

The above observations do not alter Darwin's natural selection theory and do not alter the general configuration of the species' phylogenetic tree, mainly for fungi, plants, and bacteria. For these species, there is no pure germ cell lineage; the germ cells develop from the somatic cells by a special mechanism. For animals, because the germ cell line comes directly from the Ultimate Creative Source (often called Creator/ God or Buddhahood), the trunk of the phylogenetic tree consists of two parallel components: the germ cell line is distinct from the somatic cell component. The gonadal tissue that is located in the bodily cell part (Figure 2) wraps around the germ cell line.

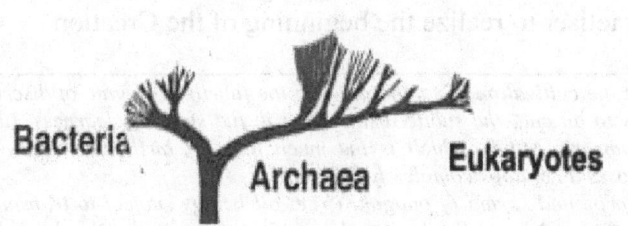

Fig 1.11 Tree of Evolution, Archaea is as small as bacteria but has nucleus

These findings may reinforce the belief in major religions (Islam, Christianity, and Buddhism) regarding the creation of human beings directly from the Ultimate (Original) Creative Source. For Christianity and Islam, it is well known that God and Allah created Adam and Eva. In Mahayana Buddhism Surangama sutra, Buddhasaid: the birth of life of all beings of the twelve (3x4: 3= present/past future x 4= directions) forms of living organisms, although separated in the ten directions of space, all attached to the same root. This mythical revelation of Buddha's Consciousness is at the moment (of Creation) that is equivalent to the very beginning of the dawn. (J of Phylogeny Evol Biol,10:11, 2022.10.250"Parents Nurture but Do not Create Children.")

At the Meditation at the deep level, when the last veil of ignorance is cleared out of the way for apparent Awareness, the meditator can see the birth of all species, as illustrated in this saying of Buddha at the end chapter of the Surangama sutra. As discussed in the paragraph of the veil of ignorance , the integration of data received by the six senses constitutes a barrier from the Original Mind to get assess to the data. The data is a whole package must be split into six separate packages. The veil of ignorance impedes humans from acknowledging the origin of species at the Creation.

In Chapter X of Surangama Sutra, Buddha reveal of the Creation:

     In the Meditation, when the aggregate of formaton is cleared, the fifth aggregate of Consciousness releases the practiser from all attraction, the Mind, are clear and transparent like crystal both within and without. This is the end of the aggregate of Consciousness which enables the practiser to realize the beginning of the Creation

> *Ananda, in the cultivation of samàdhi, when the fourth aggregate of discrimination (saüskàra) comes to an end, the subtle disturbance in the state of clearness, (that is the functioning of samsaric Mind), which is the mechanism of birth and death, suddenly explodes and exposes an outlook completely different*
> *from that of the profound karma of pudgala (, i.e. all beings subject to transmigration). This is the moment when Nirvana is about to dawn, like the cock-crow that heralds the first light of the day in the east, when the six senses are void and still and no more wander outside. Within and without there is only a profound brightness reaching the root of life of all beings of the twelve forms of birth in the ten directions of space wherein there is nothing that can be further penetrated. This contemplation of the essence of basic clinging (, i.e. the fifth aggregate of Consciousness) releases the practiser from all attraction (by old habits and new karma) and immunizes him from further transmigration in saüsàra for he has realized the identity of Mind with its self-created externals everywhere. As the nature of Consciousness now manifests clearly, he will discover its hidden depth. This is the fifth aggregate of Consciousness which conditions the practiser's Meditation.*
>
> *.As the practiser is immune against external attractions and realizes the identity of Mind, and objects, the separateness arising from the six different sense organs ceases and the Mind functions uniformly with seeing and hearing in regard to a single function which is pure and clean. In this state, all the worlds in the ten directions, together with his body and Mind, are clear and transparent like crystal both within and without. This is the end of the aggregate of Consciousness which enables the practiser to leap over and beyond the kalpa of turbid life, the main cause of which is the (first) seeming shadow of his wrong thinking*

One wonders why this simple, fundamental, and very important question for human beings has been unfortunately misled. The reason is the veil of ignorance. The knowledge has been misled by science which only focuses on the epi-phenomenon of evolution. The accomplishment of technology has tremendously improved the materialistic quality of life and has rendered people to stop believing in major religious teachings. Buddha repeatedly said Buddha's teachings are not illusional!

Since human beings are so proud of themselves, with some rational thinking, one can see that if the ancestors are monkeys/ chimpanzees, the ancestors of chimpanzees are lower-level animals; successively, the conclusion will be that the ancestors of human beings are insects or so... (Wiley, EO, 2011)

C. Conclusions.
Despite tremendous development and achievement for more than 100 years since the successful proposal of many theories with Max Planck's discovery of h constant, Albert Einstein's suggestion of light waves are quantized, Ernest Rutherford , Niels Bohr and Werner Heisenberg, Wolfgang Pauli in the design of the structure of atoms and David

Bohm's concept of implicate order of the wholeness of atoms with implication in the concept of Mind and quantum level. However, science is still limited in 5% of the Universe containing baryonic matter.

- Humans are created by the Creator. Parents or lower sentient beings only nurture children. The objection to this reality is a critical mistake due to the superficial morphologic observation with the input of the veiled Mind. Monkeys and chimpanzees do not create human beings but carry and protect human germline cells.
- The ingenious assembly and arrangement of the Creator creates the world. It is irrational that chance and accidents have to happen multiple times in Genesis to build this world as it is nowadays.
- Big Bang is still an incomplete theory for all phenomena, including logistic observation. The approach demonstrated similarities with False thought arising from Emptiness since both create a local expanding quake with acceleration to create time and space. Among the deficiencies of the Big Bang is the need for more ingenuity that is required at many levels of Creation, like forming minerals, trees, and sentient beings. Other phenomena like entanglement, interconnectedness,
- As a result, religion and theism are necessary and accompany worldly life.
- Buddha said in the Nikaya sutra (Long Discourse) pointing out th role of spirits (devas) in helping humans

> 'Ananda, just as if they had taken counsel with the Thirty-Three Gods, Sunidha and Vassak Fira are building a fortress at Pzrtaligiima. I have seen with my divine eye how thousands of devas were taking up lodging there. . .(as verse 26). Ananda, as far as the Ariyan realm extends, as far as its trade extends, this will be the chief city, Pafaliputta, scattering its seeds far and [88] wide. And Pataliputta will face three perils: from fire, from water and from internal dissension.' 1.29.

- Science is necessary for material life when humans, with the physical body, develop technology. Nevertheless, besides the physical body, the Form, the metaphysical part (Perception, Feeling, Formation, and Consciousness) is even more essential and lasts much longer. It is critical in guiding humans to everlasting happiness. Theism is necessary for the Creation of the Universe, Nature, and Species and plays an influential role in living. Religion may cause an illusional view of life with dependence on religion instead of being self-supportive and self-relying. Freud, as well as his contemporary fellows, believe that theism makes humans more dependent than to liberate them from theism.

Flower Adornment/ vn: kinh Hoa Nghiêm and Lotus/ vn: kinh Pháp Hoa are two major sutras among Buddhist sutras. Splendid, magnificent Buddha landscapes with miraculous ingenuities exceeding imagination and limitless multiple Universes. On the other hand, Surangama sutra, Buddha went into the ontologic aspect of Buddhism, laying out:
-The significance of Emptiness.
-False thought transforms the Original Mind (common to all Universes, minerals, and all sentients) into different Forms of nature, such as water, earth, mountains, and living organisms.
- The development of worldly life with inversion.
- Formation of the Consciousness by splitting the original data into six separate sources of information and the veil of ignorance, the important cause of the inversion.
   - the path of liberation from suffering.

Implication In Vegetarianism.
Vegetarianism is the practice of living depending on no-meat products, therefore abstaining from killing animals. Ovo and ovo-lacto vegetarianism allows the consumption of eggs and dairy products. Consumption of non-fertilized eggs is usually similar to milk and represents an abuse of animal welfare. However, consuming milk and dairy products is eventually indispensable for human survival and is allowable in Buddhism. Consumption of non-fertilized eggs that can carry the risk of attraction of nonphysical sentients is, therefore, not permissible.

The key point in Vegetarianism is that germline cells are strictly derived from the Creator as opposed to somatic cells in animals and vegetal cells from parents/ plants harboring vegetal buds. Somatic cells develop from germ cells, but never germ cells develop from somatic cells in animals. Therefore, animals represent the children of the Creator, while the Creator does not directly create plants, mushrooms, and bacteria. Therefore, Vegetarianism does not involve killing the children of the Creator.

Furthermore, vegetal cells can act as germ cells to create new plants, while somatic cells of Animals can not be transformed into germline cells in natural conditions. It is known that in laboratories, Somatic cells can be made to change into germline cells.

VII. KNOWLEDGE, WITH CONSCIOUSNESS VERSUS AWARENESS

Knowledge is the mental process of the acquisition of information. The information is theoretical or practical and experimental.

The acquisition follows the experience, learning/education. Furthermore, the quality of the information is characterized by the depth and the extent of details. CS/ understanding requires mindfulness with attention. With the concentration of mental energy on the process, the obtained information is rich in detail and in-depth. The extent characterizes Awareness, , i.e., the data comes from larger and more numerous areas. Contrarily, Awareness requires less attention and less mental energy. Areas of large extent accompany the information with detail and insight, and vice versa; therefore, CS and Awareness are two mental processes mutually inhibitory.

In neuroscience, CS and mindfulness, commonly accompanied by a strong attention component, use the cerebral dorsal pathway for the conduction of information much more than the ventral pathway. The cerebral ventral pathway in this information acquisition mode provides Consciousness, while the dorsal pathway focuses attention on a designated area. Attention is a prerequisite for the mechanism of CS formation.

On the contrary, the Awareness uses the ventral pathway more than the dorsal pathway. Because of the weak attention component, the Awareness covers large areas of information but fewer details and is more superficial in depth.

In daily life, such as driving, one has to use Awareness to have an overall view of the environment and mindfulness to focus attention on running the car on the street.

Mindfulness	**Awareness**
thinking	observation
anxiety	tranquility, stillness, peace
discrimination	no discrimination, equity, Indifferen
subjectivity camera-like	objectivity, artist-like
focal	global
time and extension	time and space contraction
logic	natural, as it is

requiring control /attention	natural
dreaming, abstract	reality
significane	appearance
obvious	hidden
Error and failure	accuracy and success
Past tense, difficulty in correction and planning	present tense, feasibility of correction and planning
complicated, difficult	simplified, easy
turmoil	peace
isolated	popular
peripheral, incoming information	central, inner information
temporary	permanent
many words	silence
impure	pure
terrestial power	heavenly, celestial power
recycling	preservative
samsara	liberated fron samsara
turbulent	quiet
energy consuming	energy economic
conditional reflex of Pavlov type,	Free will in morality
Function with blindness, ignorance	Function with enlightened Mind
false ego with intention, ready answere to any questions,	egolessness
Politician, university professor	Spiritual master
No morality	Morality
selective	harmonious
Mindfulness/attention, .	elightment with experience
Meditation: dorsal pathway with attention to the breathing	Meditation ventral pathway with observing the movement of abdomen and chest wall

In quantum mechanics, the attention to the particle will obliterate the Awareness. Since the observation using Consciousness is insufficient to identify the particle beyond the physical realm. Particles will only be recognized when the attention is focused on the wave, liberating the Awareness close to the original/transcendental omniscience. Omniscience is at the level of the particle to be identified.

The meditator uses the dorsal pathway to tie up the mindfulness in Meditation. The Consciousness is narrowed to the area of attention of mindfulness. The ventral pathway is clear from the CS/veil of ignorance.

VIII. VEIL OF "IGNORANCE"/ STUPIDITY (Fig 1.12,13).

A. **Veil Of "Ignorance (incomplete and deviated CS, according to Gödel's incompleteness theorems)**

The Creation started with the False thought which arose from the Emptiness. Emptiness is defined as the still, silent, homogeneous, harmonious, formless background. Since Nature tends to fill the empty and remove the redundant, the Emptiness content is not perceptible to five sensory organs but is filled with imperceptible matter and forces. As pointed out by Nagarjuna, this formless Emptiness is self-existing and, therefore, is the Ultimate Oneness that has Free Will, Ownership, and Ultimate Omniscience. The False Thought creates the deviation accompanied by obliteration of the Omniescence.

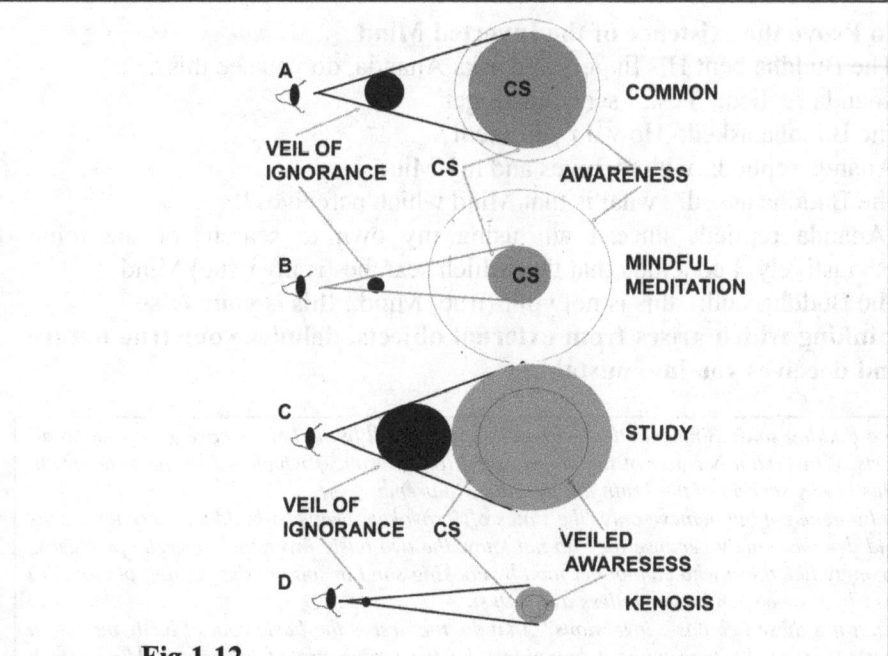

Fig 1.12
Clear circle: Awareness. Black circle: CS, Gray Circle : Understanding
A: Common
B: Meditation: attention to the restricted target: the understanding is narrower than normal and less detailed. The Awareness is
 uncovered. If the meditator changes the attention to the areas of Awareness, the Awareness immediately disappears due to the CS involved
C: Learning: increases understanding, obliterated Awareness, and ignorance develops.
D: Kenosis lack of proportional development of Awareness with narrowing awareness

Buddha said in the Surangama sutra:*Origin of Inversion*

The Buddha said: .Since the time without beginning, all living beings have given rise to all sorts of inversion because of the karmic seed (of ignorance) which is like the aksa shrub (a type of plant with large leaves covering extended area of soil).

Therefore most living beings (pratyeka-buddhas, heretics, devas and demons…) , practice wrongly, do not know the two basic inversions which are
i. the basic root of birth and death is caused by the wrong use of a Mind which people mistake for their own nature, and the
ii. attachment to causal conditions/ the Mind which obliterates the basically bright essence of Consciousness (Emptiness/Omniscience which is the fundamentally pure and clean substance of Nirvanic

To Prove the existence of the Inverted Mind
The Buddha bent His fingers and asks Ananda, do you see this?.
Ananda replied: .Yes., I see your finger
The Buddha asked: .How do you see it?.
Ananda replied: .with my eyes and my Mind.
The Buddha asked: . what is that Mind which perceives?
Ananda replied: since I am using my own to search for the mind exhaustively, I conclude that that which searches is my (true) Mind
The Buddha said: . this is not your (true) Mind.. **this is your false thinking which arises from external objects, deludes your true nature and deceives you into mistaking**

> *The Buddha said: .Since the time without beginning, all living beings have given rise to all sorts of inversion because of the karmic seed (of ignorance) which is like the aksa shrub. This is why seekers of the Truth fail to realize Supreme*
> *Enlightenment but achieve only (the states of) ÷ràvakas, pratyeka-buddhas, heretics, devas and demons, solely because they do not know the two basic inversions, thereby practising wrongly like those who cannot get food by cooking sand in spite of the passing of aeons (a very long time span) as countless as the dust.*
> *What are these two basic inversions? Ananda, the first is the basic root of birth and death caused, since the time without beginning, by the wrong use of a clinging Mind which people mistake for their own nature, and the second is their attachment to causal conditions (which screen) the basically bright essence of Consciousness which is the fundamentally pure and clean substance of Nirvanic Enlightenment. Thus they ignore this basic brightness and so transmigrate through (illusory) realms of existence without realizing the futility of their*
> *(wrong) practice.*
> **The Inverted Mind** *Probe into the false Mind Ananda, as you have enquired about the Samatha Gateway through which to escape from birth and death, I must ask you a question.. The Buddha then held up His golden hued arm and bent His fingers, saying: .Ananda, do you see this?. Ananda replied: .Yes The Buddha asked: .What do you see?. Ananda replied: .I see the Buddha raise His arm and bend His fingers, showing a shining fist that dazzles my Mind and eyes.. The Buddha asked: .How do you see it?. Ananda replied: .I and all those here use the eyes to see it..*

> The Buddha asked: .You say that I bend my fingers to show a shining fist that dazzles your Mind and eyes; now tell me, as you see my fist, what is that Mind which perceives its brightness? . Ananda replied: .As the Tathàgata asks about the Mind and since I am using my own to search for it exhaustively, I conclude that that which searches is my Mind..Thinking is unreal The Buddha said: .Hey! Ananda, this is not your Mind.. Ananda stared with astonishment, brought his two palms together,rose from his seat and asked: .If this is not my Mind, what is it?. The Buddha replied: .**Ananda, this is your false thinking which arises from external objects, deludes your true nature and deceives you into mistaking**, since the time without beginning, a thief for your own son, thereby losing

The above is an example of many paragraphs in the Buddhist sutras in which Buddha pointed out that the human common view of the world / nature does not originate from the true self, the Original Mind, but from false thinking.caused by two basic inversions that are:
- Use the CS (veil of ignorance) to view the world
- Not to use the Original Mind
    As Buddha said, false CS developed when receiving information from the external world is the culprit for the deviation of the Orginal Mind. Followings are th

B. **Five Skandhas /Or Maras**
   (vn: Ngũ Uẩn/ Ngũ Ấm).

In the Sangiti Sutta in the Long Discourses, Buddha pointedout that Human being is made up with five *pancûpadana*-skandhas: Form, Perception, Feeling, Formation and CS:
Five aggregates of grasping are:
body, feelings, perceptions, mental formations, and Consciousness.
Form is equivalent to chunk of foam, Preception as mirage, Feeeling as a buble, Formation as mental Formation (integration of data) and CS as an illusion

Significantly, the five Skandha act as formidable barriers, preventing humans from realizing their Buddhahood.
It's crucial to note that the five Skandha are not just barriers, but are referred to as five Maras. They lead to deviation and form the veil of Ignorance or stupidity.
The teaching fits very well with the current knowledge in biology and Neuroscience (please see page 80).
Ignorance is caused by the obliteration of the veil of ignorance/stupidity and is not related to intelligence. Intelligence refers to the result or the end product of integrating information in the brain and translating it into consciousness. As described later in this book, this integration process is mainly in the parietal and temporal lobes (the posterior parts of the brain), and the DMN areas. These brain parts are well developed by mathematicians and

physicists, including Einstein. Since CS is focused on narrow spaces, it differs from the Awareness which covers a large extent of the environment. Ignorance is referred to the Awareness near original/ultimate omniscience. The restriction of Awareness is the common cause of ignorance.. **This restriction comes from the predominance of CS, mindfulness, and attention.**
Information perceived in living sentient goes through five barriers or veils of ignorance: form, perception, feeling, formation, and Consciousness

the hope of clearing away Karma.

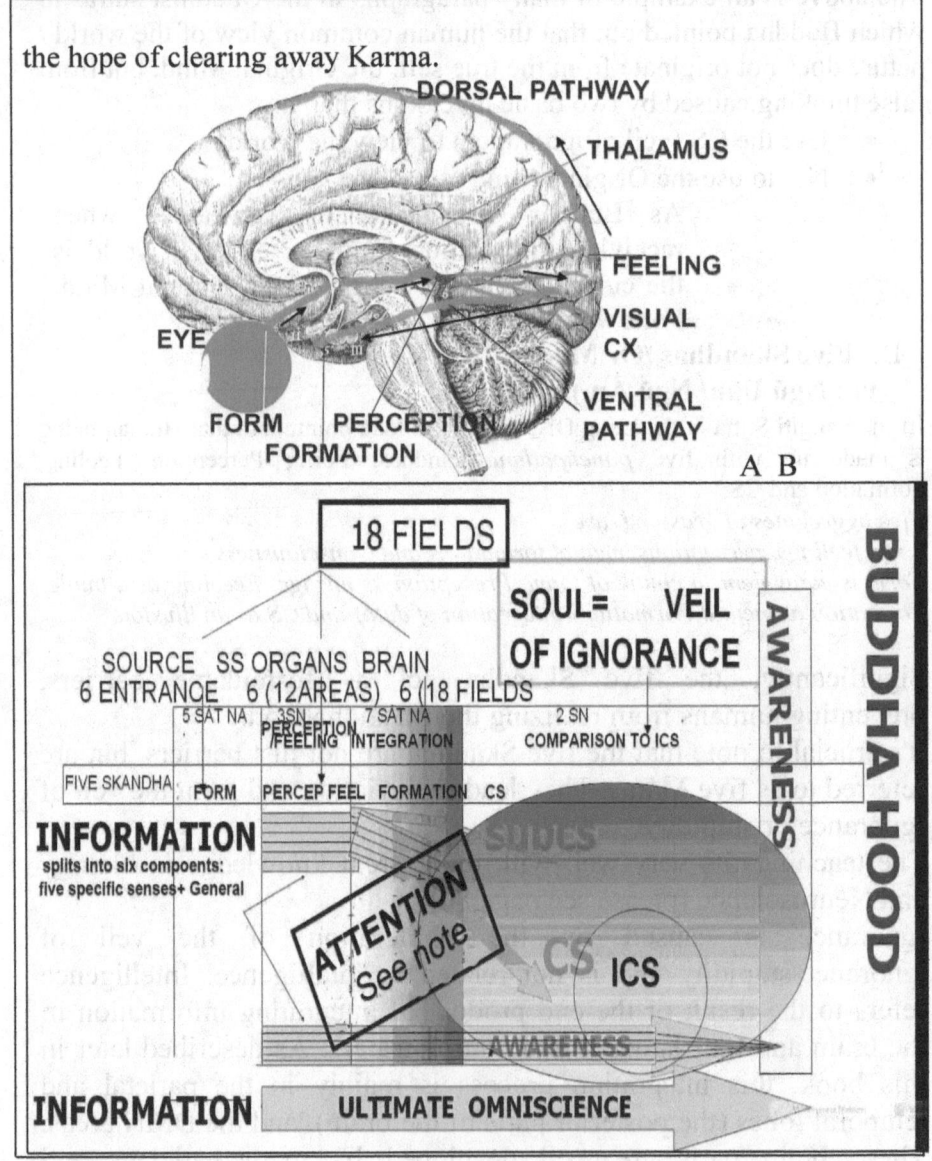

> Note: Without ATTENTION → Without mindfulness:
> There is no seeing when looking
> There is no understanding when hearing
> There is no tasting when eating
>
> **Figure** 1. 13
> A: Pathway of vision: Retina→Thakamus→ Visual cortex :
> →Dorsal Noncondtional reflex
> →Ventral pathways; d Formation of CS
>
> **Figure** 1. 13 Diagram showing the original package of input information undergoing two types of modification:
> - Splitting the original package of information into six types of information: five specific sensory organ types and a nonspecific type
> - Filtering effects by five filters
> - Milieu (forming six entrances)
> - Sensory organs (Twelve ayatana, the fusion consisting of six entrance+ six sense data)
> - Feeling: Information triggers a reflex of defense by using the dorsal pathway
> - Formation: integration of information by using the Ventral pathway (Eighteen Fields or realm, the fusion consisting of twelve ayatana+ six types of formation)
> - Consciousness formation by comparing the information in the Ventral pathway to that in the ICS
>
> Therefore, the information undergoes 18 types of filtration called eighteen realms (constituting four steps of filtration) that modify and distort the input. The last step is comparing the data stored in the ICS to label the CS. In the stage of formation, the critical factor is Attention. In the CS labeling, the connection of the ICS with the Ultimate Omniscince/Buddhahood is essential.
> In the Formation of Awareness, the original package of information also undergoes splitting into six types. Attention is also necessary, but Attention covers large areas.
>
> The brain and Attention are unnecessary for Ultimate Omniscience, and the information package is not split.
> Buddhahood is universal, present everywhere and at any time
> Without Attention, information is retained in the ICS as SUB CS or UNCS that may exert influence in the functions of viscera and Formation of CS

- Form, duration: 5 ksana=80ms
- Perception :3 ksana=50ms
- Feeling 2 ksana 2ksana=30ms. In this process, the data reach the sensory cortex that is immediately associated with the Top Down stream to the motor cortex for the unconditional reflex (for example, closing the eye when seeing a fly approaching)
- Formation; duration 7 ksana=120ms
- Consciousness ksana 32ms

(please see page 80).

By analogy the running stream of water: Form represents the Foam covering the surface of the stream, and the remaining four elements represent the stream).
The splitting of the original data from one package into six different types of information commensurate to five sensory organs and CS can be considered as one of the mechanisms of alteration of the original data. Furthermore, in the brain, the split data undergo further secondary and tertiary splitting. For example, the visual stream from the visual cortex to the inferior temporal gyrus is split to be integrated into areas of color, shape, and face recognition.

Without the body and the brain, the information comes directly to the Ultimate Omniscience/ Original Mind as it is, in a single package.
The five sensory organs split the information into six packages (5 sensory organs + general/ semantic or general Memory/ CS). Buddha also distinguished:
- Six Entrance.
- Twelve ayatana/ areas, the fusion consisting of six entrance+ six sense data (six types of Source+, six types of Organs).
- Eighteen Fields or realms, the fusion consisting of twelve ayatana/ areas + six types of formation.
Therefore, the information undergoes tremendous filtering, alteration, and modification before becoming the type of information ready to be identified by the inner CS.

These processes of alteration, modification, and filtering change, causing distortion, is the basic concept of the veil of ignorance. The information becomes individualized according to each person, society, and culture. It is said that we see things that we are not as it is.

Distortion is most remarkable with the CS due to the role of attention; with Awareness, the attention is much reduced; therefore, the information is less distorted or modified.
In Meditation, the attention is reduced to the minimum, and the information is closer to that perceived as being omniscient.

Meditation, a profound practice, offers the potential for achieving the original Omniscience and dispelling the veil of ignorance. The mechanism of this achievement will be discussed in the Chapter on Memory. In the Surangama sutra, Buddha's words intrigue, promising that as the five layers of the veil of ignorance are successively cleared, the meditator will gradually be liberated and achieve the original Omniscience (see more in Meditation).

To highlight the above mechanism of the distortion of the worldly view by the veil of ignorance/ stupidity, the following is the important excerpt from Surangama Sutra, in which Buddha explained to Ananda and the Assembly of disciples:

> *The Buddha took pity on Ananda and those in the assembly who still needed to study and. .....*
> *Excellent, Ananda, if you want to know about the innate ignorance that causes you to transmigrate in saüsàra, (you should know that)* **the roots of your birth and death are your six sense organs**. *If you want to know about Supreme Bodhi, it is these six organs that will enable you speedily to realize happiness in liberation and permanence in Nirvàna.*
>
> *.... So he bowed and asked the Buddha: How can the same six organs cause me to transmigrate in saüsàra and be happy in absolute Nirvàna?.*
> *The Buddha said: Ananda,* **both organs and their objects spring from the same source, bondage and liberation are not two different things.** *Consciousness is illusory, like a flower in the sky Ananda, your knowing originates from each phenomenon which takes on form because of your sense organs. Both form and seeing are mutually dependent, like two bundles of rushes that stand by leaning against each other. Therefore, if your intellect acts as the knower, this is the root of your ignorance (but) if it is free from seeing, it will be Nirvàna which is transcendental and pure. How then can the latter allow foreign elements to intrude?.*

Buddha explains that : both organs and their objects are identical and from the same source. Consciousness is illusional. Both form and seeing are mutually dependent. Therefore, if your intellect (Mind) acts as the knower, this is the root of your ignorance.

In summary: Buddha, in his compassion for human beings, pointed out that the source of ignorance is the six senses. Originally, the data and the six senses came from the same origin and were attached together. It is discrimination (CS) that separates them, making us more aware of the role of discrimination in our perception.

Similarly, in the Brihadaranyaka Upanishad, revered Shankara, a perfect commentator, explains: "As in the world a form is revealed as soon as the observer's eye is in touch with light, similarly the very moment that one has the Knowledge of the Supreme Self, ignorance regarding It must disappear. Hence, the effects of ignorance are impossible in the presence of the Knowledge of Brahma, like the effects of darkness in the presence of a lamp." This profound insight underscores the transformative power of self-knowledge, a power even the Gods cannot hinder.

An example of a story was written: a man with 100 sheep spends an overnight at a farm. He tied up 99 sheep to fixed posts except for one sheep because he ran out of ropes. The man asks the owner of the farm to borrow a rope, but the farm owner replies that he doesn't have one and advises the sheep keeper to pretend to tie the remaining sheep as he does for other sheep. In the morning, he is very glad that all sheep remain at the same place. He starts to untie the 99 sheep. To his surprise, the untied sheep remains as it is without getting up the journey. The man asks the farm owner how to wake up the sheep. The latter replies that you must pretend to untie that sheep as you did for the others. By following the instructions, the last sheep stands up and is ready to go. The veil of ignorance of the later sheep is no more serious the human beings who mistakenly believe that they own their bodies or assets.

KURT GODEL'S INCOMPLETENESS THEOREMS AND THE VEIL OF IGNORANCE107

Although in Science and philosophy, the incompleteness of reasoning is common and often accepted as in the saying: "humam being is imperfect," everyone, including mathematicians, finds it difficult to accept the incompleteness of this subject of logic. For a long time, the popularity of the problem of language has caused paradoxes, such as:

The paradox of lying: I am a liar (so I am not lying!), or

The barber or the doctor: I only cure people who cannot cure themselves (so does the doctor cure him?)

In Quantum Mechanics, photons or electrons originating from the same system always behave the same because there is superluminal communication even though they are thousands of miles apart. That is the EPR/Einstein-Podolsky/Rosen paradox.

The longer people live on this earth, the more paradoxes they create themselves, not to explain paradoxes but to reduce paradoxes according to the principle of entropy/disorder increasing with time.

Thus, Gödel's Incompleteness Theorem not only affirms the incompleteness of Mathematics but also extends its implications to common knowledge subjects such as beliefs in religion, philosophy, and Science. It challenges the very foundation of rationality in mathematics and raises profound questions about the limits of human understanding.

Gödel's theorem not only questions the absolute accuracy of Mathematics but also serves as a humbling reminder of the limitations of human knowledge. While mathematics is often considered the pinnacle of logical and rational thought, it too is a product of our understanding, and therefore, inherently relative. This relativity extends to other fields of knowledge such as physics, biology, and philosophy, reminding us that our understanding is always evolving and never complete.

Gödel's theorem predicts that philosophy and Science, as human endeavors, will always be in a state of flux. Philosophy, a reflection of our understanding of the world, changes with space and time, adapting to new discoveries and perspectives. Science, while striving for accuracy, is not immune to errors, as evidenced by its past mistakes in Medicine, biology, physics, and astronomy. This ever-changing nature of knowledge is a testament to our adaptability and openness to new ideas.

The shortcomings of Medicine are obvious. Inadequate understanding of diseases, such as stomach ulcers caused by the bacteria H. Pylori. The bacteria are obvious to the eyes of the research doctor but have been blind for decades....

Darwin's evolutionary tree has a serious flaw. Humans are the pinnacle of the Creator's creation, which has been mistaken in the knowledge of a researcher, making the whole world think that Humans are created and are descendants of the species of the Chimpanzee. (Kien T Mai | Canada (hilarispublisher.com) . In astronomy, the concept of the universe is constantly changing, and even the Big Bang theory is questioning the mechanism. In quantum/Quantum, the idea that the world of Quantum particles cannot be affirmed. So, the Quantum world forms a strange world separate from the macroscopic world, which was criticized by Einstein and considered incomplete by the Quantum theory. This incompleteness is clearly in harmony with Gödel's incompleteness theorem of mathematics. In Computer Science, Alan Turning (1912 – 7 June 1954), the father of modern computer science, used Gödel's theorem to predict that no matter how perfect a computer software is, it can have errors (computer glitches).

Why is that? The problem is actually simple because Science is a product of knowledge, so it is not complete, and the information is covered with distortions. The shortcomings may be small and insignificant, such as a small error in the computer while in use. However, incompleteness can be very massive, such as Science not knowing about 95% of the universe (including the Dark force). Science can only describe at most 5% of the universe. That is not to mention that Science has described it wrongly (such as about the Quantum world). Therefore, theorems 1 and 2 can be applied in the case of the universe as presented.

As presented above, CS is the veil of ignorance that limits and distorts the true Awareness of things, so CS subjects, including mathematics, are also limited and obscured by the veil of ignorance. Another way to decode the mechanism that creates the incompleteness of CS is CS's point of view, or more precisely, CS's observational position in Genesis and the Universe. CS is a dualistic view of all issues because humans live in a dualistic world. Knowledge cannot be consistent; it is biased and

Therefore, when analyzing the sentence; I am a liar, from the perspective of Monism, it is unacceptable because in Dualism, not only is the real I often overwhelmed by the false I. The real I (True Self) is the standard and only reference system that exists only in the world of Monism. In the world of Dualism, there are two I's, the true I (not lying) and the false I (lying). Therefore, the mistake is created by the True Self and the false I merging into a standard reference, becoming a self-reference (or self-judgment about oneself, thus creating a paradox. Likewise, when claiming to be I, a paradox has already been created). In the quantum world, two photons are one because they share the same CK (Simple point), and when they are separated into two, a paradox has already begun. In Gödel's Theorem, the mathematician's analytical judgment can be either from the false Self or the True Self. When the True Self is used, the result is correct. However, the True Self cannot be used regularly in all difficult problems of mathematics unless there is a rare quality of intuition.

Gödel's Theorem is inherently incomplete.

Gödel's Theorem, judging about its own category, will also be stuck in the "Liar's paradox": Gödel's Theorem itself is not as accurate and complete as it indicates. That is, the incompleteness of mathematics (or any subject of epistemology) is not complete, and it can be complete than the theorem initially indicates. For example, Gödel's Theorem says that 5% of problems are incomplete or wrong to make it easier to understand. When applying Gödel's Theorem a second time (referencing a second time), the error can be only 4%. And so on..., the error will gradually decrease. of course, always greater

than zero, not to mention that Entropy/disorder increases with time, making it impossible for the incompleteness to be eliminated. This fact is shown through scientific discoveries or the resolution of scientific problems or difficult problems with time.
In short, the incompleteness theorem's incompleteness corresponds to new discoveries in the subjects of logic, such as medicine, physics, and mathematics, every year but never solves all the mysteries of nature.

With the above concept, cultivating to become Selflessness is the ultimate method to dispel Gödel's Theorem. Slflessness is truly Paramita, transcending to the other side, no longer obscured by the veil of ignorance, that is, in the world of Oneness!

After 45 years of preaching in the earthly world duality, the Buddha said, "I do not say a word" because when speaking to disciples, the Buddha has become Anatta, so words can no longer fully express the truth, so speaking is like not speaking, what remains is only the meaning of words.
In short, the Earth/Saha/Dualistic world is a world of contradictions, paradoxes, and incompleteness. Mathematics cannot be an exception to the law of Duality, so it must be incomplete. That is also the meaning of this Saha world as a school of re-education, absolutely not the eternal home of humans, so the perfection of Goodness and Beauty can only be found in Nirvana's Heavenly and Metaphysical realms. .

C.Role Of The Attention In The Generation Of CS
Without attention, data received in the brain is not registered in the CS but recorded in the subCS and even in the Soul as karmic CS (vn: Nghiệp Thức). It is evident that attention is characterized by heightened levels of neurotransmitters, particularly norepinephrine. The role of glutamate, a key player in this complex process, highlights the received data in the cerebral cortex.

D. The World, As False Perception/Projection Of The Mind, Is Illusional Or Distorted Reality

Due to the veil of ignorance, a concept in Buddhist philosophy that refers to the lack of true understanding of the nature of reality, especially the inner CS, human perception and subsequent integration (resulting in projection) of the world become increasingly distorted in the afterlife. In the Lankavatara sutra, Mahamati once again asked the Buddha about False projections:

Questions: a) Characteristics,
            b) Circumstance of occurrence and
            c) Cause

Buddha's answer:
        a) misperception of different objects that are illusional unreal because they are created then eventually disappear, or perish
        b) The intriguing concept of attachment to illusions
        c) The mind is itself unreal, the product of the brain, Emptiness creates the Mind that projects the false images, sounds impressions. The only Truth is Emptiness

> *"Bhagavan, please explain what characterizes false projections. How do false projections arise? What constitutes a false projection? And where are false projections found?"*
> *The Buddha told Mahamati, "Projections arise when there is attachment to the misperception of different objects. Mahamati, because people are unaware that their attachments to projections of what they grasp and of the one who grasps are nothing but perceptions of their own minds, they fall prey to views of existence and nonexistence in which they are abetted by the views of followers of other paths and the habit-energy of their projections. And as they become attached to different external objects, the Mind, and what belongs to the Mind give rise to the projection of a self and what belongs to a self."*
> ......
> *Mahamati: Why on the one hand, Bhagavan, does attachment to the discrimination of the existence of an unreal object give rise to projections and on the other hand attachment to ultimate truth not give rise to projections? ....*
>
> *The Buddha said, "Because projections of ('true') existence or nonexistence do not arise. Projections do not arise when the external objects that appear as existing or not existing are seen to be nothing but perceptions of one's own Mind. By becoming aware that their projections are nothing but Mind. Thus, do they transform their body and Mind and finally see clearly all the stages and realms of self-Awareness of tathagatas and transcend views and projections regarding the five dharmas and modes of reality. This is why I say that projections arise from the attachment to things that are unreal and that once someone knows what is real they free themselves from the various projections of their own Mind."*

Another question: why does attachment to the Mind give rise to projections of illusion, and why does attachment to ultimate truth not give rise to projections?

Buddha's answer: The Mind is False, therefore its projection is false Emptiness is true, t

The essence of Buddha's teaching is that humans, once they realize that their 'normal' view is a distorted world due to the attachment to the Mind and the misconceptions from innumerable previous lives, can transform their perception through spiritual practice. This realization paves the way for them to perceive the real world.

It was erroneous to have the nihilistic conception that this world is non-existent despite the illusion due to the impermanence. *"The Einstein's moon always exists"(in this Universe)*
The following sermons highlight the difference between Brahmaism and Buddhism and the mechanism of the projection of the illusional world from the Mind:
Question:
a) Buddhda teaches that every world arises from the conditions of ignorance, desire, karma, and projection

Externalists teach that every world arises from causes, while Buddha teaches that. But the CAUSE and CONDITION are merely different words
b) Buddha teaches that multiple Buddhas can't appear in the world at the same time. But Externalist Masters are also Buddhas
Answer: a) The CAUSE of Externalists is illusional, CAUSE is Buddha is Emptiness that is non-illusional, therefore
      b) Externalist Masters are not Buddhas
"The Buddha told Mahamati, According to these members of other schools, there is something that exists that has the characteristics of neither arising nor changing. Mine transcends the categories of existence and nonexistence. It is not subject to arising or ceasing."
Emptiness, a central concept in Buddhist philosophy, refers to a state beyond existence and nonexistence. The Externalist's existing content, not being created and ceased, is not at the level of Emptiness that they hadn't yet reached. Buddha's existent content, on the other hand, is at the level of Emptiness, making the Externalist's statement inapplicable in this context.

The real entity that exists, is imperceptible to the five senses, not created nor ceased, is what one can not grasp and is Emptiness / Buddhahood/ Original Mind /Original Omniscence. The ICS, the veil of ignorance, obliterates them. On the contrary, the ICS easily recognizes illusional things/ dharma and causes the attachment to the illusional world which is readily projected.

Mahamati Bodhisattva asked the Buddha, "your proclamation of 'what neither arises nor ceases' is not unique. the Bhagavan teaches that the realms of space, nonanalytic cessation and nirvana neither arise nor cease. "...Other schools teach that every world arises from causes, while the Bhagavan teaches that every world arises from the conditions of ignorance, desire, karma, and projection. But the causes of the one and the conditions of the other are merely different words. ....the Bhagavan also teaches that whatever exists neither arises nor ceases because its existence or nonexistence cannot be determined. "Other schools also teach that the four elements are indestructible, that their essential

> nature neither arises nor ceases The Bhagavan has said it is impossible for multiple Buddhas to appear in the world at the same time. But according to the foregoing, ...there would be multiple buddhas at the same time."
>
> The Buddha told Mahamati, According to these members of other schools, there is something that exists that has the characteristics of neither arising nor changing. Mine transcends the categories of existence and nonexistence. It is not subject to arising or ceasing

In summary, Mahamati Bodhisatsava challenged twice Buddha when he asked

i. *Why, **on the one hand, Bhagavan, does attachment to the discrimination of the existence of an unreal object give rise to projections, and on the other hand, attachment to ultimate Truth does not give rise to projections?*** What Buddha replied is obvious: The Original Mind with Ultimate Truth is devoid of the veil of ignorance therefore, there is no projection, no illusion. There is only the ultimate Truth.

ii. According to Mahamati Bodhisatva, the teaching of Buddha and other schools are the same (regarding the permanence-impermanence, the cause of phenomena), then he concludes that teachers of other schools are also Buddhas. Buddha objects him because in, Buddhism Emptiness neither arises nor ceases. In other schools, illusion (a false perception that does not correspond to reality) appears to mimic Emptiness or Nirvana (that exists and is the true entity). Buddha adds: Mine does not fall prey to such categories as existence or nonexistence!.

In Surangama Sutra (proclaimed many years after Lankavatara Sutra) Buddha said:

The development of three evil causes (killing, stealing, and Carnality) due to the unenlightened Awareness will give rise to the perception of illusional form (false mountains, rivers, and the great earth as well as other phenomena)

Bodhisattsava'Pårðamaitràyaðãputra asked: 'If Tathàgata Mind, can suddenly create mountains, rivers, the great earth and other phenomena (likin in Big Bang/Falst Thought, when will the Buddha give rise to the worldly perception of mountains, rivers and the great earth (by illusional protection)?'

The Buddha said: 'No, The illusions only belong to the mundane world. Buddha is never associated with illusion represented by mountains, rivers and the great earth

.

The Buddha said:" *It is like ore which contains pure gold; once the latter is extracted, it cannot be mixed with the ore again. It is also like the ashes of burnt wood which cannot become wood again. It is the same with all Buddhas of the nirvanic enlightenment."*

> 'Thus Pårðamaitràyaðãputra, these three evil causes (killing, stealing and Carnality). succeed one another solely because of unenlightened Awareness which gives rise to the perception of form and so sees falsely mountains, rivers and the great earth as well as other phenomena which unfold in succession and, because of this very illusion, appear again and again, as on a turning wheel.'ness
>
> 'Pårðamaitràyaðãputra asked: 'If Bodhi, which is basically absolute and enlightened and is the same as the unchanging Tathàgata Mind, can suddenly create mountains, rivers, the great earth and other phenomena, when will the Buddha,,who has attained Absolute Enlightenment, again give rise to the worldly perception of mountains, rivers and the great earth (by illusional protection)?'
>
> The Buddha said: 'Pårðamaitràyaðãputra, if a man loses his way to a village by mistaking south for north, does his error come from delusion or enlightenment?'
> Pårðamaitràyaðãputra replied: 'From neither. Why? Because, since delusion has no root how can this error come from it? Since enlightenment does not beget delusion, how can it cause him to err?' (but from the veil of ignorance)
> The Buddha asked: 'If this man, while erring, suddenly,meets someone who shows him the right way, do you think,in spite of his mistake, he will lose his way again?'
> (Pårðamaitràyaðãputra replied:) 'No,
>
> 'Pårðamaitràyaðãputra, it is the same with all Buddhas in the ten directions. Delusion has no root for it has no self-nature. Fundamentally there has never been delusion and though there is some semblance of it, when one is awakened, it vanishes (for) Bodhi does not beget it. This is like a man suffering from an optical illusion which sees flowers in the sky; if he is cured, these flowers will
> disappear. But if he waits for them to appear again, do you call him stupid or intelligent?'
>
> Pårðamaitràyaðãputra replied: 'Fundamentally space has no flowers but due to defective sight they are seen as being in the void; this is already a false attitude. If in addition, they are required to appear again, this is mere folly; how then can that man be called stupid or intelligent?' The Buddha said: 'Since you have interpreted well the non-existence of flowers in the sky, why do you still ask me about the immaterial absolute Bodhi of all Buddhas creating mountains, rivers and the great earth? It is like ore which contains pure gold; once the latter is extracted, it cannot be mixed with the ore again. It is also like the ashes of burnt wood which cannot become wood again. It is the same with all Buddhas of the nirvanic enlightenment.

In summary: All projections are illusional and are not created by enlightenment but by the veil of ignorance. An example is a person with eye disease who sees flickering bright flowers in the air. After the treatment, non-existing flowers disappear and won't reappear again, (but normal and enlightened person/Buddha still see the actual world existing as it is: this world will include the authentic physical and the non-physical aspects)

As previously discussed, Emptiness, a key concept in Buddhist philosophy, shares imperceptibility with illusion. However, it's important to note that the illusion is non-existent, created and ceases. Emptiness, on the other hand, refers to the lack of inherent existence in all phenomena, a concept that is central to understanding the nature of reality in Buddhism.

In the Diamond Sutra, a key Buddhist text, Buddha said: 'If one sees me in forms, If one seeks me in sounds, He practices a deviant way, And cannot see the Tathàgata.' This quote underscores the Buddhist teaching that ultimate reality, or the Tathàgata, cannot be found in the external world, but is to be realized within oneself through spiritual practice.

E. Dependent Origination In The Twelve Links

"Cause And Effect" Or "Karmic Causality" (Fig 1. 4) The twelve links of cause and effect are one of the discoveries of Buddha shortly after the achievement of full enlightenment and characteristic of Buddhism in explaining both

-the samsara/birth and death cycle and

- the cause of sufferance in current life.

For many Buddhist philosophers, the Dependent Origination or Paticca Samuppàda is a profound concept. The circle of twelve components, with its endless succession, invites us to contemplate its cyclical nature, sparking intrigue and deep thought.

The circle, indeed, has no end or beginning. But it is also true that at the Creation (of the Universe/Multiverse), there is an entry into this circle at the component of Ignorance. *It is critical to recognize that Ignorance is caused by a False thought that develops within Emptiness as a noumenon (self-nature)*. In the circle of twelve links, the successive progression in a clockwise direction is caused by the so-called Causal relationship (Duyên Hệ/Patthana) with the involvement of multiple Dharma components. This characterization of the Causal relationship adds a layer of complexity to the principle of Dependent Origination, challenging us to delve deeper into its understanding.

The continuity of the links in the circle of cause and effect in the cycle of reincarnation creates the phenomenon of "repeated origination," making it impossible for people to connect Cause and Effect directly. It makes people feel that "Effect" comes naturally without any cause. In the Bible, when Jesus answered the question of why the man who was born blind, Lazarus, was blind, It was not because he or his parents sinned but that the works of God might be revealed in him (note: The works of God are the Circle of Cause and Effect).

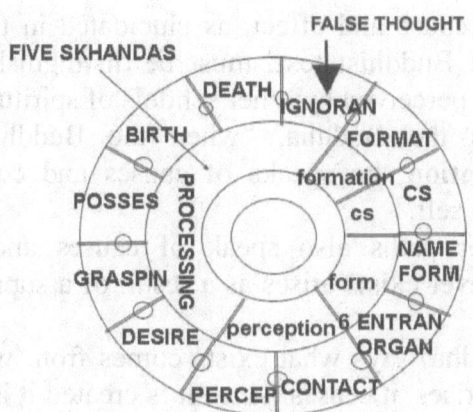

Fig 1.14 : twelve links of dependent origination
1. Ignorance (the Soul with Karma: at creation with further addition of karma in successive lives). *Karma is the recording of bad deeds or memories*
2. Formation (corresponding to the ensoulment in the embryo. In the Creation, Formation represents the process of development of the emotion and CS in the living being). This is the neuroscientific process of integration of data. It should not ve misunderstood as a motor activity of the physical parts of the body
3. CS (creation of CS + Emotion). After Ensoulment the Karmic CS is embodied in the Inner CS, to form the template for the manifestation of the emotion/ Three poisons:Greed-Anger- Stupidity
4. Generation of Name (conforming with the concept of Mind creating Form).
5. Generation of Form (Note Formation generates CS, Name, and Form at the same time in the concept of Patthàna/vn: Lý Duyên Hệ). In the realm of Formlessness, (Sentient beings have no Form), there is no formation of the Form but still, the Formation of the internal mental state consisting of 52 types of Dvipannàsa Cetasikà, used as reference mental states (similar to the Inner CS) cause of the Formation, then Consciousness, then Form...)
In the Dependent origination or Paticca samuppàda, steps of Formation, CS, and Name-Form are successive).
6. Contact and,
7. Perception for the process of Perception.
8. Desire and,
9. Grasping and,
10. Possession constituting the process of express integration and,
11. Birth and,
12. Death: the last five steps result in the development of the past cause for the future outcome.

This cycle of cause and effect, as elucidated in the Lavantara sutra, a significant Buddhist text, must be distinguished from the cause and effect as perceived by other schools of spirituality.

Mahamati asks the Buddha, "when the Buddha speaks of **dependent origination**, he speaks of causes and conditions and does not speak of a self.

Followers of other paths also speak of causes and conditions, namely, that whatever exists arises as a result of a supreme deity or force. .

Bhagavan (Buddha) says what exists comes from what does not exist, and once it arises, it ceases (since it is created it is perishable). Mahamati argues that ignorance is the condition of Memory and so on up to old age and death (in the twelve links of cause and effect), is a teaching of no causes, not a teaching of causes. For other school, the teachings would appear to be superior, because according to other schools, the cause does not arise from conditions but gives rise to what exists

The Buddha told Mahamati, "I do not teach that there are no causes, nor do I confuse causes and conditions, rather 'because this exists, that exists. The Awareness that these are nothing but perceptions of one's own Mind. Mahamati, as long as people and are unaware that these are nothing but perceptions of their own Mind, it is they who mistake the existence or nonexistence of external objects. I have always taught that things arise due to the conjunction of causes and conditions
not that they arise without a cause."

---

*Mahamat iasked the Buddha, "Bhagavan, when the Buddha speaks of **dependent origination**, he speaks of causes and conditions and does not speak of a self. Followers of other paths also speak of causes and conditions, namely, that whatever exists arises as a result of a supreme deity or force, or time, or minute particles.*

Bhagavan says what exists comes from what does not exist, and once it arises, it ceases (since it is created it is perishable). According to the Bhagavan, (Mahamati argues that) ignorance is the condition of Memory and so on up to old age and death (in the twelve links of cause and effect), is a teaching of no causes, not a teaching of causes. The teaching established by the Bhagavan goes like this: 'Because this exists, that exists.' It does not acknowledge a gradual existence.

The teaching of other schools would appear to be superior, because according to other schools, the cause does not arise from conditions but gives rise to what exists.

The Buddha told Mahamati, "I do not teach that there are no causes, nor do I confuse causes and conditions, rather 'because this exists, that exists. The Awareness that these are nothing but perceptions of one's own Mind. Mahamati, as long as people and are unaware that these are nothing but perceptions of their own Mind, it is they who mistake the existence or nonexistence of external objects. I have always taught that things arise due to the conjunction of causes and conditions not that they arise without a cause."

(Meaning: This paragraph is very difficult to understand in common sense and is the main argument for Buddhism's superiority to other schools. In the twelve links of cause and effect, both causes and effects are illusional, coming from the projection of the Mind; therefore, they can not be considered as the primary cause of the phenomena, but the relationship is only based on the principle of relative correlation. The CAUSE coming from Deities called by externalists is illusional because it does not truly exist. The ultimate CAUSE in Buddhism is Emptiness, a concept that can enlighten our understanding In the twelve links of dependent Origination, the CAUSE is a product of the veil of ignorance and is illusional because the CAUSE will eventually perish. The SELF isEmptiness is not part of the twelve links. Ignorance is not the problem of memory but represents the illusion. The teachings of the Externalist Master are easier to understand than Buddha's because the Externalist teachings represent the common errors due to the common veil of ignorance. DO NOT confuse the CAUSE (in the twelve links) that only represents the CAUSE of the concerned CONDITION, not as the original CAUSE

F. Four Noble Truth
This is the first Sermon given in the Deer Park after attaining the Buddhahood
The sermon is the fundamental doctrine of Buddhism and encloses the Dependent Origination in the twelve links of karmic causality. The Four Noble Truth can be divided into two groups:

i. - Life consists entirely of sufferance
    - The cause of sufferance can be found: the Ignorance
ii. - The cessation of sufferance is only found Nirvana
    - The path leading to Nirvana: To practice the Eight-fold

Noble Truths—Buddha taught: "Whoever accepts the four dogmas, and practises the eighfold Noble Path will put an end to births and deaths.

G. Eightfold Noble Path (see page 87)

H. THE APPLICATION OF EMPTINESS NATURE IN LIFE:122
Emptiness Nature, a fundamental concept in Buddhist philosophy, refers to the lack of inherent existence in all phenomena. It is not a state of nothingness, but rather a state of interdependence and interconnectedness.
The continuity of the rings in the circle of cause and effect, in their intricate and continuous rotation, gives rise to the phenomenon of

"repeated cause and effect." This complexity can make it seem as if "Effect" emerges without a discernible Cause. In the Bible, when Jesus responded to the query about why Lazarus was born blind:
It is not because of the person or the parents who sinned; but it is so that the works of God can be revealed in the person (note: The work of God is the circle of cause and effect)
The Path of Purification (Visuddhimagga) said: No one creates the karma, nor does anyone receive the retribution,
There is only the phenomenon of flow: seeing it differently is not right

The Law of Cause and Effect expresses karma summarized as follows,
• The cycle of 12 causes and effects, an inescapable cycle that continues until the end of karma. Therefore, Cause and Effect is not just difficult to relate to each other but unavoidable.
• Dependent Origination: It's a direct, immediate connection between Cause and Effect, without any intermediary, emphasizing the immediacy and directness of this relationship.
• True Nature Dependent Origination: a concept that explores the fundamental nature of all things in Creation, revealing their interconnectedness and the inevitability of change.
• Six Great Causes and Effects: These are the fundamental elements that shape our existence. For instance, our Consciousness determines our perception, Space provides the context for our experiences, and the Four Great Elements (earth, water, fire, and air) form the basis of our physical being.

Dependent Origination: The secondary factors that contribute to Cause and Effect.. for example, a grain of rice needs water, fertilizer, and care to grow into a rice plant
• Store CS: The stored Karma can produce results when convenient according to the cycle of causes and conditions, for example, according to the law of Dependent Origination, the 7th Consciousness (manas) Dependent Origination or**IX. Interrelationship Between Memory-Consciousness- Emotion - Karma- Veil Ofgnorance And False Ego 102**
Emptiness, the profound and ultimate origin of the Creation, constitutes the True Ego, self-nature/noumenon/Buddhahood/Holy Spirit. Its depth and complexity invite us to delve deeper into its implications.

It is a False thought, a manifestation of the self-nature of Emptiness, that triggers the Creation, born out of Ignorance, and commences with the cataclysmic and transformative event known as the Big Bang.

Post-Creation, the generation of the Mind and Dharma is devoid of self-nature. The term "Dharma" encompasses both the Mind and the Dharma itself. In this context, "Dharma is the primordial Dharma which is nearly indistinguishable from Emptiness or at the interface of Emptiness-Creation.

The interaction of the Mind and Dharma creates the Consciousness System (CS), also referred to as the Mind, the motor activity, and the Emotion. Emotion represents a blend of motor activities and the CS. Emotion is a fusion of motor activities comprising facial expressions, body gestures, and internal organ reactions.

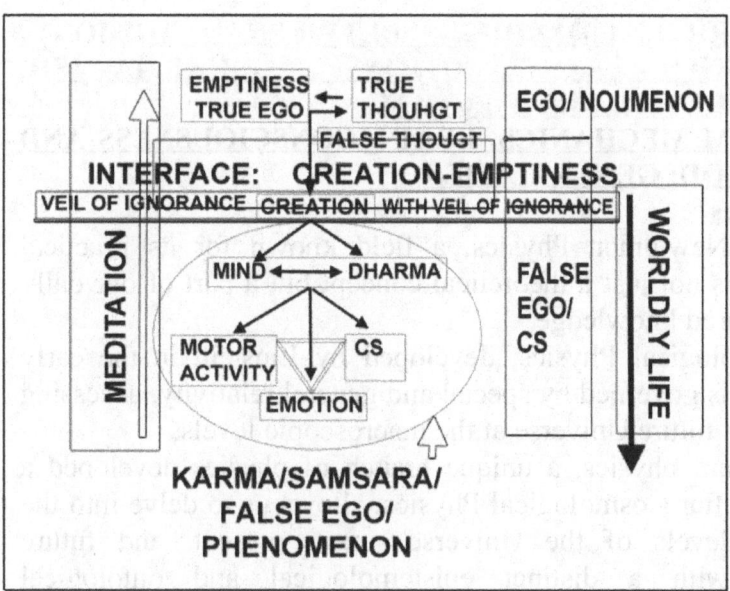

The Mind, serving as the repository of memories, is the essence of the CS. These memories/ MM, are the archives in the Brain and the Soul, housing past, present, and future data received by the five sensory organs. The past data is preserved as Alaya CS. Among all the data, the implicit MM stands out as the most crucial, shaping the cycle of birth and death. While explicit and sensory MM have their roles, they are significantly less important.

The false ego, a complex interplay of Karma and the Veil of ignorance, is a pivotal concept. It is symbolized by the Mind-Dharma combination, which encompasses the CS, Emotion, and motor activities.

The morality with Four Immeasurable Virtuous Mind, a beacon of hope, executed in the creation is always more or less tainted by the illusion (projection of the true nature of the world through the Mind) and False ego. As a result, Arahats, who are still tainted with some degrees of ignorance, remain distant from Bodhisatsava and Buddha.

Without spiritual practice, the ignorance keeps accumulating, and the veil becomes thicker and thicker, serving as the source of turmoil in the Creation until the ending of this Universe. However, with dedicated spiritual practice, this ignorance can be dispelled, offering a path to enlightenment and peace.

X. INTERRELATIONSHIP BETWEEN MEMORY- CONSCIOUSNESS- EMOTION -KARMA- VEIL OFGNORANCE AND FALSE EGO124
X. QUANTUM MECHANICS, BRAIN, CONSCIOUSNESS AND BUDDHAHOOD: GENERALITIES125

A. Generalities

Classical or Newtonian Physics, a field known for its practical applications, is not just a theoretical concept but a part of our daily living and shared knowledge.

Cosmological Physics, developed by Einstein in the early 20th century, is governed by special and general relativity, accessing the earlier and future Universe at the macroscopic levels.

Quantum physics, a unique branch of physics developed a few decades after Cosmological Physics, allows us to delve into the microscopic levels of the Universe's past, present, and future phenomena with a distinct epistemological and ontological awareness and understanding.

As a result, Quantum Mechanics/ QM is not only interested by physicists in exploring and improving physical life but also by those concerned with the metaphysical aspects of life, such as Consciousness, supernatural phenomena, and spirituality.

The characteristics of QM are the quantum particles, such as photons, neutrinos, and electrons ....that constitute the Universe's building blocks and do not have real entities defined by their mass and velocity. A quantum particle has no definite location at any moment but is represented as a probability expressed in a wave function. This is in contrast with the macroscopic Form, in which the location is expressed in a linear function.

Heisenberg's profound statement upon receiving the Nobel Prize in 1932 resonates even today:' *The atom has no immediate and direct physical properties at all.'* This revelation extends beyond the atom, suggesting that if the building blocks of our Universe are not physical, then the same must hold true for the whole in some way. The interaction between QM and CS is equally intriguing, as Bohr's words echo: 'If quantum mechanics hasn't profoundly shocked you, you haven't understood it yet.' This shock factor extends to spirituality, where metaphysical phenomena can surprise even the most skeptical, unless they choose to ignore the supernatural realm.

QM shares some characteristics of Emptiness, a philosophical concept that denotes imperceptibility and interconnectedness. However, most quantum particles lack the pervasiveness of Emptiness, with the exception of Neutrino.

CS is the direct product of the Brain but requires the presence of the Original Omniscience/ OM of Emptiness.

Nevertheless, OM is permanently obliterated or inhibited by the CS. Briefly, CS is characterized by:

· Imperceptibility like OM,
· Capability of receiving data and integrating data to transform it into thinking (thought)
· Pervasiveness and interconnectedness when free outside the Brain (nervous system). (in the Brain and nervous system, the pervasiveness and interconnectedness of CS are limited within the Brain)

Emptiness is the origin of the Creation; therefore, Emptines is also called Creator, equivalent to God/Buddhahood in religions, and is the non-Duality/ Oneness/ Ultimate Pole. Due to its intrinsic nature arises the False Thought/ Big Bang; Emptiness gives birth to the Duality consisting of:

i. Force with subsequent development into Form (quantum particles/ QM, perceptible Matter/ Baryonic Matter, Dark Matter, secondary Force (like electromagnetic, strong, weak forces and gravity) and Dark Force.

   ii. OM with secondary development of CS by the intermediate of the Brain.

The concept of OM is not uniquely addressed by the Buddhist sutra but also by Kant (see below) and other scientists (does the Universe have cosmological memory? Does this imply cosmic Consciousness? by Walter Christensen, from the Department of Physics Pomona , CA, USA; The Quantum Hologram and Nature of Consciousness by Edgar Mitchell , Apollo Astronaut; How Consciousness Become the Physical Universe by Menas Kafatos , computational Physics CA,USA).

B. CAN QM BE RELATED TO A PROTOTYPE OF CS?

In QM, there is a problem with the observation using the CS. Since the Brain (of the macroscopic world) creates CS, the measurement of the natural wave function (that belong to a microscopic dimension) of the particle location is paradoxical. To consider the Brain as a quantum computer, the problem of QM in expressing the measurement has to be resolved. There are two methods for fixing the problem:

i.   **The method that *cannot be* applied to the brain model**. This method uses the concept of Unitary Evolution or Quantum Coherence, in which the particle has the duality of the existence of particle and wave function. The accurate measurement is impossible at any moment , sine the measurement is expressed in a wave function. As a result, there is a "superposition" on the result

ii.   **A method that *can be* applied to the brain model**. This method uses the concept of Objective Reduction or Quantum Decoherence. In which the wave function collapses into a straight line that gives a real/affirmative measurement. Obviously, this method is "artificial" due to the interference of the observer, as in the case of the ERP paradox experiment. In terms of Physics, the observation is not continuous anymore at the level of QM/

In this context, the concept of 'STOP GAP' refers to a point of discontinuity in observation. This 'STOP GAP' represents the gap between the object and the observer in the macroscopic world. Philosophically, closing this gap signifies the fusion of the object and the observer. This idea of fusion, however, is only a philosophical concept in our 'reality', as these two entities always remain distinct despite originating from the same Emptiness. In simpler terms, the fusion can only occur if these two entities return back to Emptiness.

This means that only the OM/Original Mind/Ultimate Omniscience can reveal the true value as it is. The CS is the veil of ignorance that makes the knowledge incomplete. This incompleteness is mirrored in Enstein's claim about a Hidden Variable in QM or his unhappiness about Copenhague interpretation of QM. To Einstein, The moon always exists regardless of the observer's attention. In Copenhague's concept, the observation alters the" reality," or at least makes the reality fuzzy at the microscopic level; the recent confirmation of the EPR paradox demonstrates the deterministic fate of the particles when leaving their source is consistent with the concept of Emptiness interconnectivity, and homogeneity. The EPR, therefore, confirms the deterministic fate of each entity in the cosmos. It does not refer to the uncertainty concept of Copenhague's interpretation of QM, which belongs to the concept of CS created by the Brain.

C. LIMITS OF THE CONSCIOUSNESS

As previously stated, and as in the preceding paragraph, CS is not the product representative of reality but the product of illusion. The observations from CS are product filtered by the five layers of the membrane, the veil of Ignorance (Form, Perception, Feeling, Formation, and CS) forming the CS of Parikalpita Misconception of the true nature of phenomena (distorted CS) and Paratantrasvabhāvastu. Vikalpah pratyaya udbhavah (common sense CS). So, the observation result is a reflection of the observer, not from reality, the truth of the observed object. Physicists know that too. This speaks of the inconsistency of CS when compared to UO. UO of Buddhism/Holy Spirit includes all things as they are, in the creation of the saying in Buddhism: Small is not outside, and Big is inside.

It is worth noting that the Buddha often showed His disciples mercy for their limited distorted CS that was obscured by the veil of ignorance from thousands of lifetimes. With the concept of Ultimate Omniscience (UO), the uncertain world of Heisenberg's Quantum Mechanics may cease to exist. UO, being all-encompassing and infinite, resolves the problem of Subject and Object. It provides a comprehensive understanding of reality, in stark contrast to the limitations of common sense and distorted CS.

D. CAN THE BRAIN BE A QUANTUM COMPUTER? 129

The Diagram of microtubule as a computer. (H1.15)

The CS is generated from the Brain after receiving information from five senses and general CS. Since the development of artificial intelligence (AI), many scientists have thought that the Brain may represent a mechanism of a system of computers that receives and transforms information into CS. As a result it is interesting to investigate the mechanism of this computer system in the Brain.

The physicist with much interest in this issue is Sir Roger Penrose, who is considered to have made the greatest contribution to physics after Einstein, being Knighted, and the 2020 Nobel Prize in Physics for the Black Hole. He studied the CS in collaboration with Dr. Hameroff, an anesthetic specialist, and psychologist at the Arizona Research Institute. The two believe that microtubules (MT) in neurons play a crucial role in the brain's computer system. These MTs, which typically have dimensions of 25nmx50 micrometers and are formed by combining tubulin protein, are connected together throughout the neurons, forming a complex computer network. This network, when combined with other elements in the brain, forms Orchestrated OR/Orch OR, a phenomenon according to Penrose that explains the generation of CS in the brain.

MT typically have dimensions of 25nmx50 micrometers, formed by combining tubulin protein that forms the wall of the MT. MT/Microtubules usually combine a set of two or three. MT is abundant in every cel. In the cell there are three types of microfilaments that form the skeleton for the cell (the supporting system responsible for the shape of the cell). Three different forms of microfilaments are the smallest actin (for involuntary movement), the largest myosin (only in muscles for voluntary movement), and the intermediate filaments (fibers that only serve as supporting mechanical skeleton, not for the movement of cells). In addition to these microfilaments, MTs also play an important role in arranging chromosomes in cell division. MT also functions in the cilia of the respiratory cell, moving the secretory mucin or the motility of the sperm. The tubular shape of MT suggests the transporting role in of nutrients including ions/electrons in cells,

There are many MTs in neurons. The polarization on the wall of the MT is suspicious to be the mechanism that forms the underlying mechanism for computing. MT has long been known to have a function of linking to synapses/neural connections because in experiments, MT damage affects the function of synapses. MT referral to CS was suggested by Sherrington (1957), and was developed by Hameroff (1980) and suspected to be a cellular/molecular automata. The wall structure of the MT may correspond to the function of making a network that connects the synapses (not the connections of neural cells/1011 neurons) from the spines to the end of the axon (103 spines per neuron). Can computing in cells be done by MT? Sir Penrose estimated that the number of tubulins (108 per neuron, flashing at 107 /sec) that is sufficient to do the work (high capacity MT-based computing inside neurons could account for the organization of synaptic regulation, learning, and memory and may act as the substrate for CS).

MTs are placed in the nerve cells but separate from each other. Penrose used the concept of gravity, which is called the gravitational computer phenomenon, similar to the gravity force of objects connected together.

Again, in the Brain, only some parts of the cortex are involved in the formation of CS; the other parts, such as motor and accessory motor cortices, only perform unconditional and conditional reflexes without CS. These areas, which Penrose refers to as 'Zombies' (without soul), do not contribute to the generation of CS. Thus, according to Penrose, CS generation only occurs when computing phenomena are orderly and arranged accordingly.

- To date, the comments of Hameroff and Penrose, which propose a novel theory about the relationship between microtubules and consciousness, have received little agreement by biologists. This lack of consensus underscores the ongoing debate and the need for further research in this area.

As discussed above, the electrical activity to create changes of the computing type does not show the integrity of the information as it is. Reduction process of particle and waves phenomenon is known through the phenomenon of harmony reduction (Orch OR). Penrose also uses the concept of proto-consciousness (almost similar to Priori Origin/Transcendental Consciousness) to supplement the CS created from the MT. Penrose argues that CS is beyond the framework of computing and belongs to another theory of physics that has not yet been discovered. Penrose's idea is consistent with Godel's mathematical theorem (1931), saying that no consistent system of axioms whose theorems can be listed by an effective procedure (i.e., an algorithm) is capable of proving all truths about the arithmetic of natural numbers. It is the Truth that the Word cannot fully describe idea, or that the Word cannot fully describe religion. Penrose called the above type of CS the Precursor of CS that needs the Brain to create the CS

Penrose posits that this type of CS is original to living beings. The original CS, or priori CS, is a unique concept that is almost equivalent to UO. This idea does not challenge the notion that CS outside the Brain is generally more intelligent than when it is attached to the Brain. UO, in its perfect acquisition of information and transformation into thought, is a fascinating concept to contemplate (Of course, CS outside the Brain does not take action or sound).

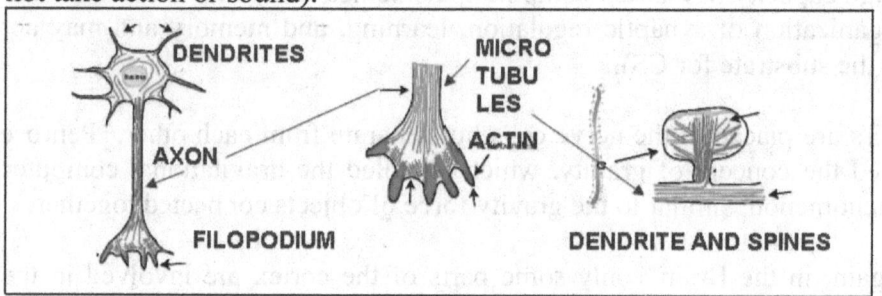

H1.15 Neural cells, beard, axis and tail with actin fibers and MT microtubules

• In organisms with a cell expressing CS, the destruction of MT does not impair CS, a finding that challenges existing theories.

• If MT indeed plays a role in the formation of CS, then the MT in neurons should not only be abundant but also exhibit dynamic changes in shape. Surprisingly, this is not the case, raising intriguing questions for future research.

• In the non-sensory cerebral cortex, the abundance of MT is comparable to that in the sensory cerebral cortex. With the notion of MT as amicrocomputer, this observation challenges our understanding of MT's function and underscores the need for further research to unravel this mystery.

If MT is not a microscopic quantum computer, it is possible that MT only serves as a supporting and communication function in the cell. The information transmitted from the arrival sites in the spine moves along the MT to the MT of the axon to eventually move to the MT of the axon terminals. The number of spine coves is greater when compared to the number of MT in the axon. This shows that each MT tube conveys many neurons. As seen above the lining, the soul is guided to go through the body and axis. The conductive material in MT can be Ca^{++} cations. When the spine of the synapse is stimulated, Ca enters the cell, partly to regenerate into cAMP or cGMP, commonly referred to as memory molecules, the other part transferred by MT to retreat to release neural connectors out of retreat (Cold Spring Harb Perspect Biol 2017;9:a025817 J Exp Bot.

70:387–96, 2019, Cells, Synapses, and Neurotransmitters Joseph Feher, in
Quante Hum Physiol,Sec Ed,2012, Cell Biosci (2020) 10:26, J Neural Transm (2014) 121:799–817, Neurochem Int. January 2008 ; 52(1-2): 142–154, Cell Mol Neurobiol, August 2022 DOI:10.1007/s10571-021-01064-9, Theoretical Biology and Medical Modelling 2010, 7:34)

In summary, QM reveals areas of limit of human CS.
Therefore, the paradigm of the Brain containing microtubules/MT as an assembly of Quantum Computers is unable to solve some problems, such as:
 - In unicellular organisms with evidence of CS, the destruction of MT does not alter CS.
 The absence of CS in parts of the brain and body without CS, despite the presence of MT, has not been explained well.
 - In OBE or NDE, CS exists in the absence of the Brain: example of CS outside the Brain.

As a result, in the formation of CS from data received in the Brain, MTs in neurons and other types of cells only subserve the role of transmission of electrical impulse (via cations Ca++ and Mg++), connecting different parts of the cellWhen presynaptic tic data enteto the postsynaptic sp,iitand likely proceeds to tmicrotubule (he) MT in the axon to reach the next synapseHowever, sincence the total number of spines exceeds the number of microtubules per section of the axon, the transmission he data the in MT is not specific to the spineIn this context, the concept of the 'soul' is introduced. . T'he s'ois proposed to guidedes the clear transmission from the spine to the axon termi,nasnittisobelieved to carryhas the imprint of toriginal dataint.

E. Paradigm Of Soul As A Quantum Computer (Fig1.16)

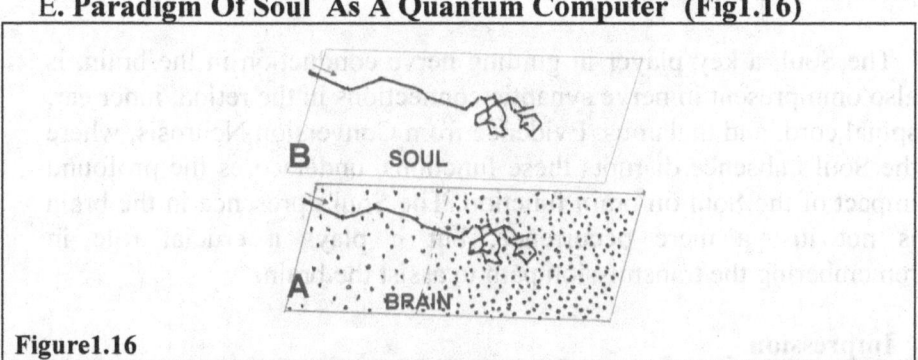

Figure1.16

> Paradigm with the Brain: A: Pathway of transmission of data and integration in the sensory cortex (dots: synapse site)
> B: the imprint of Pathway of transmission of data and integration onto the Soul.

In the Chapter on CS, CS is created in the Brain can be summarized as follows:

    i. Input data to specific sensory cortices and HIPPO

    ii. Interconnections with associated cortices for characterization (for example, visual cortex V1 associated with V2..6, temporal cortex for color, shape face emotion...).

    iii. ACC is the cortex of error detection for recognition of the new data as similar data previously stored (designated as in the Inner CS/ ICS cortex/ Default Mode Network): the coming data is labeled as such.

    iv. Incoming data dissimilar to data in the ICS is labeled with a new name.

In the above process, the brain plays the role of registering and integrate the data in the DMN, retrieving and remembering it.

    The hypothesis of the Soul diagram (as illustrated on pages 271-272) consists of neutrinos disposed of in specific and orderly patterns in a piece of Dark Matter. Neutrinos are combined with electrons in the brain, and electrons are attached to positively ionized atoms like Ca^{++} or Mg^{++}. When synapses are activated with coming data, the Neutrino-electrons are detached from the synapses. This event served as an imprinting process of the data into the Soul. As a result, the Soul guides the path of transmission and integration in the cortex. Of note, because the Soul is always sitting in the OM, the Omthe final step of formation of the CS

The data quality appeared as CS reflects the content of the Inner CS that obliterates the OM. Clearing the Inner CS will result in the perception of data as it is

The Soul, a key player in guiding nerve conduction in the brain, is also omnipresent in nerve synaptic connections in the retina, inner ear, spinal cord, and thalamus. Evidence from Conversion Neurosis, where the Soul's absence disrupts these functions, underscores the profound impact of the Soul on brain function. The Soul's presence in the brain is not just a mere occurrence, but it plays a crucial role in remembering the transmission path to assist the brain.

Impression

As Buddha pointed out to his disciples, Human beings have been obliterated from the Ultimate Omniecence for countless lives by the veil of ignorance/ stupidity; misconception is common. Void or Emptiness is commonly misunderstood as nothingness per se. Once Emptiness is fully known as the source of energy, Ultimate Omniscience, and all types of Forms, one can imagine that the miraculous power in the Universe can be conceived. The form (living sentients, nature like mountains or rivers are non illusional as long as the deviated Mind does not attach them with ignorance. Clearing the veil of ignorance is the radical treatment of the distorted, illusional view.

For religious practice, Emptiness/ the Ultimate Pole/ The Oneness create the Duality comprising Omniscience and Ignorance. Ignorance is the root of the Dharma Seals: Impermance, Sufferance, and False Ego. They are intercorrelated, but there is no causal relationship. The most essential of the above three Dharma Seals are:

1) The sufferance, highlighted by the Four Noble Truths (*sufferings, origins, cessations, and path of cessation of suffering), The Path is the EightfoldNoble Path (* **Right View, Right Thought, Right Speech, Right Action, Right Livelihood, Right Effort, Right Mindfulness, and Right Meditation**) *(translated into Six virtues of perfection* **Ultimate Charity**/ *Dana-paramita,* **Ultimate Discipline**/ *Sila-paramita,* **Ultimate Patience**/ *Ksanti-paramita, Ultimate Devotion/ Virya paramita,* **Meditation** */Dhyana-paramita , and* **Ultimate Omniscience**/*Prajnaparamita (note: Ultimate is equivalent to Egoless).*

The ignorance, or Stupidity (erring or deviating from Mindfulness, but not related to intelligence), is manifested as the veil of ignorance caused by the presence of the Five Skandhas. The Five Skandhas, which include Form, Perception, Feeling, Formation, and Consciousness, are the components that make up an individual's experience of the world. They act as a barrier or filter, shaping our perception and understanding of reality. When we are not mindful, these Skandhas can lead to a distorted view of the world, creating the veil of ignorance.

Physical component: Form,
Non-physical components:
Perception (receiving information from sensory organs).
Feeling: the preliminary reaction to the data using the dorsal pathway in the brain.

Formation: integration of the information in the brain ready for labeling CS.
CS formation of CS based on the pre-existing CS.
Engaging in spiritual practice, such as mindfulness meditation, has the profound ability to dissolve these physical and non-physical barriers, unveiling the ultimate omniscience. This transformative power of spiritual practice is not only inspiring but also motivating, encouraging us to embark on our own spiritual journey.

3) Absence of Egolessness (False Ego)

Einstein was not wrong to point out that: *"Science without religion is lame; religion without science is blind"*. The blindness should be understood as an inability to understand the revelation in Buddhism. As living sentients of the Creation, human beings do not need to reject this world, although it is impermanent, because the living creature's lifespan is much shorter than that of nature. The problem is the failure to recognize its wholeness, which includes the physical and the metaphysical realms.

CHAPTER II:
SUMMARY OF THE EMBRYOLOGY, NEUROANATOMY and PHYSIOLOGY Of THE BRAIN

I. Highlights of Important Stages in the Development of the Brain

Aims:

Highlights of the early stages of the embryology (Fig 2.1)

This chapter aims to highlight the successive stages in the embryo's development and the brain's formation.

This may shed light on the ensoulment and possible mechanism development of neuronal connections necessary for different brain and body functions. Furthermore, the ensoulment may open the door to the understanding of the mechanism of the formation of the store Consciousness, a type of Memory from many previous lives before the ensoulment into the present embryo. This mechanism may be essential for understanding cases of some individuals having the capability of remembering their previous autobiography and the role of Karma in past lives influencing the outcome of the present lives.

A) Highlights of the early stages of the embryology (Fig 2.1)

- Embryology is a branch of science studying the miraculous development of living organisms, including humans. This review of the development of the central nervous system is necessary for understanding the mechanisms of formation of various functions such as Memory, cognition, and the relationship between the body and the environment.
- Day one: The day after fertilization, the egg (ovum in the mother) starts the first cell division into two and then into four and eight cells.
- Before day 4th, the egg is called a blastula, composed of up to eight cells (with identical chromosomes) with identical cellular and genetic features without cellular differentiation. Blastula is a compact block of cells. Afterward, two empty and superimposed spaces will be developed, separated by a layer of cells.
- Days 15-17: An embryonic disc appears in the cell layer separating the two cavities. The upper cavity will become an amniotic cavity associated with the placenta.

-The upper surface of the embryonic disc is destined to be the ectoderm, the precursor of the skin.

-The lower layer and the lower cavity will form organs with a cavity, namely the stomach, small and large bowels, gallbladder, urinary bladder, and peritoneal cavity.

- Between the ectoderm and the endoderm will develop the mesoderm, precursor for the mesenchymal /or interstitial tissue including bones,

- muscles, blood vessels, spleen, and major solid organs like heart, lungs, liver, pancreas, and kidneys.
- On the 17th day, a longitudinal streak called the primitive streak which rapidly becomes longer and larger and becomes the neural plate.
- The neural plate bends longitudinally and downward to become the neural tube with subsequent development into the spinal cord, hindbrain, midbrain, and Forebrain on the 20th day, as shown in the Figures.
- The shared development of the nervous system and the skin from the ectoderm has the ultimate meaning: these are structures communicating with the exterior environment.
- Days 20-30 develop the Diencephalon, then the Forebrain, and finally the Telencephalon. The diencephalon is composed of the Thalamus, Hypothalamus, and adjacent nuclei. The telencephalon will become the neocortex.
- The Right hemisphere develops a little earlier than the Left side and is larger under the effect of testosterone. Neurons start appearing during this period. As a result, there was a concept that abortion was permissible before this period without violating Pro-life tendencies.

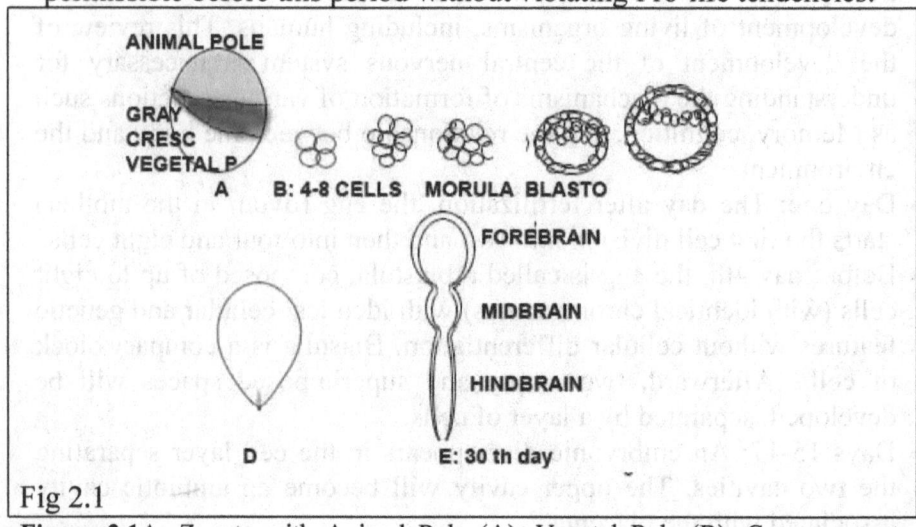

Fig 2.1

Figure 2.1A: Zygote with Animal Pole (A), Vegetal Pole (B), Gray crescent, Front (V)
2.1B: 8 cell stage=Morula to stage of Blastula, 1C: Gastrula with embryonic disc
2.1D: After day 17, Primitive embryonic streak appears and expands until day 25 with the formation of the neural groove
2.1E: day 30; formation of midbrain and Forebrain,

II. BRIEF REVIEW OF THE CENTRAL NERVOUS SYSTEM
A) Generality

The brain architecture consists of many modules of gray matter specialized in a few specific functions. Because Memory formation is due to the connection between neurons' dendrites, cognition is the result of the integration of information at different parts of the cortex and Thalamus. Therefore, a review of important areas of the cortex, Thalamus, hippocampus, and many subcortical gray structures is essential for understanding the mechanism of different functions of the central nervous system and possibly the metaphysical aspect of the brain. The central nervous system consists of
• Telencephalon represents the last and the most developed part in the phylogeny evolution and comprises two hemispheres, basal ganglia (BG) and olfactory bulb.
• Forebrain or procephalon is the most rostral part of the brain. In humans, it represents a small portion below the hemisphere and in front of the Diencephalon
• Diencephalon composed of Epithalamus (Pineal gland and Habenular nuclei), Thalamus, Sub Thalamus (including Zona Incerta and Reticular nuclei), Meta Thalamus (Geniculate bodies, Mammillary bodies. The development of the Diencephalon is crucial in the formation of the central nervous system with the involvement of many recently discovered genes (Martinez-Ferret 2012)
• Midbrain=mesencephalon composed of the superior and inferior Colliculi (ICS and SC= auditory and visual relay stations) Tegmentum, (including the part number of nuclei of the ascending activating reticular system=ARAS) and brain stem including the Substantia Nigra (SN) producing DOPA. Lack of DOPA results in the development of Parkinson's disease.
• Pons =connecting the Right and Left Cerebellum, nervous fibers connecting the brain and spinal cord, and nuclei of the ARAS
• Medulla. Note: Brain stem=Midbrain+Pons+Medulla

In comparison to other body organs, understanding the brain regarding the structure and the mechanism formation of different functions in the control of the body's physical and metaphysical aspects has come late. The brain is somewhat comparable to a black box of an airplane and perhaps part of the investigation in discovering the miracle of nature hidden in animals as well as in human beings. The common knowledge about the brain is limited to certain functions, like parts of the cerebral cortex for sensory reception, movement commandment, and some basic neuroendocrine control. It is surprising to reveal the brain contains large number of structures with quasi-defined and specific roles in receiving, storing incoming information, integrating, interpreting, and transforming into Memory cognition and conditioned reflexes. To perform different activities, there are areas of the brain playing the role as manager for activating, inhibiting and harmonizing the activities. Particularly, important functions of the brain are usually managed by more than one center for backing up in case of failure of one of these centers. Examples of centers for wakefulness and sleep include vlPO, MnPO for sleep, Basal Forebrain (BF), Lateral Hypothalamic area, SupraChiasmatic Nucleus, MCH/Melanin Concentrating Hormone, ARAS...for wake.

. 1) Cortex 140

The cortex is the last developed structure of the central nervous system. The Prefrontal cortex is the most developed portion of the brain. In addition, the brain is composed of specialized structures forming different modules in a pattern comparable to the government with various departments. Each department is responsible for certain specific functions, in collaboration with the others, to collect adequate information and acquire organization and harmony before the cognitive and executive orders are issued. Data is first received in the Thalamus as a relay information department/center. The information is sent to respective and associative modules (cortex) for integration, then to a center of Inner Cognition to become cognitive information. From there, conditioned reflexes are generated, and executive orders are issued and implemented. Until now, the essential components, physical components, or elementary particles of consciousness have been unknown; however, we know the neural correlate of cognition.. The division of the cortex into many areas was previously studied by K Broadman, a German anatomist based on the cytohistoarchitecture; different regions of the cortex were distinguished, characterized, and modified according to the function of these areas in many subsequent studies. The cortex, regardless of location, is composed of 6 distinct layers. Neurons of these layers are interconnected, and connected to other cortical areas, to the contralateral side through the corpus callosum, to subcortical gray nuclei, and to the spinal cord. The connections are frequently bidirectional. Neurons with long axons (usually connected to the spinal cord without a relay, are usually pyramidal and of large sizes.

. THE BRAIN CONTRIBUTES TO THE CREATION OF INNER CONSCIOUSNESS or INNER MIND Original Mind/ Buddhahood, but not the brain that creates the CS 137

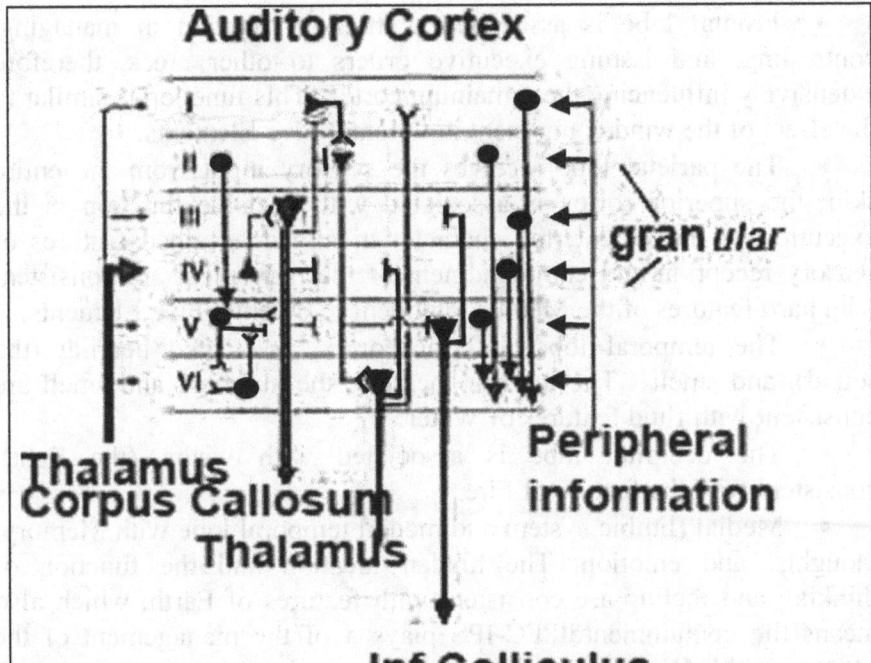

Fig 2.2 Architecture of the cortex composed of six layers, interconnected together, for the purpose of mutual control by intermediate microcircuit. Particularly, despite all six receiving input from the Thalamus and contralateral cortices, layer 4 (granular) receives a major part of the input from the Thalamus, playing the role of information processing. **Layer 4 is most often demonstrated in Sensory, Entorhinal, and Retrosplenial cortices.** Layer 3 sends outputs to the contralateral cortex. Layers 3, 5, and 6 send outputs to the Thalamus

2. Overview of the function140(Fig 2.3)

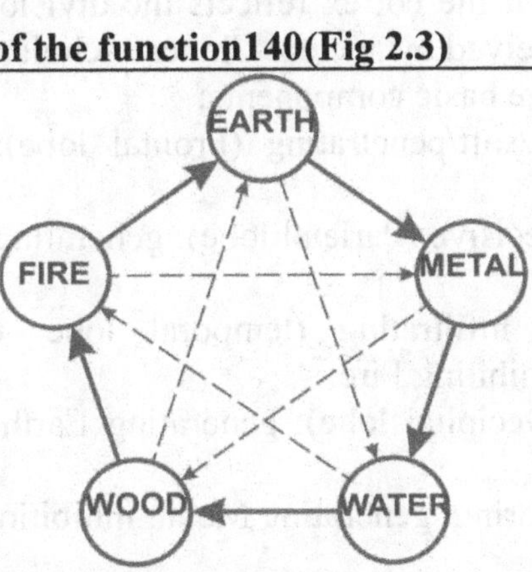

Fig 2.3 Five components with continuous arrows representing generation and discontinuous arrows representing inhibition

- Frontal lobe is associated with high function in managing, controlling, and issuing executive orders to other areas, therefore extensively influencing the remaining cortex. This function is similar to the effect of the wind component in Oriental Five Elements.
- The parietal lobe receives the sensory input from the entire skin; the superior cortex is associated with decisive function in the execution of movement; the characteristically determined features of sensory reception and commandment of the movement are consistent with hard features of the Metal component in Oriental Five Elements.
- The temporal lobe, inferior cortex, associated hearing (the sound), and smell ` The inferior location, the audition, and smell are consistent with fluid features of Water.
- The occipital lobe is associated with vision (the light), consistent with the feature of Fire.
- Medial (limbic system and medial temporal lobe with Memory, thoughts, and emotion. The hidden location and the function of thinking and feeling are consistent with features of Earth, which also means the containmentdlPFC-IPS plays a of the management of the Memories (MM)

   -Superior and Inferior Temporal Gyri for semantic MM
   -Medial Cortices: PFC, Temporal Lobe with Limbic system composed of Cingulate Cortex Hippocampus, Amygdala for emotion, MM.
   - Note: anterior Cingulate Cortex plays the role of error detection: comparison of new information and information in the ICS

The above division of the cortex reflects the division of nature into 5 components, as conceived by ancient Chinese scholars, representing the harmony between five basic components:

 a) Wind/Wood/soft/penetrating (Frontal lobe): generating Fire, inhibiting Earth;

 b) Metal /hard/decisive (Parietal lobe): generating Water, inhibiting Wood;

 c) Water/low, infiltrating (temporal lobe, exterior surface): generating Wood, inhibiting Fire;

 d) Fire/Light (Occipital lobe): generating Earth inhibiting Metal; and

 e) Earth /storing: generating Metal, inhibiting Water.

3. Mechanism of Connections

There are numerous and intricate connections between different layers, areas, and subcortical nuclei in the brain. These connections, often bi-directional, direct, or indirect, are facilitated by relay or intermediate structures. The relationship between the cortex and the Thalamus, the most prominent connection in the brain, is of significant importance. Two theories have been proposed to explain this relationship, underscoring the importance of understanding these complex brain structures.

i. Hypothesis of Protcortex: The cortex develops under the influence of incoming input to the Thalamus. This theory has significant implications for our understanding of brain development, as it suggests that the Thalamus plays a crucial role in shaping the structure and function of the cortex. It is based on the fact that the Thalamus develops before that of the cortex.

ii. Hypothesis Protomap: the neo-cortex develops as a result of the primordial pattern of brain tissue associated with different specific functions in the early development of the brain, consisting mainly of uncommitted or non-differentiated neurons. The hypothesis has gained some popularity since studies in the 2000's. It is critical to note that control of motor function and the perception of sensory inputs are executed in the cortex in the superior part of the brain. However, for important functions like Memory, cognition, and thought, the related cortices are in the inferior and deeper part of the brain (a phenomenon similar to the fact that animals and humans keep their daily earnings and foods in safe and hidden places).

It is highly likely that the development of the subcortical structures and the overlying cortex is concomitant to the phylogenetic evolution. The development of axons and dendrites of subcortical neurons in gray nuclei can both dictate and find the appropriate destination in the cortex. The debate between the two theories of Protocortex and Protomap is ongoing, akin to the age-old question of the egg and the chicken: neither precedes the other, highlighting the dynamic and evolving nature of our understanding of the brain.

B) Cellular component of the brain
1) **Neurons (Fig 2.4)**

Neurons are the most characteristic cells of the nervous system. Neurons have three parts: the cell body, the axon, and the dendrites. The cell body makes up the gray matter, whereas the axons surrounded by the myelin sheets make up the white matter. The neuron cell body shows numerous short cytoplasmic extensions, called dendrites, terminated by multiple thin, usually bushy, spiny extensions.

. These thin spiny cytoplasmic represent the sites of contact with cytoplasmic thin terminal extensions of axons from other neurons to form the synapses. In addition, neurons show a long and sizeable cytoplasmic prolongation of different lengths called axons. Axons frequently have many collateral branches; thin, bushy, cytoplasmic terminal extensions terminate each. These spines and thin cytoplasmic extensions aim to increase the number of synapses.

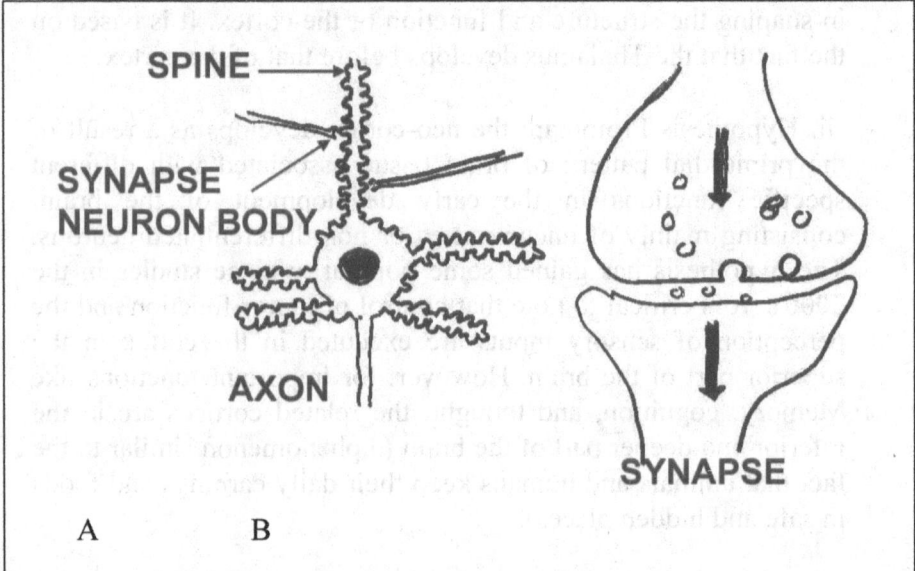

Fig 2.4
A. Neuron with spiny cytoplasmic extension from the body to form synapses with other c neurons
B. Showing the synapse. Note vesicles containing neurotransmitters to communicate with cytoplasmic extension from other neurons

Important notes:
-Each neuron can have synapses with one or many neurons (each human neuron has hundreds of thousands of synapses)

- Each synapse is specific for each neuronal transmission. The significance is that neural transmission does not depend on the specific neuron but only on the neuronal transmission when it is first established. As a result, neural transmission has its pathway of connection. Therefore, once established, a neural transmission involving multiple synapses will "remember to choose its synapses to connect with. This phenomenon is similar to the electrical wiring in a large building with many switches, light bulbs, and machines.... Each button is specific for each system of light bulbs and devices.

In the brain, numerous pathways of connection through many synapses within the cortices and nuclei have been described. However, the mystery is how to activate these pathways of extensions of axons from other neurons to form the synapses. In addition, neurons show a long and sizeable cytoplasmic prolongation of different lengths called axons. Axons frequently have many collateral branches; thin, bushy, cytoplasmic terminal extensions terminate each. These spines and thin cytoplasmic extensions aim to increase the number of synapses. Stimulation and inhibition have yet to be unlocked in many instances. Researchers are aware that many nuclei (or centers) are moderating, activating, and inhibiting other centers from making different functions of the brain work by increasing, decreasing, and turning on or off the connections. The PFC, a key player in this intricate system, likely plays a vital role in these controls, enlightening us about the complex regulatory mechanisms of the brain.

The synapse connection to establish the MM and the CS is, of course, not random but highly selective. This selectivity, a key aspect of neural transmission, will be discussed in detail later in the chapter on MM and CS, engaging the reader in the intricacies of synaptic specificity.

2. Glial cells(Fig 2.5)

Astrocytes are the most populated cells in the brain and account for 2/3 volume of the brain. They are characterized by the presence of many cytoplasmic processes. For a long time, they are believed to play the role of supporting, a role similar to that of the fibroblasts in the non-nervous tissue. In animals, one astrocyte has cell processes covering up to 140,000 synapses of 300-600 dendrites from 4-6 neurons. The astrocytes are larger in humans, and each can have contact with 2,000,000 synapses. Recent studies have expanded the function of astrocytes in the metabolism of glucose, facilitating the action of Ca^{++}, glutamate necessary for the transmission of the neural pathway of conduction in the chain reaction involving NMDAR, AMPAR, CREB, GlunRs Gliotransmitters, Ephedrin, Cholinergic, Nicotine, Interleukin.... From these findings, astrocytes maintain, support, and facilitate the neuronal conduction of information about pain, itch, and other functions of the nervous system. The pain and itch sensory conduction from the spinal cord is modulated by astrocytes (Goldman 2010, Zhang 2003). On the other hand, reactive proliferation in the injuries of the nervous system can also contribute to the increase of the sensory input., Astrocytes (and especially microglia) produce several cytokinin-like IL-1beta, TNF 1, 6,18, Astrocyte-derived CCL2 and CXCL1, CCR2, and CXCR1 (ji 2019). With these proinflammatory, cytokinin glial cells take part in the development of degenerative neuronal disease and depressive disorders (Barone 2017, Haroon 2017, Maydych 2019, Ndayisaba 2019 Phillips 2019, Kim 2019 Rana 2018).

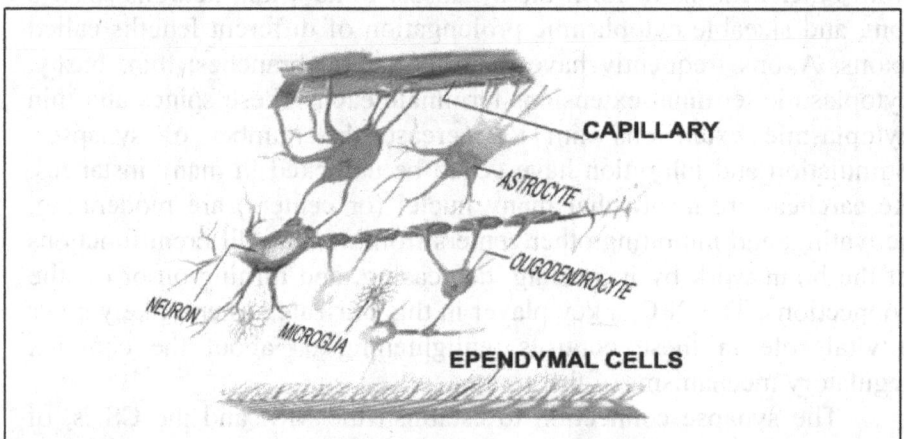

Fig 2.5. showing the interaction of astrocytes and microglial cells and neurons in the Creation of the pro-inflammatory reaction

Astrocytes are the most populated cells in the brain and account for 2/3 volume of the brain. They are characterized by the presence of many cytoplasmic processes. For a long time, they are believed to play the role of supporting, a role similar to that of the fibroblasts in the non-nervous tissue. In animals, one astrocyte has cell processes covering up to 140,000 synapses of 300-600 dendrites from 4-6 neurons. The astrocytes are larger in humans, and each can have contact with 2,000,000 synapses. Recent studies have expanded the function of astrocytes in the metabolism of glucose, facilitating the action of Ca^{++}, glutamate necessary for the transmission of the neural pathway of conduction in the chain reaction involving NMDAR, AMPAR, CREB, GlunRs Gliotransmitters, Ephedrin, Cholinergic, Nicotine, Interleukin.... From these findings, astrocytes maintain, support, and facilitate the neuronal conduction of information about pain, itch, and other functions of the nervous system. The pain and itch sensory conduction from the spinal cord is modulated by astrocytes (Goldman 2010, Zhang 2003). On the other hand, reactive proliferation in the injuries of the nervous system can also contribute to the increase of the sensory input., Astrocytes (and especially microglia) produce several cytokinin-like IL-1beta, TNF 1, 6,18, Astrocyte-derived CCL2 and CXCL1, CCR2, and CXCR1 (ji 2019). With these proinflammatory, cytokinin glial cells take part in the development of degenerative neuronal disease and depressive disorders (Barone 2017, Haroon 2017, Maydych 2019, Ndayisaba 2019 Phillips 2019, Kim 2019 Rana 2018 C) **Neural connections.**

1. Overview 146

Neural connection, a crucial feature of the nervous system, involves the transmission of electrical impulses between two neurons. The direction of this communication designates one neuron as pre-synaptic and the other as post-synaptic. Each synapse, the point of connection, is unique and specific to an individual electrical impulse. This impulse carries a specific piece of information, underscoring the function and importance of the synapse in neural communication. Given that a synapse is a narrow space between the cellular membranes of two neurons, a mechanism is required to transmit the information across this space..

2. Development

Neural connections, a unique and essential feature of the nervous system, play a crucial role in transmitting electrical potentials from one nerve whisker to the next. This transmission creates an electric current, which is essentially a stimulus, or information. The significance of these neural connections, as revealed by research, underscores the remarkable mechanism of the nervous system.

3. The formation of neural connections 146

Done during the embryogenesis into the development of an adult organism. In an adult organism, the neural cells no longer reproduce or reproduce very little, the neural connections are still made thank to the neuroplasticity mechanism to adapt to the need to integrate more information. The dendrites develop to form spines, mushroom-like structures , and filopodia

a) the formation of neural connections-muscles.

The most crucial function of the nervous system is to efficiently control movement, acting as a mediator between the automatic and voluntary mechanisms and the motor organs/muscles. The finger-shaped dendrites, when touching the muscles, initiate the Acetylcholine production process, a key step in muscle movement. Each type of muscle, whether it reacts quickly or slowly, is connected to each type of adaptive nerve, demonstrating the system's efficiency in controlling a wide range of movements.

b) In the central nervous system, Acetylcholine is replaced by Glutamate with the corresponding MNDA receptor

- In the fetus, the intricate process of CNS development is guided by the enigmatic guidepost neurons, found in the developing olfactory tract connecting the olfactory bulb to the Entorhinal cortex EC. These neurons play a crucial role in directing filopodia to the CNS, a phenomenon unique to the fetal stage.

-Guidepost neurons and Axon guidance to help filopodia find their adaptive target
Wnt proteins, a group of signaling molecules, are not just required but are instrumental in the production of the central nervous system in the fetus. Their role in CNS development is a fascinating area of study that continues to enlighten researchers in the field.
-Netrins, Slits, Ephedrins, Somaphorins, and CÁM/cell adhesion molecules are the elements that attract or repel filopodia
-Glia cells, a type of non-neuronal cell in the central nervous system, are required for the development and maintenance of the CNS.
NEUREXIN-NEUROLIGIN "handshaking" Neurexin/NRX is a cell membrane molecule. The inner part of the cell activates microvesicles to secrete neurotransmitters. The outer part of the cell shakes hands with Neuroligin to transmit information to the next post-mitotic nerve cell. The above research brought the 2013 Nobel Prize to James E Rothman (US) at Yale University, Randy W Schekman (US) at UC Berkeley and Thomas Südhof (Germany)

4. two modes of connections
modes of connections: (Kandel ER. 2006)

a) Chemical synapses 146
Fig 2.6): the most common neurotransmitters are adrenaline, noradrenaline, acetylcholine, glutamine, DOPA, serotonin histamine, galanin, and orexin....These neurotransmitters are synthesized by neurons in nuclei that are usually capable of producing a few neurotransmitters specific to the function of nuclei.

The neural transmission consists of creating the high voltage to be transmitted to the low voltage. The neuron has to pump the Ca^{++} cation or other cations like Zn^{++} from outside of the cell membrane into the inside /cytoplasm of the cells. This can be done by opening the cell windows. The cell windows consist of receptor NMDAR (N-methyl-D-aspartate) or AMPAR (DL-alpha-amino-3-hydroxy-5-methylisoxazole-propionic acid). In an inactive state, these receptors are blocked by Mg^{++} cation, in the presence of neurotransmitters like Glutamate (the most common neurotransmitter), Mg++ is removed, and Ca++ (or other cation like Na+) can enter into the cytoplasm, creating the negative charge on the outside of the cell membrane. This process requires a chain of reactions in the cytoplasm to restore the initial state of NMDA or AMPA. cAMP (called Memory molecule) plays a critical role in this chain reaction (Barco A, Marie H 2011)

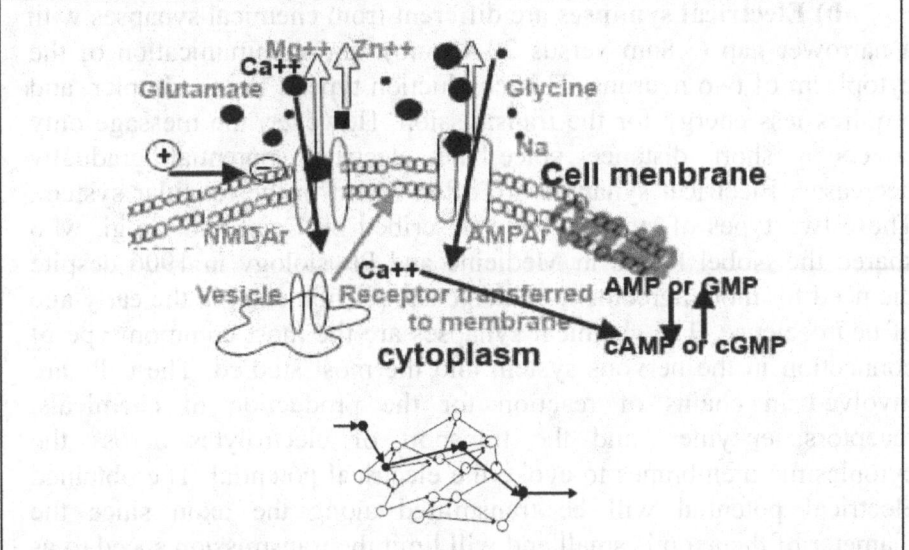

Fig 2.6: Cell membrane (+)>>>(-)Glutamate attached to Receptors NMDA or AMPA (Medina 2011, Ghasemi2011)
NMDAR (for Ca++ hay AMPAR for Na+ when combined with Glutamate or Glycine, kids out Mg++ and open the window for Ca++ entering the cell.
Intracytoplasmic Ca++ activates the gene to make new proteins copy AMPAR or NMPAR to replace receptors losing Mg++. Receptors without Mg++ hay Zn++ are retrieved and transported in the cytoplasm for recycling by attaching Mg++ or Zn++.
Note: vesicles release AMPAR to be attached to cell membranes. AMP or GMP is transformed to cAMP or cGMP in the presence of Ca++

The change in the electrical potential of the cell membranes is to render the neuron at the level of firing (to communicate with other neurons). The electrical potential of the membrane, when the neuron starts firing (producing electrical potential/ electrical current), is
-55mV (a phenomenon called Excitatory Post Synaptic Potentia/EPSP) (normal electrical potential is -70mV). For example, at –40mV, the neuron becomes refractory/unresponsive to stimulation for firing, a phenomenon called Inhibitory Post Synaptic Potential/IPSP).

Furthermore, to make the cell membrane at the firing level, one excitation is not powerful enough. There must be multiple excitation excitations in the same or adjacent synapses/ a chorus of excitation!)

- The electrical current travels along the axon to reach the dendrites and initiates the release of neurohormone to activate the next post-synaptic membrane.

b) Electrical synapses are different from chemical synapses with a narrower gap (3.8nm versus 20-40 nm), and communication of the cytoplasm of two neurons. This conduction type is faster, simpler, and requires less energy for the transmission. However, the message only covers a short distance since the electrical potential gradually decreases. Electrical synapses are often found in the reticular system. These two types of synapse were described by Cajal and Golgi, who shared the Nobel Prizes in Medicine and Physiology in 1906 despite the need for more agreement on this basic phenomenon in the early age of neuroscience. The chemical synapses are the most common type of connection in the nervous system and the most studied. The cells are involved in chains of reaction for the production of chemicals, receptors, enzymes, and the transport of electrolytes across the cytoplasmic membranes to evoke the electrical potential. The obtained electrical potential will be transmitted along the axon since the diameter of the axon is small and will limit the transmission speed to as slow as 1mm per second. The increase of the rate and the amount of data of conduction are consistent with the enlargement of axons by getting wrapped around the axon with many layers of myelin sheets secreted by oligodendrocytes in the central nervous system or Schwann's cells in the peripheral nervous system. In addition, neurons in the PFC and motor cortex may show strong coupling to the network in the same areas. In cases of the motor cortex, the coupling of neurons is involved in the coordination of the movement rather than in increasing the intensity of the movement. In the PFC, the coupling to the network contributes to subtle changes in the interaction between neurons in learning Memory and consolidation

3. New type of neuronal connections Recently, McFadden (McFadden 2020) proposed a hypothesis of connection based on the electromagnetic field to contribute to the formation of cognition. The basis is that electromagnetic is eventually created when there is an electrical current. The paper was published in a new journal. The hypothesis is to circumvent the current problem in understanding many mysterious phenomena of cognition such as mediumship, out-of-body experience, near-death experience, peduncular hallucinosis, sleepwalking, and dissociated identity disorders....

D. Neuroinflammation

Neuroinflammation usually alters the central nervous system. This phenomenon is beneficial in repair but damaging when the phenomenon represents the over-reactivity. Chemicals involved are chemokines (CCL2, CCL5, CXCL1). (CCL= geneCC chemokine in chromosome 17, playing the role of chemotaxis attracting inflammatory cells into the sites of interest. Monocytes, T lymphocytes, and endothelial cells produce these chemicals. Other chemicals are messengers (NO and prostaglandins) and reactive oxygen species (ROS). Monocytes are smaller than macrophages which only exist in the tissue, not in the blood, and account for up to 10% of cells in the CNS. Neuroinflammation and inflammation outside the CNS eventually increase (or damage) the Blood Brain Barrier, a semipermeable wall isolating the CNS from the rest of the body to preserve the integrity of the CNS from harmful chemicals and organisms. Over-reactivity to traumatic injuries can expand far beyond the areas of damage. Aging is often associated with an increase in neuroinflammation levels and a decrease in the anti-inflammatory mechanism. Therefore there is an enhancement of neurodegenerative diseases. Neurodegenerative disease like Alzheimer's due to the production of amyloid materials and neurofibrillary tangles is associated with demonstrated inflammatory markers. Parkinson's disease, with a deposit of synucleoproteins, is frequently associated with a history of inflammatory illness of the intestines, constipation, and Dysbiosis with bad bacteria that may be released from the bowel and cross the Blood Brain Barrier and infect the nervous tissue. Multiple Sclerosis: Neuroinflammation alters the Blood Brain Barrier; peripheral activated T lymphocytes can cross the BBB and enter the nervous tissue. Capoxone can control the T cell reactivity to prevent the relapse of MS and PD (Guzman-Martinez 2019, Disabato 2016). Stress and insomnia increase neuroinflammation. On the other hand, Meditation and physical activity inhibit genes involved in neuroinflammation (Seo 2019, Buric 2017, Rozich 2020, Chen 2021, Huang 2021, Cossu 2021, Gong 2022, Guzman-Martinez 201 DiSabato 2019 Trimble MR 2015).

E. INNER CONSCIOUSNESS
1. Concept of ICS (Fig 2.7)

Hippocampus represents the initial step in the consolidation of Memories (MM). Synaptic connections in the MM are essential in the formation of Consciousness. Example: face, name..., mental status used in Pavlov's conditioned reflexes. Cortices of Inner Consciousness

are the cortical areas that store Memory. Inner Consciousness is used as Model (Comparative) Consciousness to compare with the newly integrated data. Inner Consciousness stores different, distinct, and meaningful/ semantic MM of general and non-specific types in life. After death, the Inner Consciousness will be carried by the Soul and pasted into the Brain of the individual in the next life. In addition, in life the ICS is continuously updated with new information coming from the five senses and concurrent cognition. As a result, the ICS can be regarded as an encyclopedia for consultation regarding the information in life. Components of information similar to that in the encyclopedia are labeled as such. (Example: the incoming information on an orange will be matched up with that of the orange in the encyclopedia. If the incoming data of an orange from Vietnam has a skin of green color distinct from Sunkist Orange in California, the green color of the orange will be added up in the ICS. The above process of updating the ICS can be considered as the mechanism of learning: The newborn baby has instinctive reflexes for survival, such as motility of Inner organs for basal metabolism such that of gut, heart, lung, and kidneys.... Other fundamental reflexes like eating, sitting, walking, and phonation develop without needing learning. Following skills in living, like language, emotion, and social interaction, are obtained and added to the ICS through MM and MM consolidation with the help of learning, teaching, and training. The portion of the brain responsible for discerning the similarities and differences between the incoming and internal information is likely the anterior cingulate cortex/ACC. ACC will be discussed later in the book..
(Raichle ME 2015)

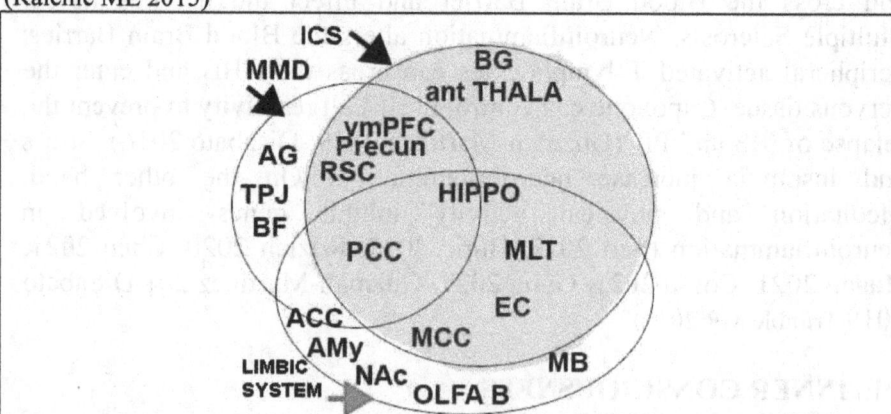

Fig 2.7 Cortices of Inner Consciousness, Default Mode Network and Cortices of the Limbic system share many common areas of cortex

2. Hippocampus

As a part of the Limbic system, the hippocampus (HIPPO) is considered an essential center of MM with widespread connections to the entire cortex, the adjacent thalamic nuclei, and the opposite HIPPO. After being processed in different cortices, the information integrated and converted into pieces of Consciousness which are sent to specific sensory cortices for specific sensory information. Non-specific or general information such as time, place, name of things/ objects, and general meanings like love and hate are sent to the HIPPO.

In the sensory cortices, the information is stored in neural connections known as synapses. Consolidation may be necessary. However, the links appear relatively stable unless disturbed by learning and retrieval. The type of information/ MM, also called implicit MM, related to procedures (involving movements) that are sent to the Basal Ganglia and the Cerebellum is also stable in preservation for retrieval. For the Hippo, with its small size, the incoming information/ Consciousness can not be stored in this small organ for an extended period. The information is therefore categorized as semantic and non-semantic types. The explicit, semantic MM is sent to the middle Temporal Lobe. The implicit MM, such as speaking, eating, dressing, walking, driving, and performing surgical procedures... are stored in the cerebellum and the basal ganglia.

According to the standard theory of MM consolidation, the HIPPO stores and retrieves recent general MM and sends recent MM to the frontal cortex (ventromedial PFC/vmPFC) for consolidation. The

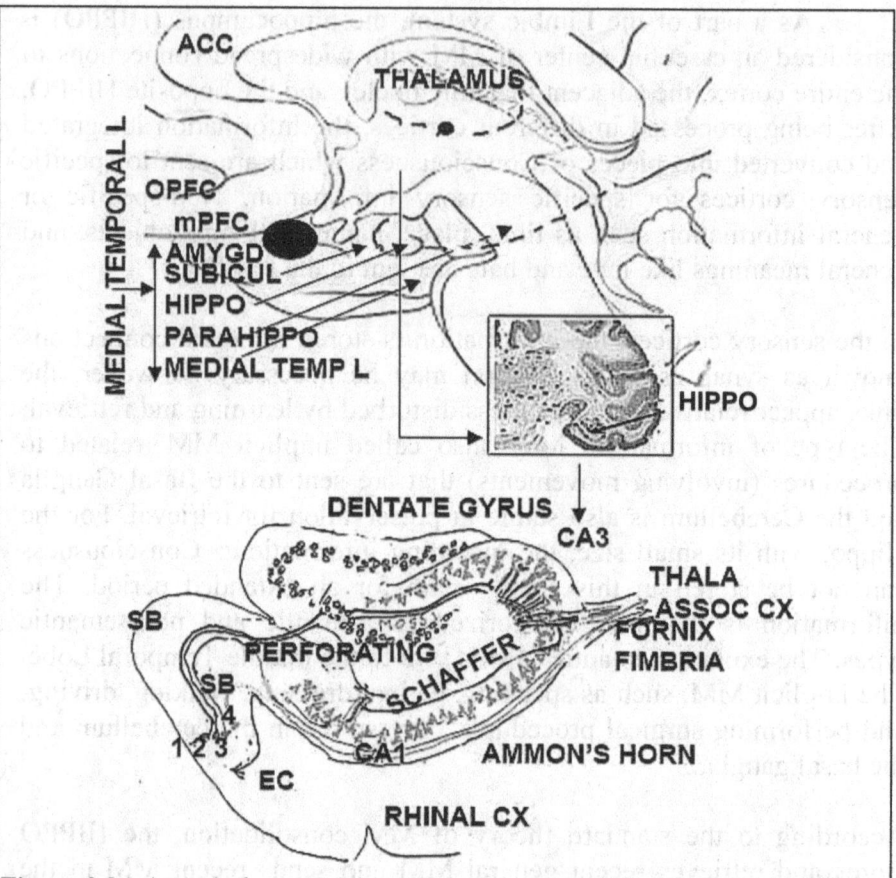

Figure 2.8A (upper) shows Orbitary PFC, ventromedial PFC Amygdala, Hippo , Mammillary body, and ParaHippoo with different areas of the Anterior Thanamus ATN
Inset: cross-section of the HIPPO
Fig 2.9B
The Hippocampus has such a shape to accommodate the interconnection between various regions to fulfill the function of encoding and storing Memories and Consciousness. -Entorhinal Cortex/ EC= Entorhinal Cortex connecting to Hippo and the cortex
-Subiculum/ SB (Supportive structure =betweem EC and Hippo CA1), connecting with CA3

Hippo associated with functions:
Dorsal Hippo: CA3 dedicated to the MM of place with Place cells.
Ventral Hippo: CA1 linked with amygdala/fear

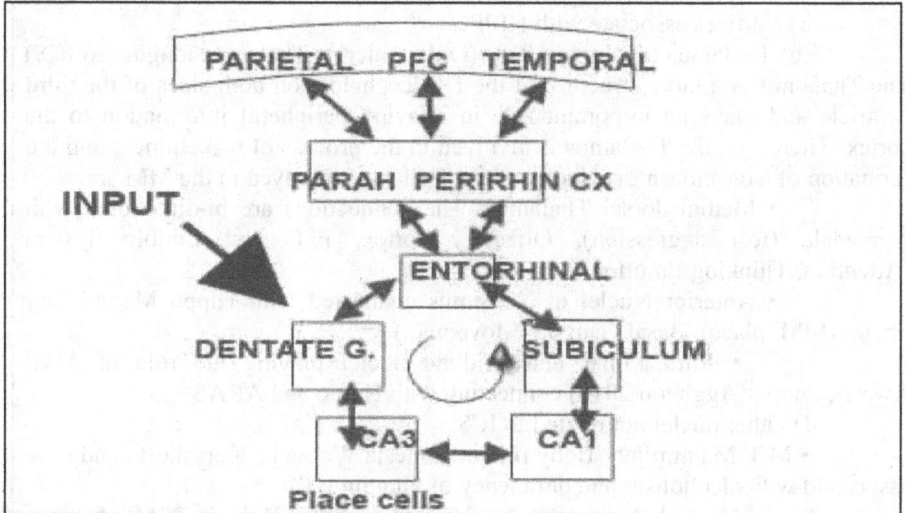

Fig 2.9C Peripheral information coming to Hippo/Dentate Gyrus and CA3, transferred to CA1 then toSubiculum >> Entorhinal cortex >> Para Hippo/Perirhinal >> later to the sensory cortex

HIPPO does not play the role of retrieval of remote MM. However, further investigation leads to a newer concept of MM management. The Hippo retains traces of explicit MM after sending away the information (to vmPFC). Therefore, the traces of MM in the Hippo sub-serve are the keys to the retrieval of different pieces of MM. This concept of MM traces is supported in the following experiment. In mice trained in search for food, genes ZIF268 in the HIPPO have been demonstrated to produce protein transcription factors facilitating the retrieval of long-term MM. About 400 genes of this type increase activity during sleep (when MM are sent up to the vmPFC for consolidation and to the Posterior Cingulate Cortex/PCC). Tetracaine injection into the HIPPO will abolish the capability of MM retrieval. In training or learning, HIPPO often repeats this process of retrieval and consolidation many times. Of course, intra-cytoplasmic chemical reactions in the recycling of receptors for Glutamate are necessary for the consolidation of MM.
...

Further reading
Note: a) Stress increases Glutamate in NMDA associated with Matrix MetalloProteinase/ MMP-9 of CA1 of the Hippo with the decrease of Nestin-3. Nestin 3 in CA1 can modulate stress and emotion(van der Kooij 2014). Protein Kinase C/ PKC in HIPPO alters the synaptic connections with subsequent modification of behavior (Hains 2009)

  b) Cortices associate with HIPPO
   c) Thalamus (Crabtree 1018) (Only Anterior Thalamus longing to ICS)
The Thalamus is a large structure of the Diencephalon, on both sides of the third ventricle and plays an important role in relaying peripheral information to the cortex. Therefore, the Thalamus is involved in the process of wakefulness and the formation of Consciousness. Nuclei of the Thalamus involved in the MM are:
   • Medial dorsal Thalamus: The connections are bi-directional with Amygdala (fear aggression), Olfactory cortex, PFC and Limbic System (Attention, Thinking Emotion)
   • Anterior Nuclei of Thalamus connected with Hippo Mammillary bodies (MM, place), Basal Ganglia (Movement)
   • Intralaminar and Midline nuclei playing the role of MM, Awakefulness, (Aggleton 2010) connecting with Hippo and ARAS
  d) Other nuclei not related to ICS
   • MB/ Mammillary Body role in amnesia Wernicke Korsakoff syndrome associated with alcoholism and deficiency of vitamin B12
   • Ventral Tegmental Nucleus of Gudden Role in MM of space (Dillinghan 2015) role in amnesia Wernicke Korsakoff syndrome, sleep apnea, Alzheimer disease,Schizo
   • Amygdala
   • N Reuniens interconnected with vmPFC, Hippo is involved in playing the role of short episodic MM of place. In remembering the route to return home, the connections between NRe, vmPFC, and CA1 of Hippo in critical (Jin 2005). Connection with the Amygdala, Hippo is also essential to avoid dangerous places.

3. PFC
  - social interaction, emotion, empathy,
  - The Hippo sends recent MM to the vmPFC during NREM sleep of stage 3-4 (Stage of sleep without Consciousness during Slow Wave Sleep/SWS) evidenced by Sharp Waves Ripple picked up in EEG. The MM is consolidated in the vmPFC. Later on, after 2-3 years, the MM is sent to the PCC, probably for more stable consolidation and for a longer duration, usually known as autobiographic MM.
  - vmPFC corresponds to the Third eye. The third eye is the imaginary location 1-2 cm behind the point on the Forehead midline, between and slightly above the eyebrows. The role of vmPFC is commonly known as playing a function in management behavior and expression of emotions in social interaction. In addition, due to its location in the front of essential nuclei involved in the Consciousness, vmPFC may play the role of the Third eye: communication with the outside spiritual world. This function includes communication and receiving information from estranged Souls.
  - dmPFC, located between vmPFC and ACC. Role in high-level mental status, like the feeling of the self, theory of Mind (guessing mental status of the others (Baetens), mental value,

sympathy, empathy. And role in the retrieval of long-term MM (Li 2019) and autobiographic MM (Nawa 2020)

4. Limbic System and Medial Temporal Lobe/ MTL

Broca named the limbic system to designate the structure in the medial side of each cerebral hemisphere due to its shape molding around the Corpus Callosum and the Thalamus. It was believed to manage the expression of affection and sentiment. In addition, PCC of the limbic system is mainly for the storage of autobiographic MM,

a) MidCingulate Cortex/ MCC (DOPA):
  Cognitive role in monitoring social activity with decision, reward-related information
  Perception of chronic pain, fear, and obsessive-compulsive disorder (Anterior MCC)
  pMCC: PTSD and PSP

b) PCC, RSC: component of DMN PCC has connections with vmPFC and Precuneus/, Insula, Hippo /MTL. PCC shows decreased fMRI activity and increased functional connectivity during mental attention and retrieval of MM. vmPFC and PCC are important cortices for the storage and retrieval of MM. PCC is considered the cortex for autobiography. RSC, the cortex behind and adjacent to PCC. RSC may share this role of autobiography. RSC is distinct from the PCC by dysgranular layer 4 (fewer granular cells than other granular 4, playing the role of information processing). Besides the role of episodic explicit MM (Todd 2015, Mitchell 2017,Vann 2009 Rolls 2019 Thompson 2010, Mitchell 2018, Dixon 2018, Cunningham 2017, Aggleton 2009)., RSC likely represents the storage of the store MM , , i.e., Karma)

c) Anterior Cingulate Cortex/ ACC, well known as the cortex for error detection in addition to the role of pain detection and place. As a result, ACC is involved in the formation of the CS. The role of differentiating between right and wrong information is consistent with the necessary task of the cortex to detect information similar to that already present in the ICS and the new information not present in the ICS. The abundant connection between ACC and Hippo supports this function of the ACC. The alteration of the retrieval of the MM accompanies the lesion of ACC.

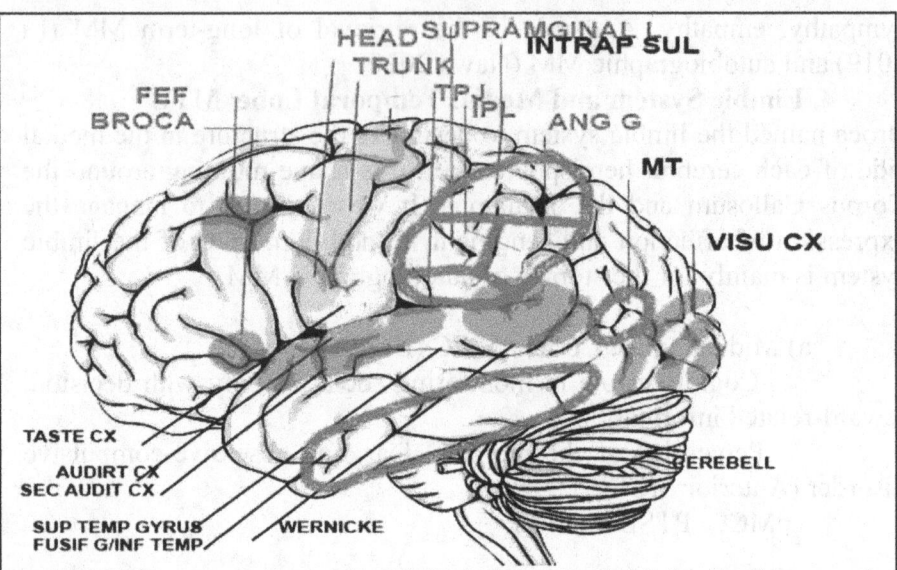

Fig 2.10 PPC, a component of the DMN, is the cortex behind and adjacent to the Tactile sensory cortex, which also plays a role in processing the episodic spatial MM mental imagery (Cabeza 2008, Hutchinson 2009). PPC has extensive connections with most areas of PFC, MTL IPS: Intraparietal Sulcus between Superior and Inferior Posterior Parietal cortices, a central executive network component for managing different brain areas.
 b) TPJ Temporoparietal junction related to OBE and Theory of Mind
 c) Angular Gyrus: DMN, OBE, retrieval of MM, attention (Right AG)
 d) IPL Inferior Parietal Lobule: composed of - Angular gyrus - Supramarginal gyrus SGM: episodic MM, anosognosia, construction apraxia, dressing apraxia, contralateral sensory neglect, contralateral hemianopia, or lower quadrantanopia. (Harrison 2009, Wagner 2005, Berryhill 2007, Berryhill 2007).

d) **Precuneus** (episodic MM), a DMN component, has decreased activity in thinking, attention, Meditation, sleep and Anesthesia (Conil 2021, Cunningham 201).

- Considered as a hub for CS in environment-driven tasks as opposed to the Insula involved in volition-driven CS.
- Role in mental visuospatial imagery with a neural connection of Precuneus with Zona Incerta (undetermined function, ? motor limbic integration, visceral function, pain), Pulvinar, Superior Colliculi (visual relay center), Medial Dorsal, Intralaminar of Thalamus (relay center for sensory input), PFC, and the relationship of Precuneus with IntraParietal Sulcus/ IPS
- Role in episodic explicit MM

The location is consistent with that of Chakra 12, Yoga practitioners believe Chakra 12 to be the open window of exit of the Soul in case of Out of Body Experience

- Response inhibition

e) Medial Temporal Lobe (MTL), including Hippocampus, Amygdala (fear), and Entorhinal Cortex/ EC (Memory, navigation, and time)

MTL site of storage of core episodic MM (Reber 2017, Liu2021, Puttaer 2021, Mars 2012).

Amygdala: Connections with Hippo, VTA LC Trigeminal Nucleus, TRN/ Thalamus (Dorsomedial N.), vmPFC role in MM for fear, anger, and aggression. The essential nucleus in the arc of reflex of Pavlov. Information coming to the Hippo is processed in the Dentate Gyrus and sent through the Subiculum to the PFC.

f) Posterior Parietal Cortex/ PPC and the cortex **of the :**
  -**TemporoParietal Junction TPJ (Fig 2.10) role:**
  Theory of Mind, Out-of-body experiences ,Temporal order judgment
  -**Angular Gyrus/ AG** role: language, number processing and spatial cognition, Memory retrieval, attention, and theory of Mind
  -**Inferior Parietal** Lobule/ IPL sensorimotor integration.
  -**Supramarginal Gyrus phonological** processing and emotional responses,

Empathy and sympathy (Right TPJ)/ language cognition, processing, and comprehension of written and (Left TPJ) spoken language

F. Polyvagal Theory (Fig 2,11)
(Poly=many, Vagal=wandering)
is proposed in 1994 by Stephen Porges, Ph.D. in Chicago. The widespread innervation of the Vagal nerve in the body covering the skin, viscera, and glands of the upper respiratory tract of salivary, lacrymal, and mucous types deserves a particular study. The theory suggests the mechanism of reflexes in response to stress and stimuli upon the above sites. The nerve forms the Parasympathetic system, having actions to counter-balance those of the Sympathetic system. The Sympathetic system tends to make the body react to fight and flight. The Parasympathetic system is responsible for an opposite reaction for tranquility and health with rest and digestion. The third state for social engagement involving immobilization/ shutdown is undertaken by the Dorsal Vagal system (The Ventral Vagal system is for rest and digestion)

i. Dorsal Vagal Complex/ DVC, in continuity with the RAS including the PeriAqueductal Gray/ PAG, responsible for immobility/ shut down status usually seen in small insects becoming frozen in danger.

ii. **Ventral Vagal Complex/ VVC** (Ambigus Nucleus and Nucleus of Solitary Tract) is responsible for tranquility and health with rest and digestion.

The reaction using a mechanism of DVV is only used when the sympathetic response and the VVC fail to protect the living organism from the insults.

Comments and Objections

Clinicians and Neuroscientists oppose the theory due *to:*

The evidence of DVC in low invertebrate insects can *not be demonstrated*

The proof of bradycardia in stress and Respiratory *Sinusal arrhythmia= RSA is difficult to demonstrate.*

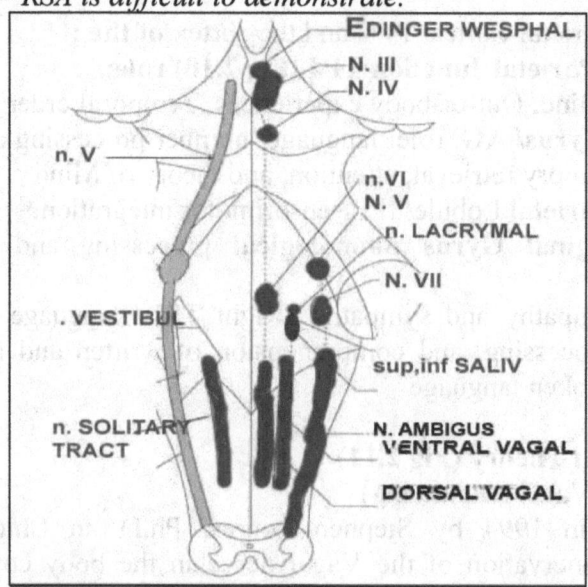

Fig 2.11: Vagal Nuclei

The theory tends to attribute emotion's bodily and faci*al expression as being controlled by the X nerve, disregarding other mechanisms proposed by neuroscientists with the involvement of NAc, VTA, Amygdala, and PAG...*

Polygonal theory plays an additional role in exp*ressing sentiment and emotion.*

Applications:

Although the theory is not as critical as suggested by Porges, the theory is conveniently used to explain:

• Inhibition of the Sympathetic system (Porges, 2011; Deuchars et al., 2018)

• Control of the hypothalamic-pituitary-adrenal axis (Porges, 2011),

• Neuroinflammation (Pavlov and Tracey, 2012), brain-gut interactions (Bonaz et al., 2018; Fülling et al., 2019)
• Neurogenesis (Laborde 2018)
• Epigenetic (Follesa et al., 2007; Biggio et al., 2009; O'leary et al., 2018
• Mechanism of regulation of Emotion(Geisler et al., 2010; Kok and Fredrickson, 2010; Kok et al., 2013, Zilioli et al., 2015; Dang et al., 2021, Geisler et al., 2010, Williams et al., 2015)
• Mechanism of regulation of social interaction (Kashdan and Rottenberg, 2010; Colzato et al., 2018, Kemp et al., 2012; Geisler et al., 2013; Kok et al., 2013 Williams et al., 2019; Eggenberger et al., 2020),
• Maintaining well-being and adaptation to the environment (Werner et al., 2015; Young, Benton, 2018, Dedoncker et al., 2021 Hillebrand et al., 2013; Jandackova et al., 2016; Fang et al., 2020 Richardson et al., 2016; De Brito, 2020)),
• Sneezing: overaction of the whole body to stimuli to the nasal mucosa. To prevent such inappropriate, a simple external pressing over the nose ala (lateral wall of the nostril) to decrease the itchy sensation of the nose. The sneezing overreaction likely decreases if the pressure is applied late.
• Ingestion of a hypotonic solution stimulates the secretion of VIP/ Vasoactive intestinal peptides, which in turn promote the Vagus nerve and then the osmolarity center in the Brain
• Vertigo of vestibular origin is commonly accompanied by high GI motility, abdominal pain, vomiting, and diarrhea. Crystals in the posterior semicircular canal cause Benign Paroxysmal Positional Vertigo syndrome. The vagal ending in the Posterior Semicircular Canal is stimulated and causes an increase in GI motility.
• In gastroenteritis, with or without accompanying other viral symptoms like sore throat, sneezing, fever, vomiting, and diarrhea. The vagal endings in the skin exposed to the cold are stimulated, leading to the overreaction of the vagus, including the GI and the vestibular system, causing vomiting, diarrhea, abdominal pain, and dizziness
The Vietnamese and Chinese commonly use procedures like Coin rubbing (in the upper back) or Moxibulism, Suction Cupping, or Spa-Sauna to activate the Vagus nerve. As a result, the processes cause cutaneous vasodilation.
• Abdominal pain in Children, stress causes the activation of the Vagus nerve
• GENIAL model (Genomics—Environment—Vagus Nerve—social Interaction—Allostatic regulation—Longevity

• Is the model proposed by Kemps linking Emotion, Physical Health, and Longevity and Social relationships > Behavior+Psychology+Physiology Health

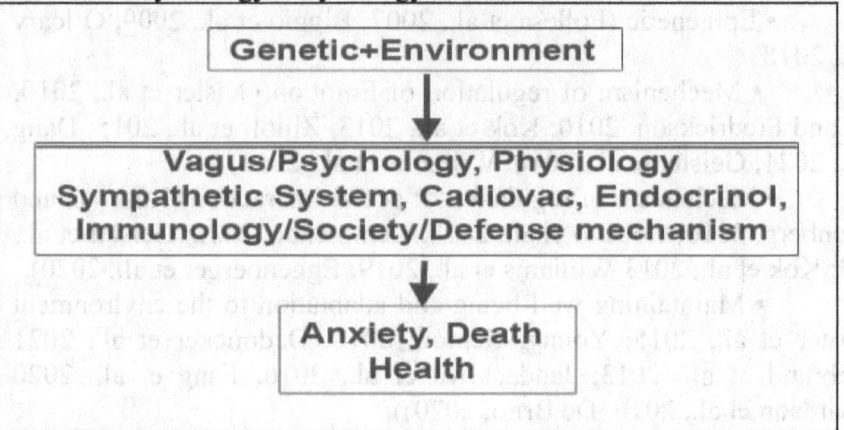

In this model pathway G-E-N-I-A-L, the Vagus plays a critical role in handling the stress linked with behavior toward health, medical treatment, psychological factors, i.e. moods, emotions, personality attachment, Physiologic factors, i.e. sympathetic/ parasympathetic cardiovasc neuroendocrine immune with results of wellbeing/ longevity/ premature death/ loneliness.

Chapter III
MEMORY
I. Introduction

The evolution of living organisms can be considered as the result of adaptation to the environment according to Darwin's theory of species evolution (1859). The adaptation involves the physical aspect of the species and the non-physical part, including the behavior, the language, and the other metaphysical aspects. The evolution of these latter parts of the species is related to the nervous system, particularly the brain. It is well known that the phylogenetic evolution of the central nervous system is characteristic. The higher the phylogenetic development, the more complicated the Telencephalon. Parallel to Darwinism related to a living organism in general, Neural Darwinism is a theory proposing that the brain function results from selective development from cortical groups of neurons forming primary repertoire (a part composed of many items). This means that the brain's function is a result of the selective development of groups of neurons, influenced by the natural selection of the animal body, species development, and Consciousness adaptation. The Proto cortex theory proposes that the evolution and transformation of the brain depend on genetic mutations that take many hundreds or thousands of years to accomplish the necessary adaptation. In addition, epigenetic factors are also crucial for the mechanism of somatic transformation. The epigenetic role is based on the modified gene expression secondary to chemical modifications of proteins like histones, a molecule that provides support for chromosomes containing genes.

This journey of understanding is in line with the direction of Darwin's theory, as opposed to the Promap theory, which proposes that the cerebral cortex is composed of different zones of a specific function. This zone dictates the organization of the subcortical gray nuclei. The study of human Memory has been a subject of interest for philosophers and psychologists, given the elusive nature of its mechanism of formation and retrieval. However, with the progress of biology, chemistry, comparative neuroanatomy, and neuroscience, our understanding of the function of different brain parts has significantly illuminated the mysterious roles of various areas of the gray matter, including the cerebral cortices. This progress has not only informed us but also enlightened us about the intricate processes involved in memory formation and retrieval.

In brief, the mechanism of neural conduction and connection through neural synapses in laboratory invertebrate animals has significantly clarified the Memory formation and retrieval mechanism. On the other hand, clinical studies of neurological deficiencies in patients in parts of the brain related to Memory storage and retrieval have led scientists to identify the brain as the site on which Memory is based. The formation of MM is similar to registering events in a notebook; the writing in the notebooks is usually reviewed for addition, clarification, and modification. For MM, the incoming data is transcribed into a format suitable for stable storage and even transferred to more permanent places. Peripheral inputs to the Thalamus are sent to sensory cortices (specific sensory organs) for storage. Non-specific/.general data are sent to Hippocampus for triage and then uploaded to vmPFC, Cingulate cortex, and Medial Temporal Lobe. for definite storage

The Soul plays a profound role in the formation and retrieval of the MM. MM can be conceived as the interplace between the physical (body) and metaphysical (Soul) realms. It is reasonably believed that MM is not stored only in the Brain (DMN) but also in the Soul, which ultimately controls the retrieval of MM. However, since the Soul has not been recognized in neuroscience as an entity, the above process has yet to be studied and has created a gap in the mechanism of some MM disorders. This contemplative aspect of the Soul's role in memory formation invites us to think deeply about the complexities of memory and the gaps in our current understanding.

II. Definition of Memory
A. Features and relationship with Consciousness
Memory is very close to our lives, however challenging to characterize.
1. Far Eastern concept

A living being is a whole entity with visible and invisible parts. The above division is artificial to serve for examination investigation and subjective understanding. This division can be represented by the classical Chinese symbol of the Utmost Extreme/ Supreme Ultimate or Ultimate Pole/ vn:Thái cực/ Taijitu, which creates the Dualist Negative and Positive (Yin and Yang). In this diagram, the separation is usually:

- Unequal: the two parts have different names Yin/ Yang, and

- Incomplete in each part, there is a dot representing remnant and also the seed for future change in the opposite part:
. Minor Yang in Great Yin
. Minor Yin in Great Yang

Traditionally, for the living being, the positive part represents the physical body, the negative part, the metaphysical aspect called the Soul, and the Mind composed of Memory, Consciousness and Karma.

The Minor Yin likely represents the parts of Memory in the forms of neural connections.
The minor Yang represents the Memoirs / Memorials
   According to Buddha, each living being is composed of five Maras/ Aggregates of Grasping: Form, Feeling, Perception, Mental formation, and Consciousness (vn: Ngũ Uẩn: Sắc Thọ Tưởng Hành Thức).
   • The Body or Yang part represents the Form.
   • The Soul/ Mind or Negative part represents the remaining four of five Maras.

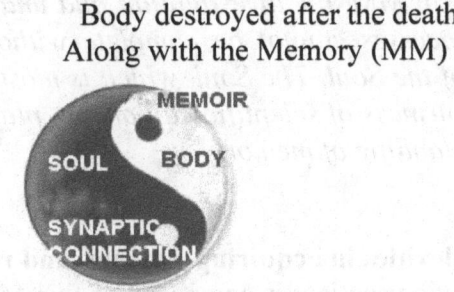

Fig 3.1
the MM + Memoirs is a part of SOUL/ MIND

After death, the Yang Part/ body disintegrates. However, something remains after the body leaves the earthly world. The remaining part is the memoirs of that individual representing the minor Yin. The Soul/ Mind leaving the body carries traces of neural connection or equivalent.

2. In Buddhism,
All sentient beings are composed of five components, called five Maras/ Skandhas or Aggregates, used to interact with the environment
    i. Form representing the physical part of the body . The physical part consists of structures that can be determined by:
      - Matter or Material, Dust/ vn: Sắc trần/ skt: Gunas,

Ruparammana
- The five sensory organs: vision, touching, tasting, hearing, electromagnetic energy and -neuronal structures consisting of neurons-and synapses

ii, iii, iv, v. Four Maras representing the metaphysical component. The metaphysical part is the part that cannot be detected by the five senses and the electromagnetic energy (see below for listing)

Memory (MM) is not a component of Maras, but rather the acquired data. It serves as the crucial interphase between each individual's brain and the non-brain parts, including the body, environment, and society. In the brain, memory plays a pivotal role as an intermediate component between the physical and metaphysical forms. It is often understood as an informational processing system, with a sensory processor responsible for receiving data and transcribing it into signals recognizable to the brain, thereby facilitating the formation of consciousness.

Important note:
It is crucial to note that due to the presence of the metaphysical component in MM, the investigation and understanding mechanism of MM processes cannot be complete without acknowledging the existence of the Soul. The Soul, which is mostly beyond the realm of the consciousness of scientific equipment, plays a significant role in our understanding of memory.

B. Complexities in acquiring, storing, and retrieving the MM

Since the living sentient is composed of five Maras/Skandhas, the data must penetrate these five components so that the data is recognizable. Kant said: *We do not know self-nature/ 'noumena.' We only know reality in terms of how our active minds structure/ organize/ form our experiences of Mind-independent reality.*

1. Form: Detecting object or problem. This stage is dictated by the nature of the objective that reveals its nature to the subject, influenced by the environment and the readiness status of the subject. The latter component comprises gaining information in the form of learning for entertainment, knowledge in studying, and practicing. In other words, knowledge only relies on what the object and environment reveal to the subject.

2. Perception: Acquiring information through the five Sensory organs. Sophisticated equipment can improve perception but is still

limited by Quantum mechanics' fuzziness when reaching the quantum level..
As Heisenberg and Bohr pointed out, speed and momentum can not be pinned down momentaneously. As a result, Artificial Intelligence/ AI must have the upper limit of Consciousness.

3. Feeling: after traversing sensory organs, the information follows the dorsal stream (Top-Down stream), going through motor areas of the cerebral cortex to elicit the reflex of reaction for defense

4. Formation: The integration of information in various areas of the brain to convert into forms compatible with Consciousness formation.

5. CS identification

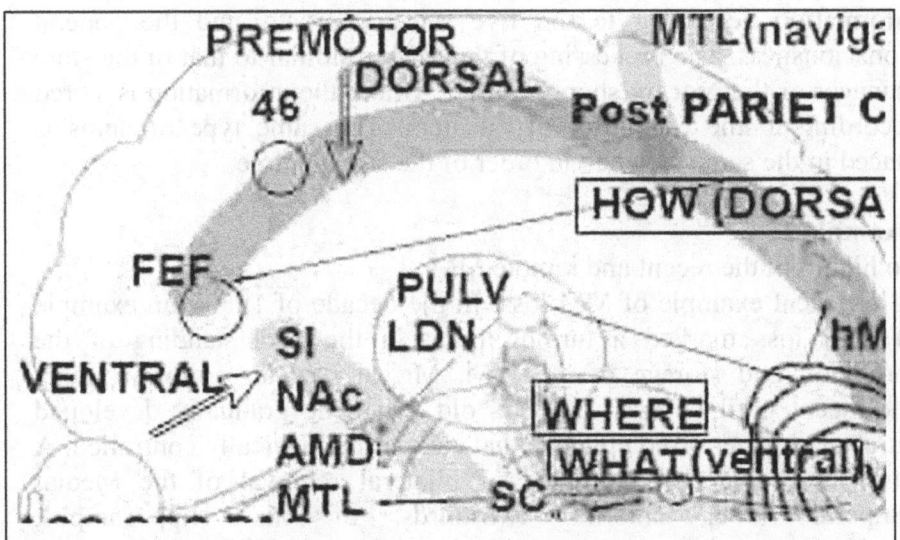

Fig 3.1 Dorsal stream- *V1 >motor cortex>Frontal Eye Field:* for eye movement (HOW?) for defense reflex. hMT: The extrastriate area MT [V5 (hMT+) in humans] located in the "dorsal pathway" of the primate brain is specialized in the processing of visuospatial coding of movement direction for visual motion information , i.e., in the control of visually guided hand movements. Ventral steam to Inferior Temp Lobe for CS (WHAT) and to SC (WHERE)

Process of recognition of data. The significance of this process is often neglected in studies of MM or Consciousness. It is very obvious that the recognition of data can only occur if the data is directly or indirectly in contact with the recognition system. As pointed out by Buddha, the recognition system is not the sensory organs or the brain but the Original Mind/Ultimate Omniscience/UO. In living sentients, data reach the UO through the sensory organs and the brain (indirect contact through the **so-called veil of ignorance**). The direct contact creates the recognition **as the data is**. (see also chapter I, paragraph Emptiness generating

III. Different types of Memory
Examples of Memory
Examples
Originally, the information existed as a whole package. For Ultimate Omniscience, Original Mind/ Buddhahood/ Holy Spirit, the information is perceived as it is. However, in living sentient, for intercommunication using the brain, the information must be first received in the brain. In the brain, the information is integrated, shared with previous information, and connected with motor areas for motor execution and Consciousness areas for emotional expression. In addition, the original package must be split into six different types of information according to the five sensory organs and the general Consciousness. The processing of the MM is similar to that of the store manager or the grocery shopper: In the Brain, the information is stored according to the categories of the item. The same type of items is placed in the same area and in order of the storage time.

Example 1:
To highlight the recent and remote MM
A historical example of MM loss: In the decade of 1950s, an example of MM loss marked a turning point in the understanding of the formation and storage of the MM. Mr. H was an ordinary young adolescent until he was 13 years old, when he gradually developed intractable generalized epilepsy that can not be medically controlled. A surgical intervention consisted of bilateral removal of the Medial temporal lobe, anterior two-thirds of his Hippocampus, parahippocampal cortices, entorhinal cortices, pyriform cortices, and amygdala. The remaining hippocampal tissue and anterolateral temporal lobe became atrophic. After the surgical procedure, the grand mal was under partial control, but there was a severe loss of capability to form new long-term MM (antegrade MM). As a result, Mr. HM is trapped in the present time and is subject to a book entitled "Permanent Present Tense" edited by Suzanne Corkin

Temporally graded retrograde amnesia (for events in the year preceding the surgery) (loss of HIPPO, key in retrieving MM)

Loss of capability to form new semantic MM (stored in MTL)

No loss of implicit MM regarding procedure remained intact (stored in BG)

No loss of long-term MM of events before the surgery (stored in PCC)
No loss of short-term MM (not stored yet)

Example 2: Second example,
To highlight sensory MM, explicit episodic, and semantic MM
On day N, while learning to worship the Buddha, I see Mr X making an offering to the Buddha at Huong Pagoda:
Day N: explicit episodic
I see: explicit
Mr X: MM of Vision
Huong Pagoda: Place, episodic MM
Worshiping Buddha: Procedural MM
Making Offerings to Buddha: Semantic MM
A week later: Remote MM

Information from four sensory organs (skin, eyes, ears, and eyes) comes from the thalamus to the specific cortices. For the smell, data reach the olfactory bulb, a small organ, the information bypasses the Thalamus to reach EC directly. This MM is quite stable. Therefore, consolidation is not often necessary. Thinking and other non-specific information are integrated and form explicit (episodic and semantic) MM. All information is then referred to the ACC for comparison: Information identical to those in ICS is labeled as such.

Non-specific info corresponding to explicit MM is then sent to the Hippocampus to become the recent explicit MM. In a period of a few days to a week, the MM is either sent to MTL for semantic MM or to vmPFC.

From the above two examples, MM is categorized (Fig3.2)
A. Specific MM of Five Sensory Organ MM is related to sensory cortices
· Vision: Occipital cortex V1-V6→ Inferior Temporal Lobe for Face cells and TEO and TE cortices and →hMT for the motor control
· Hearing: Superior Temporal Cortex
· Touch: Parietal
· Smelling: Olfactory bulb
· Tasting low Parietal cortex
Because the cortices for receiving and storing the information are large enough, the information does not need consolidation.

B. Non-specific/ General MM,

This type of information, whether from any sensory organs (ear, eyes...) arrives in the brain (Thalamus as the first relay in the brain) and will not end up in the sensory cortex but come at the Hippo, MTL, BG, and Cerebellum, depending on the type of MM related information:

a) **Implicit MM** of procedural type: to Cerebellum or Basal Ganglia, This type of MM is very stable and does not need consolidation. The MM is also part of Inner Consciousness and can also be stored in the Soul and will be imprinted into the new baby in the next life (please refer to pge 263 for an explanation). A specific Soul may not be necessary to retrieve the implicit MM in case of multiple Souls in DID. For example, routine procedural habits like walking, eating, shopping, and cooking. The evidence can be seen in cases of DID: The retrieval of the implicit MM is feasible in the presence of the Soul, regardless of the principal or the second Soul. However, specific professional procedures like music playing or surgical skills likely require a specific Soul attached to those skills.

b) **Explicit MM**
· Anterograde / recent MM

The information from the above cortex of Inner Referral CS or MM/ Default CS or MM is sent to the Hippocampus (Dentate and CA3 of Hippo). Since the Hippo is a small area with limited storage capacity, the information can not be kept for a long period of time. According to the standard consolidation theory, The information must be destroyed after being sent to the appropriate cortex for consolidation. However, this theory can not explain Hippo's role in retrieving long-term MM (LTM) . For example, the injection of lidocaine in the Hippo will abolish the capability of the animal to retrieve the LTM. Thanks to further investigation, it turns out that the Hippo plays more of a role than that of a store of short-term MM. Furthermore, in the explicit MM, there is a component of information called semantic MM, as seen in the following example.

Example 3:
To further highlight the semantic MM
  This morning, she gave me an orange ball-like shape with rough skin and a particularly good smell.
  This morning: episodic
  Smell: semantic
In 1972, Canadian psychologist Tulwing categorized two types of explicit/declarative MM as two separate stores of MM in the brain.

- Episodic: life events involving time and place relationships and on perspective, experiences, and specific events stored in MTL.
- Semantic: factual and conceptual knowledge expressed in words, language, names of colors, the sounds of letters, the capitals of countries, and other essentials. The semantic MM can not be separated from the episodic MM since the two above types of MM are often closely interrelated, with the episodic MM being the prominent declarative front of the event. The semantic MM characterizes the facts, context, and figures. The semantic MM is stored in the Inferior prefrontal cortex (vmPFC) and the posterior temporal areas, PCC, and other regions.

c. **Autobiographic MM (Self-Memory system)** is the system of auto-recording the events of life consisting mainly of semantic MM (Coway 2000), having a role comparable to the black box in an airplane.

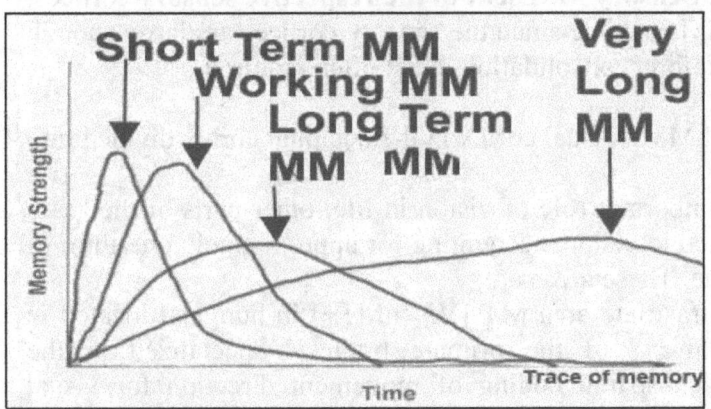

Fig 3.2 Types of MM and Time

IV. Formation of MM
Consists of stages of
 i. encoding,
 ii. storage,
 iii. Consolidation,
 iv. retrieval with reconsolidation and
 v. transfer to the next life

A. Encoding of MM
The incoming data is first split into five types of information for the five sensory organs, and the general type of information is received and transformed into the electrical potential activating cytoplasmic reaction in the neuron. The reaction is then transmitted along the axon-axon

terminal-dendrites, to generate a similar electrical potential at the dendrite of the next neuron.

The key feature of this encoding is the synaptic connections between the dendrite-terminals of the axon. The encoding involved numerous scientific discoveries in neuroscience in the past century has been reviewed in Chapter II. The encoding consists of connections between axon-dendrites that are specific for each piece of information (see Fig 1.1,2,3).

B. Storage (Fig 3.3)
The information coming into the brain comprises many types and subtypes: specific sensory, general, implicit, and explicit episodic or semantic subtypes. Each type or subtype will be stored separately in the cortex or subcortical nuclei.

1. **Specific Sensory MM sent to the respective sensory** cortices. cortices. The MM is stable since the sensory cortices are large enough for storage. Therefore, consolidation is not often required

a) Visual MM: Visual cortex in the occipital cortex divided into V1-V6

Because of the important role of vision in life, other parts of the brain are also implicated in vision, accounting for approximately one-third of the cerebral tissue. These areas are:

hMT+ The extrastriate area MT [V5 (hMT+) in humans] located in the "dorsal pathway" of the primate brain is specialized in the processing of visuospatial coding of movement direction for visual motion information, i.e.,

in the control of visually guided hand movements

TE and TEO cortex are the terminals of the inferior temporal gyrus related to the Visual Ventral pathway. They play the role of categorization based on the similarities of the objects. The tasks are done quickly, accurately, and without Consciousness (for example, car versus truck, bird versus chicken. TEO is organized topographically according to the retina for recognizing visual features. TE is more specialized in MM of the whole object. These areas do not have a role in visual acuity.

b) Auditory MM

The auditory cortex also plays a role in the spatial and temporal frame of reference in the perception of sound.

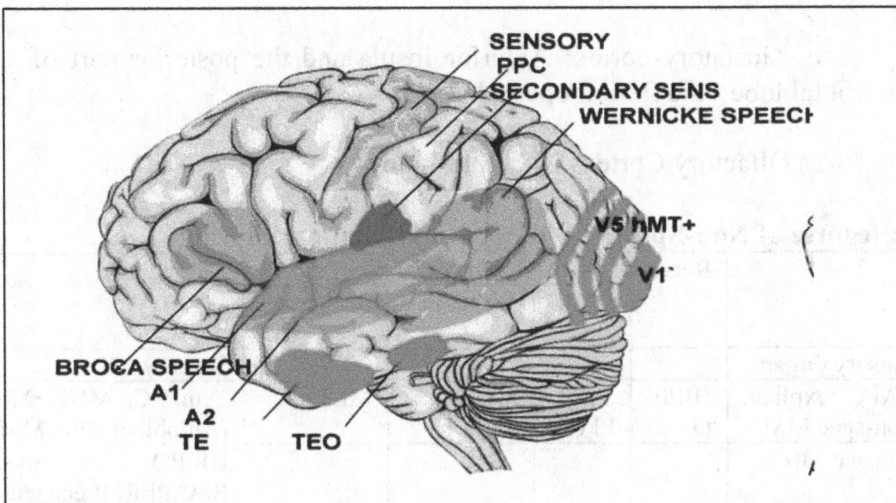

Fig 3.3 Cortical storage of MM Composed of primary A1 and secondary A2 in the Temporal superior gyrus

c) Sensory Cortex in Parietal Cortex, with secondary sensory cortex (Posterior Parietal Cortex)

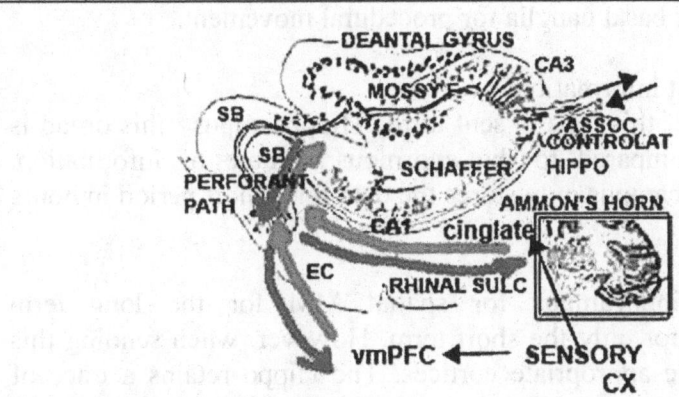

Figure 3.4 A showing Orbitary PFC, ventromedial PFC Amygdala, Hippo , Mammillary body, and ParaHippo with different areas of the Anterior Thalamus Nucleus/ ATN.

        B: -Entorhinal Cortex/ EC= Entorhinal Cortex connecting to Hippo and the cortex.

-Subiculum/ SB (Supportive structure =betweem EC and Hippo CA1), connecting with CA3

The Hippocampus has such a shape to accommodate the interconnection between various regions to fulfill the function of encoding and storing Memories and Consciousness.

d) Gustatory cortex: Anterior Insula and the posterior part of the frontal lobe =the frontal operculum

e) Olfactory Cortex=Olfactory bulb

Categorie of Non-Specific Memory and Memory Storing

	Recent	Remote			Long trrm
		Ẽxplicit	Virbatim	Semantic	
Sensory Organ					Sênsory CX
MM explicit, Non-spec MM	HIPPO	Vm PFC	MLT	MLT	vmPFC, MLT→PCC (Autobiographic MM)
Space MM					HIPPO place), RSC,PER, Precuneus
Pricedures/ implicit					BG, CEREBELLUM

2. Non-specific MM (information)

a) Implicit MM is stored in either the cerebellum for movement for equilibrium or basal ganglia for procedural movement.

b) Explicit information.

All information of this type is sent to the Hippocampus> this organ is of small size compared to the enormous volume of information. Therefore, Hippocampus only stores the data for a short period in hours or a week.
As a result,

- Hippocampus for spatial MM for the long term but explicit MM for only the short term. However, when sending this information to the appropriate cortices. The Hippo retains a trace of MM, which sub-serves as a clue for retrieval of the respective MM. The concept of MM Trace develops following the fact that Hippo is necessary for retrieving long-term explicit MM despite the absence of capacity to retain long-term MM. Injection of lidocaine into the Hippocampus abolishes the ability for retrieval of long-term MM
- MTL : Long term explicit episodic MM
- vmPFC, PCC , Long term explicit semantic MM
- In addition, the Hippocampus PPC, Precuneus, and Retrosplenial Cortex are additional areas of MM for spatial, episodic MM storage.

3. Default Mode Network (DMN), Limbic System, and site of Inner Consciousness (Fig 3.4,5,6,7)

The DMN is a system connecting various and distant cortex areas. The cortices are: vmPFC, dmPFC, PCC, Precuneus, Retrosplenial Cortex, Posterior inferior Parietal Lobe, Angular gyrus, TPJ, Lateral and Anterior Temporal Lobe, Entorhinal Cortex/ EC, [Hippo (?) ACC (?) + Subcortical BF, Thalamus Ant MedDorsal]. In 2001, Raichle first discovered the network,

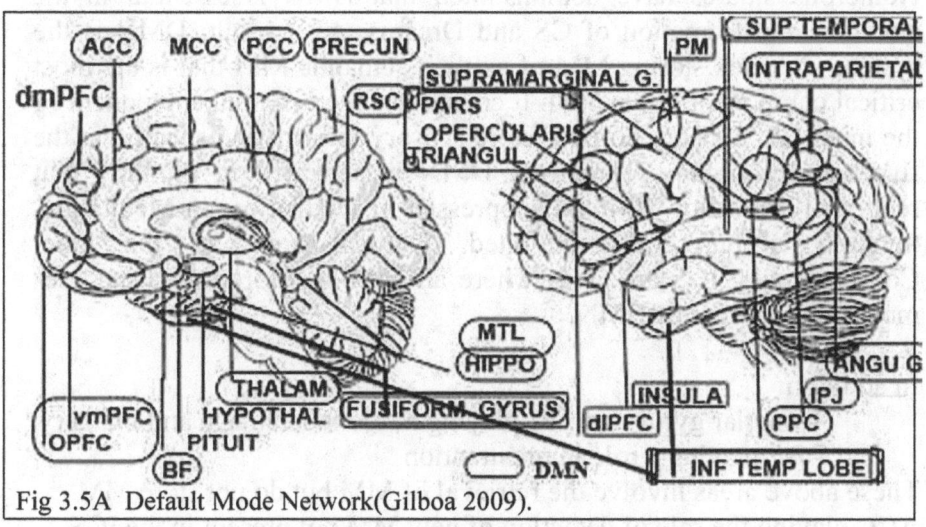
Fig 3.5A Default Mode Network(Gilboa 2009).

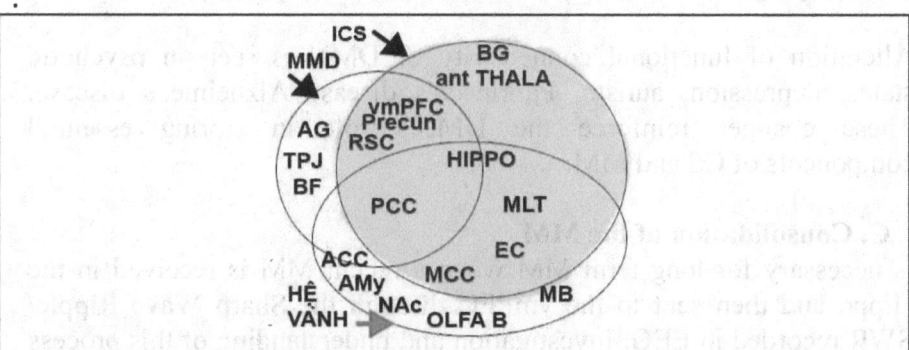
Fig 3. 5B Default Mode Network/ Limbic System and IC. The Limbic system, except for the Amygdala, is involved in the retrieval of MM but not storing MM. NAc: Nucleus Accumbens, MB: Mammillar bodies.

DMN's cortical display standard features are changes in blood flow and oxygen, associated glucose consumption, and increased functional connectivity during thinking processes and attention, with increased blood flow in wakefulness. The changes are not affected by the method of examination with fMRI, blood oxygenation-level-dependent

(BOLD) at rest (rest-stated=rsBOLD or at work (task-based=tbBOLD). The fact that increased functional connectivity is occasionally associated with decreased blood flow is difficult to understand. A hypothesis is proposed that the increased blood flow and energy consumption in brain tissue connecting various areas of DMN accounts for the decrease in the DMN itself. (Logothetis, 2001, Greicius 2003 Andrews-Hanna 2010 Parker 2019, Vishnubhotla 2021 zhang 2021 Prestel 2021 Razlighi 2018)

Numerous studies have demonstrated that DMN is essential in the attention and formation of CS and Dreams. As a result, DMN is the system of cortex storing MM of explicit semantic MM that is the most critical component of the MM. It constitutes the essential component of the inner CS. Sensory cortices store sensory information specific to the five sensory organs. Therefore, DMN is activated in thinking that requires retrieval of MM and suppressed in Meditation. In Meditation, thinking is purposedly inhibited. Cortical areas of the Inner Consciousness/ ICS are areas where all MM are stored. ICS includes most cortical areas of DMN.

In addition:
-Angular gyrus and TPJ playing a role in language and imagery
- BF playing a role in preattention
These above areas involve the retrieval of MM but do not store MM -
ACC playing the role of detection of new MM not present in the ICS

Alteration of functional connectivity of DMN is seen in psychotic states, depression, autism, Parkinson's disease Alzheimer's disease. These changes reinforce the DMN's role in storing essential components of CS and MM.

C. Consolidation of the MM
is necessary for long-term MM when a recent MM is received in the Hippo and then sent to the vmPFC through the Sharp Wave Ripple/ SWR recorded in EEG. Investigation and understanding of this process have been the topic of a large number of studies published in literature over the last three decades

1. Standard Memory Consolidation/ SMC.
Since MM arrives at the sensory cortices, BG, and Cerebellum is stable, this SMC only applies to the explicit MM.

The theory was first proposed by Marr in 1971 (Marr 1971), based on works by Russell, Scoville and Gottingen, Muller & Pilzecker in decades1990 (Russell 1946, Scoville 1957). The neural conduction activated by the input in the form of information arrives at the Dentate Gyrus CA3 of Hippo, then sent to CA1 (Place cells for spatial MM, responsible for forming the navigation map). In 1 to 2 weeks, the MM is called short-term MM. In Laboratory animals, hippcampectomy or lidocaine injection into the Hippo will abolish the short-term MM.

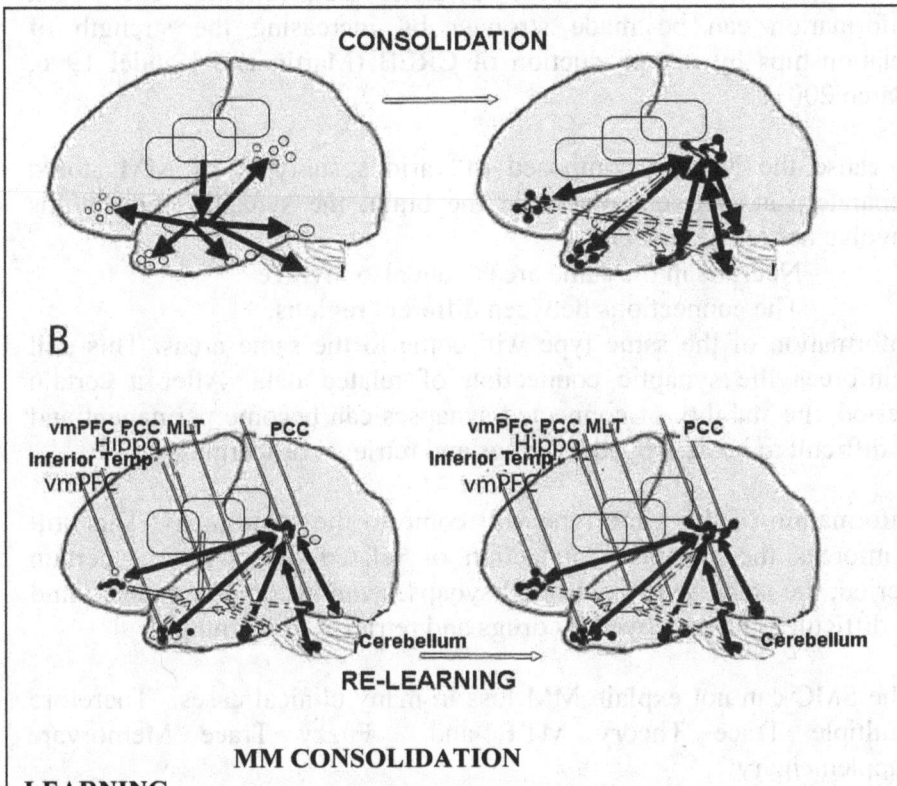

MM CONSOLIDATION
LEARNING
Fig. 3.6 A,B: CONSOLIDATION: information input to the Thalamus
 - implicit MM Middle Temporal Lobe (MTL), to Basal Ganglia, Cerebellum and MTL
 - Explicit MM to Thalmus →HIPPO→vmPFC for consolidation then →PCC for definite storing (solid node) HIPPO only retains the MM traces to facilitate the retrieval of MM
Fig.3.6 B: RELEARNING: with MM traces, the MM is reconnected to the HIPPO.

Afterward, the MM is sent to the vmPFC to become the long-term MM. The process is automatically and unconsciously performed during NREM sleep, usually in stages 3-4/ slow wave sleep/ SWS associated with delta and alpha waves (Winocur 2011), when there is no Consciousness because of acetylcholine. The activity can be seen in EEG as Sharp Wave Ripples/ SWP arising from CA1. As pointed out in Chapter II, this consolidation process requires protein production by the neurons involved and chains of chemical reactions involving cAMP and CREB...to establish the synaptic connections. The weak information can be made stronger by increasing the strength of relationships by overproduction of CREB (Martin and Kandel 1996, Barco 2001)

Because the MM is composed of various subtypes of MM stored separately at different places in the brain, the synaptic connections involve not only between:
- Neurons in the same areas, but also involve
- The connections between different regions,

Information of the same type will come to the same areas. This still reinforces the synaptic connection of related data. After a certain period, the stability of connected synapses can become permanent and is difficult to be destroyed by drugs and retrieval or learning.

Information of the same type will come to the same areas. This still reinforces the synaptic connection of related data. After a certain period, the stability of connected synapses can become permanent and is difficult to be destroyed by drugs and retrieval or learning.

The SMC can not explain MM loss in many clinical cases. Therefore Multiple Trace Theory MTT and Fuzzy Trace Memoryare supplementary

2. Multiple Trace Theory MTT

In addition to the HM case, other cases with limited lesions of the MTL show no significant alteration of the MM. On the contrary, extensive lesions of MTL, especially involving the inferior PFC, the inferior Temporal Lobe, and the Occipital Lobe, show alterations of long-term MM, including autobiographic MM (Bayley 2005). These changes demonstrate the MM consolidated outside the Hippo

Reasons to add the MTT to the theory of Standard Memory Consolidation/ SMC are:
- SMC can not adequately explain the different roles of Hippo in the long-term MM

- In the case of the destruction of Hippo, long-term explicit and semantic MM are not entirely altered
- Contextual and spatial MM have severely altered the loss of formation of new explicit MM
- In human beings and mice, navigation older than one year is independent of the Hippo, eventually located in the ACC. However, this contains fewer details than

From HIPP, MM is sent up to vmPFC, PCC...(Hebscher 2019, Nadel 2011, Moscowitch 2005, Rolls 2020)
in the case of the presence of the Hippo. The fact suggests that the Hippo contains traces of MM of the Map.

The complete retrieval of explicit semantic MM requires the integrity of the Hippo and other cortical areas , i.e., vmPFC and PCC (demonstrated by the technique of IEG /Immediate Early Gene c-Fos, an expensive Technique) (Denny et al. 2014, Rugg and Vilberg 2013; Wheeler et al. 2013). However, retrieval of long-term semantic MM requires the involvement of the vmPFC rather than the Hippo.

Therefore the Hippocampus and the Neocortex continue to interact in the case of episodic memory. The reactivation of an episodic memory trace creates a new memory trace. Thus, an episodic memory is represented by multiple traces in HIPPO.

Semantic memory can be stabilized in the neocortex and in aome cases, influenced by the hippocampectomy.

3. Fuzzy Trace Memory

To solve the above problem, and based on the similarity between explicit MM with that of the navigation map of O'Keefe (O'Keefe 1978), Nadel (Nadel 1997) and Moscovitch (Moscovitch 1992, 1995, 2006) proposed to replace the SMC with this MTT and Fuzzy Trace MM. In FTT, the explicit episodic MM can be divided into two subtypes (Brainerd & Reyna, 2004, 2015; Reyna & Brainerd, 2011).
- Verbatim MM having characteristic typical features of the explicit MM. (**Verbatim MM trace**)
- Gist MM are faint traces because the MM traces only consist of semantic forms of the objects or events. (**Gist MM trace**)
  Verbatim MM is lost in aging, but Gist MM retention is relatively intact. This represents cases when the aged person may vaguely remember the events

Comments
The Trace Theories can account for the role of the HIPPO in long-term episodic MM and some cases of semantic MM. In addition, the function of traces may be more essential than being thought. They may play the role of the keys to getting access to long-term MM stored in PCC. As **a result MM in this current life are always associated with MM trace. In contrast the MM of previous lives are not associated with any traces in the HIPPO. This may account for most people's inability to recall previous life MM.** Some persons, especially some children, can retrieve previous life MM. Some meditators at high Dhyana (3 or 4) can retrieve many previous lives MM. This can only occur with the aid of the soul playing the role 4) on the role of the soul in retrieving MM)

Further reading
- The explicit episodic MM in the MTL can become semantic MM after a certain period: for example, the place of meeting one's wife becomes, with time, a special place difficult to forget. As a result, the Hippo is necessary for this conversion, and the distinction between episodic and semantic MM is only sometimes clear-cut.
- In humans, the lesion of vmPFC is not followed by loss of MM as in the case of animals. Still, it is usually followed by alteration of MM associated with unintentional fabulation covering up the lacunae of the MM, probably as a secondary mechanism to maintain the social interaction.

In learning, the old MM of the same type as the recently learned MM is often prone to be erased, consistent with retroactive inhibition. This MM is often prone to be erased because the old and retrieved MM becomes unstable if not reconsolidated again. Therefore, relearning reconsolidates the MM. As a result, the neurons need a certain time to produce proteins necessary for synaptic connections. The consolidation involves neurons within local areas but also neurons interconnecting areas.

- Time of Consolidation: fMRI shows that the consolidation begins as soon as 24 hours in humans, and reconsolidation can be seen many years later or during the entire life.
- The consolidation usually does not begin right after receiving the MM. Consider cases of taking notes in a notebook; the notes frequently have to be reviewed, corrected, erased, and updated appropriately. Therefore, it is advantageous for the consolidation to be delayed for a certain time in the Hippo. The MM will eventually be consolidated after being sent to MTL or vmPFC and integrated with other MMs of similar types to update the CS.

D. Retrieval of the MM, Reconsolidation, and Reorganization.
1. Retrieval of MM consists of reactivation of synaptic connections of related types of MM. The retrieval is a process opposite to the consolidation of the MM: The information is subcategorized into different subtypes of MM. In the recovery of MM, the various subtypes

of MM are reconnected to re-establish the original MM. Reactivated synapses become labile. Evidence for this ability can be demonstrated in cases of loss MM of specific household infrequent tools recently used, then misplaced in another place in the house. Therefore reconsolidation of the reactivated MM is necessary to maintain the long-term MM. This process is similar to the new MM consolidation involving the Hippo, MTL, PCC, Cerebellum BG and sensory cortices.

2. Factors influencing the retrieval, role of testing, reorganization, relearning

a) Anxiety decreases the retrieval but increases the consolidation

b) Repeated retrievals facilitate retrieval. Similarly, testing increases the retrieval of MM and is considered an efficient consolidation method. This method is superior to that of relearning because relearning needs the involvement of reactivation, updating new information, reorganization, reconsolidation (, i.e. more steps than the testing), and of course, superior to no-activity(Bjork, 1994, 1999; Bjork & Bjork, 1992).. In addition, testing has the advantage of reorganizing the MM component in a more proper structure. There are various methods of enhancing the retrieval (Moreira 2019 Karpicke 2014, Bridge 2014, Gao 2016):

· Free recall: advantage for retrieval of semantic and autobiographic MM

· Cue recall: filling up the gaps in sentences

· Recognition recall: Replying to questions

In the past several decades, studies on the method of multiple-choice questions and fill-in-the-gap have demonstrated benefits (Lalame 2013 ,Delaney, 2010; Karpicke, Lehman, & Aue, 2014 Carpenter, 2009, 2011; Pyc & Rawson, 2010).).

The implicit MM, the most stable MM with the overview of the stored information and images, is helpful but not adequate for retrieval of the whole MM. For the images, Precuneus and dlPFC contributing to imagery play the role of retrieval MM, (Blalock 2016, Melrose 2019) Euston 2012, Pastötter 2019, Wiklund-Hörnqvist 2014 Fletcher 1996)

c) Feeling of knowing corresponds to the state of consolidation with difficulty in the retrieval of MM

d) Inaccurate MM: For example, in a sentence composed of "bed, rest, awake, snooze, snore," when asked to repeat, 50% of replies include the term "sleep". This phenomenon represents the poor MM often seen in young people

An additional source of MM may also cause inaccurate MM

e) Feeling of loss of MM before the MM is lost: due to the partial loss of specific components of the explicit episodic MM

f) Reorganization and relearning. Reactivated MMs need to be reconsolidated and reorganized to add up new MM

3. Right And Left Hippo Differential Role in Retrieval of MM.

For the mice's Hippo, the optogenetic inhibition of CA3, whether on the R or L Hippo, will damage the short-term MM. For spatial MM, the Left CA3-CA1 stores more MM than the Right (Shipton 2014). In human beings, the L Hippo is more specialized in Episodic MM and the Right Spatial MM (Ezzati 2016)

4. Role of the Soul in the storing and retrieval of MM

In the phenomenon of reincarnation, the law of Karma/cause and effect/ seeds and fruits, autobiographic MM are consolidated, stored and then transferred to the store MM in the next lives. The examples are MM of parts of their previous lives. This process of memorizing previous life can only be conceived with the concept of the existence of the Soul. MM plays the intermediate role between the past and the present lives and interfaces between the physical and metaphysical realms. Each individual (humans and animals) has their Soul in this concept. The body represents the physical part, while the Soul represents the metaphysical part. The combined physical and metaphysical elements constitute the whole entity of the individual.

In cases of multiple personalities (also called Divided Identity Disorder/DID), one individual (human or animal) has two or more two Souls. As a result, each patient of DID has two (or more than two) identities having individual Souls. Each identity has separate MM of explicit episodic and semantic MM but shares implicit MM such as walking, reading, dressing, and driving ....The two Souls are independent in identity but share the same body and brain. In addition, the Soul communicates the Autobiographic (semantic and episodic MM) with the brain: in other words, MM coexists in the brain and the Souls.

The hypothesis of the mechanism of amnesia in cases of DID is the specific role of the Souls in retrieving the individual episodic and semantic MM. For the retrieval of implicit MM, the role of the Soul is non-specific. Therefore, each identity has a separate explicit MM but shares the same implicit MM in the brain. In this hypothesis, the Soul plays the key or facilitating factor in retrieving the MM, probably in the reactivation of the synaptic connections.

The underlying mechanism of the Soul in storing and retrieving MM is quite complex and is described in chapters of CS and Soul

E. MM of the Subconsciousness and Unconsciousness
1. Definition

Sub-CS and Un-CS are two parts of the CS of the present life composed of particular forms of information received and stored in the brain, unlike the standard type of information of the CS. The retrieval of MM in the Sub- and Un-CS is difficult.

Sub-CS and Un-CS can be manifest in CS, a slip of the tongue Freudian slip of tongue /, uncontrolled reactions, dreams, imagination, creativity, intuition, reaction formation, rationalization by involuntarily exerting on the Manas Consciousness (the 7th CS), emotional and psychosomatic illness.

Part of the sub-CS and Un-CS can be carried over to the next lives and become Store CS.

a) Sub-Consciousness consists of information lacking full or sufficient attention; sub-cs is divided into Subliminal Sub-CS and Supraliminal Sub-CS, just below and above the level of Awareness with some difficulty in retrieval.

b) Un-Consciousness; The information is stored in portions of Ego, Super-Ego, and Id. It should be differentiated from store MM, which contains the MM from previous lives, whereas Un-CS only includes the information of the present life
The MM can be almost and only retrieved in dreams
The difficulty in the retrieval of the information in Sub-CS and Un-CS may also be related to the problem of consolidation, organization, and the absence of MM traces

2. Repression on the MM
According to Freud, painful or unnecessary MM can be voluntarily repressed (Anderson 2000). Repeated repression may render the retrieval more difficult. Furthermore, the non-voluntary recovery of a particular MM can also be suppressed. This phenomenon may be due to the non-specific repression inadvertently associated with voluntary repression (Taubenfeld 2019). As a result, the repression of MM influences not only the inadvertent MM but also other unrelated MM.

In the vision repression of the imagery, the visual image of perception may become unconscious if it does not match the similar image in the imagery process within the brain (Dijsktra 2022, 2019,2017 Taubenfeld 2019). Lesions of the Temporal lobe not only alter the formation of the cognition related to the visual information but also influence the storing, consolidation, and retrieval of the related information. Furthermore, the repression of the retrieval of certain information influences the retrieval of related and unrelated information.

Note: *The whole information comprises the information in the CS (in living being)in actual life, CS in a previous life (in Inner CS), and knowledge in the form of sub-CS and un-CS that the CS and the brain can not control. The Sub-CS, Un-CS, and Store-CS constitute the uncontrolled parts of reality for all living beings. Only the fully enlightened human, i.e, Buddha or similar Levels, can recognize and control these parts.*

F. MM of the Fetus and Newborns
The brain starts its development after the third intrauterine month. The MM of the fetus can be demonstrated after the 33rd intrauterine week. In studies in Netherlands with vibroacoustic" sound devices, the sound can be transmitted through the abdominal walls of the mother; the fetus shows evidence consistent with the development of long-term MM (Spencer 2002, Krueger 2014, 2015,2019 Granier-Deferre 2011). The evidence is the change of fetal heart rates. This development extends until the newborn period (Gonzalez 2008).

G. Cleansing of Karma of the Actual Life by the Meditation
The Karma is the store MM mainly composed of semantic MM that can be good or bad and MM in the Sub-CS and Un-CS. These MM are stored in the DMN of the Inner CS, shared with the Soul, and carried over the next lives. According to the Eastern law of Karma and of the cycle of birth and death, the Karma in the form of MM passes from the past through the present into the future not in straight lines as in common thinking but in a circle or, more precisely in helical or spiral and entangled lines. In this line of transmission, Karma is crucial in the law of cause and effect or result/law of causation/causality: good or bad or indifferent deeds, words, and thoughts all bring their results of the same type on the doer in the future. One reaps what one sows. As a result, the present effect reflects the past cause.

Buddha said in Surangama Sutra:
....since desire and love are tied so closely together, no disengagement is possible and the result is an endless succession of the births of

parents, children and grandchildren. This comes mainly from (sexual) desire which is stimulated by love..Since passion cannot be destroyed, living beings born from wombs, eggs, humidity and by transformation tend to use their strength to kill each other for food.

Cleansing bad karma is similar to deleting unwanted files on a computer. This mechanism can be exemplified in the following paragraph from the Vimalakirti sutra when Vimalakirti explained to Upali (top Buddha's disciple regarding rules of law/ Sila and Vinaya – skt) the sin that is the delusional thought and the sin cleansing that is based on the understanding about the sin committing

Vimakakirti advises that Upali should go about wiping out the doubts and remorse caused by the sins, because the offense by its nature does not exist. As the Buddha has taught us, when the Mind pure, the living being will be pure. Deluded thoughts are defilement. All phenomena are born and pass into extinction, never enduring, All phenomena are the product of deluded vision, like dreams. One who understands this is called a keeper of the precepts, this is called well liberated.'

> Upali, do not make the offense these monks have committed even worse than it is ! You should go about wiping out their doubts and remorse at once and not trouble their minds further! "'Why do I say this? Because their offense by its nature does not exist either inside them, outside, or in between. As the Buddha has taught us, when the Mind is defiled, the living being will be defiled. When the Mind is pure, the living being will be pure. As the Mind is, so will be the offense or defilement. The same is true of all things, for none escape the realm of Suchness. "'Now, Upali, if one gains emancipation from delusion through an understanding of the nature of the Mind, does any defilement remain?'
> "'No,' I replied.
> "Vimalakirti said, 'In the same way, when all living beings gain an understanding of the nature of the Mind, then no defilement exists. Ah, Upali, deluded thoughts are defilement. Where there are no deluded thoughts, that is purity. Topsy-turvy thinking is defilement. Where there is no topsy-turvy thinking, that is purity. Belief in the self is defilement. Where there is no such belief, that is purity. "'Upali, all phenomena are born and pass into extinction, never enduring, like phantoms, like lightning. They do not wait for one another or linger for an instant. All phenomena are the product of deluded vision, like dreams, like flames, like the moon in the water or an image in a mirror, born of deluded thoughts. One who understands this is called a keeper of the precepts, one who understands this is called well liberated.'

Free Will causing bad Karma is not Free Will, because Free Will is designed to be only given to human beings for spiritual advancement, by acting according to Four Immeasurable Minds of Vitue.

As discussed in the previous paragraph, when MM is reactivated for retrieval, the synaptic junctions become labile and, if not reconsolidated, will eventually be removed by neuroplasticity (destruction and repair). In most people, the MM retrieval is only feasible for MM of the actual life stored in the Inner CS, particularly in the PCC or DMN.

The purpose of Meditation is to restrict the data in the focus of mindfulness/attention.
The MM retrieved from the ICS may burden the attention. In this case, the IPS will eventually discard the data from ICS. Data discarded by the IPS become unattended and will not be re-recorded in the DMN as usual. However, some MM may be recorded at the SubCS or UnCS level. The significant consequence is the MM is cleansed to various extents. The study of cases of Hemineglect syndrome provides the evidence. Data from the Hemineglect space (usually the Left hemispace) is unattended and is not recognized by the patients. Once data are recorded in the Sub or UnCS levels, it can still cause secondary effects)

However, in many religions, it is believed that an enlightened Master can remove or put it out of the way from life for the disciple's spiritual advancement. A disciple in Buddhism progresses in the path of spirituality along with the gradual cleansing of Karma since Karma constitutes the obstacle in the spiritual path.

It is worth noting that when retrieved, MM from the ICS (also known as the veil of ignorance) is discarded by the IPS and therefore, is unknown by the meditator. As a result, the MM cleansed in the Meditation is anonymous and is not experienced by the meditator.
The process mainly cleanses the karma of the current life and likely does not cleanse the karma of previous lives because retrieval of MM is possible for the current stored MM.

Karma of the actual life is stored in the form of MM in the Inner CS. The Inner CS is comparable to an encyclopedia for reference. Data, represented by information obtained from the five sensory organs, must be referred to as the ICS to become the CS that living beings use daily. Information is similar to the existing CS is labeled as it is. New information that can not match anything in the ICS will be updated in a process known as learning.

In Laṅkāvatāra Sutra, there are three forms of CS known as Svanhavalakshana-traya.
    i. Parikalpita or wrong discrimination
    ii. Paratantra consistent with common sense

iii. Parinishpanna: correct view.

Therefore the brain plays a double role as a) a machine for action for the spirit, the filter with distortion of the input from the outside world andb) storing and processing each individual's activities. The activities include all activities from the body, speech, and Mind. The activities may directly or indirectly, intentionally or unintentionally, or may not result in any physical or mental effects in the past and current lives. Unintentional activities are not conscious and are stored in the Sub-CS or Un-CS.

Repenting or confessing one's sin is also a way of retrieving bad MM. However, they differ from karma cleansing meditation since the latter method covers much of karma. The concerned persons do not even know some karmas they created.

H. Storage of Karma
-Karma of actual life is stored in the ICS, particularly in the PCC. All semantic components of activities and information are first processed in the Hippo, then transferred to the MTL and vmPFC for consolidation and stored as long-term MM.
The MM is later transferred to the PCC which is considered a site of autobiographic MM storage. Autobiographic MM can be retrieved in the presence of the specific individual Soul into the present time. After death, the autobiographic MM/ Karma of actual life is carried with the Soul departing from the body and transferred to the next life brain, probably in the RSC.

Karma of previous lives, contained in the Soul, is transcribed or pasted into the developing brain, probably in RSC.

In summary, the Karma of this life and previous lives are stored in the Inner CS.

Furthermore, since the Creation is itself illusional, created by the false thought arising from Emptiness, the cleansing of Karma can only be done from the Mind (by understanding how the sin is committed), as illustrated in the following section of Vimalikirti Sutra

Comments
It is logical in all major religions to consider that all sins/karma originate from its noumeno (self-nature), the Emptiness.

is non-discriminative and is not Right or False. Therefore, the ultimate solution of the cleansing of karma is enlightenment, not to point out the karma. Ironically, to attain enlightenment, karma has to be cleansed
It is also noteworthy to mention here that along with karma, merit points

I. Generation of Karma

The body comprises five skandhas (Panca/vn: Uẩn). The Five Skandhas are generated through multiple steps after Creation leading to the Universe. Before the beginning of the Creation, the Emptiness/Formlessness/Nothingness (thể Không, bình đẳng, không phân biệt, tĩnh lặng) is homogeneous, tranquil, and non-discriminative The enlightened state of Buddhahood become deviated into unenlightened with next departure from the Emptiness/the immaterial Nature: Then arise a sudden transient false/misleading/ delusional, ignorant /stupid thought (Vọng Niệm) (Avidya) with the CS generated from the unenlightened state. Therefore, CS is impermanent/ ephemeral (Anitya) with the generation of false Ego.

In the Creation, originally pure and homogeneous water mixes up with Earth, becomes impure, and is far from the Emptiness of Nature.
The impurity leads to turbulence with mobility consistent with a feature of the Wind. Turbulence with the Wind causes the Fire. Subsequently, water and fire generate vapor and rain, exposing the Earth above the water level and making rocks and mountains, englobing fire inside. These conditions are sufficient for plants to grow.

Due to ignorance, humans and other living beings mistakenly believe in owning theirs. In fact, their body belongs to the Creator/ Ultimate source of Creation or Buddhahood. Therefore the body is originally egoless (Creator, not them, who owns their body). Along with this mistake, humans continue making more mistakes and deviations from the Buddhahood. Furthermore, the CS that initially deviated from the Awareness Paramita (Prajna-paramita) becomes more and more discriminative/ egoist/ subjective. In Suragama Sutra, Buddha pointed out to Ananda and the General attending audience that the ultimate mechanism of cognition of six Dharma coming from the five sensory organs, and thoughts (Manas CS) does not rely either on the Dharma, nor from the sense organs, the brain behind the sense organs, the thinking but from the Buddhahood/ immaterial Nature. The Immaterial or Emptiness Nature contains every form of the Creation since the Creation is born from the Emptiness. This is the principle in Vajracchedika-Prajna-Paramita Sutra/ vn: Sắc tức thị Không, Không tức thị sắc Rupam eva sunyata, sunyataiva rupam Matter is immaterial, the immaterial is matter. In the above meaning the material word is present but essentially delusional because it is composed of the Emptiness and vice versa: From the Emptiness arises the Form/ the Material.

In summary, Buddhahood is empty, and because it is empty, it contains everything in the homogeneous and harmonious state. However, everything in the Universe is delusional, impermanent/ epiphenomenal, unlike Buddhahood, which is permanent without death or birth. In addition, because Buddhahood is permanent and unchangeable, therefore, Buddhahood has no space and Time (Time=0, Space=0, unchangeable Awareness). As a result, Buddhahood contains forms of all Time and anywhere in this Universe and other universes if existing. Any inventions up to this date and in the future are always within the Buddhahood. The original CS/Cognition based on Buddhahood is always perfect. It is mundane to say that Buddha is also an ideal Prophet.

Free Will and Libet's Experiment
(Please see Free Will in Chapter XI, pge)

J. Mechanism of Karma Cleansing
Glia is composed of Astrocytes, microglial cells, and oligodendrocytes, and plays an essential role in support, nutrition, and neuroplasticity (destruction and repair) (destruction and repair) in the central nervous system.

The astrocytes account for about 2/3 of the nervous tissue of the brain. They play the role of supporting, by producing glial filaments, Astrocytes, with their numerous and prolonged cell processes, make up a cytoplasmic network covering the neurons... A single astrocyte can form a network covering up to 140.000 synapses from 4-6 neurons with 300.000 dendrites in small living organisms (Gao 2010, Bushong 2002, Oberheim 2009, Halassa 2007). In Humans, one astrocyte can cover up to 2 million synapses.

A more critical role is in the metabolism of the neuron: regulation of the metabolism of glucose in neurons, the vital nutrient along with oxygen, and **neuroplasticity (destruction and repair) (destruction and repair).**

This latter role consists of the removal of unused synapses and the regeneration of new synapses.

K. Role of the Ultimate Omniscience (vn: Trí Huệ Bát Nhã) (Fig 37,.8,9

As revealed by the Buddha, the Creation was initiated from a False Thought (vn: Vọng Niệm) arising from Emptiness. In some particular circumstances, Emptiness can be represented by the Implicate Order conceived by David Bohm, a theoretical Physicist. From the original stillness, homogeneity, balance, and non-discrimintation, the Emptiness demonstrates the imperishable and unmeasurable force and the Ultimate Omniscience. Unmeasurable power accounts for all forces of the Universe, including the Dark Force causing the Universe's accelerating expansion. The Ultimate Omniescience accounts for the orderly, organized, magnificent, ingenious formation of all Forms. The Ultimate Omniescience was personalized as God in major religions. Since it was conceived by scientists (physicists and mathematicians), the Universe could not have been created as it is today by a random combination of elementary particles (photons, bosons quarks...) in a period of 14 billion years after the Big Bang. Mathematically this Universe may require at least 1.5 times the Universe's age. As a result, the alternative mechanisms of Creation are chance or the Spiritual power of God. The chance mechanism does not seem sufficient to be responsible for multiple complicated formations of structures in nature. For example, Rudolf Virchows and Louis Pasteur believe that life can only be created from life (living cells only give birth to living cells).

> **In the practice of meditation,** the recall of MM in Inner consciousness is a unique process. Thanks to the ACC, which is tasked with finding the source of information when peripheral information is limited or cut off, this MM is not recorded as CS due to the absence of an attention mechanism. Consequently, it is eliminated by the IPS, which is not used for attention in meditation. This results in the erasure of MM in the inner consciousness, unknown to the meditator. This process is distinct from the mechanism of repentance or confession, as individuals focus on recalling MM in the hope of clearing away Karma.

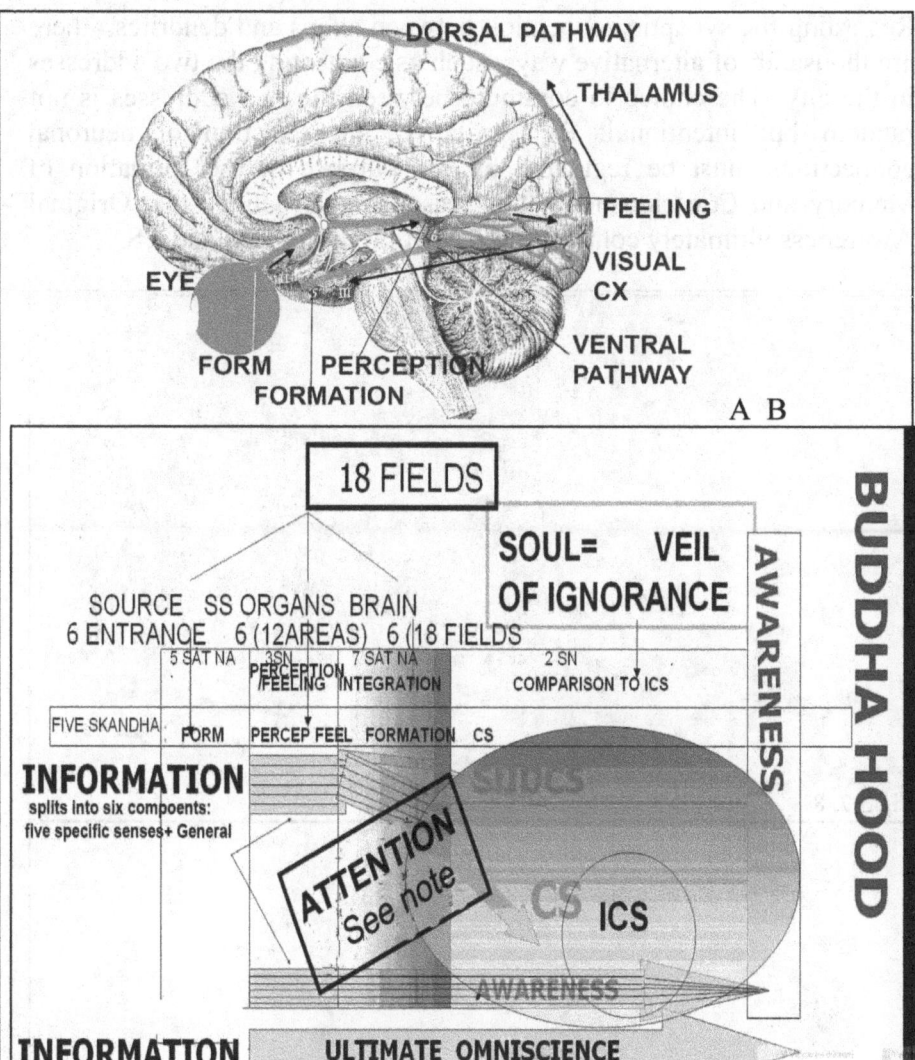

Figure 1. 13
A: Pathway of vision: Retina→Thakamus→ Visual cortex :
→Dorsal Nonconditional reflex
→Ventral pathways; d Formation of CS
B: Diagram showing the original package of input information undergoing two types of modification:
Splitting the original package of information into six types of information: five specific sensory
Note: Fields used in this diagram is equivalent to Filters

Regarding the synaptic connection between axons and dendrites, there are thousands of alternative ways, such as connecting the two addresses in the city. The choice to commute between the two addresses is not random but intentional. By similarity, the selection of neuronal connections must be regulated by the purpose of the formation of Memory and Consciousness. It is reasonable to believe that Original Awareness ultimately controls the formation of the MM and CS.

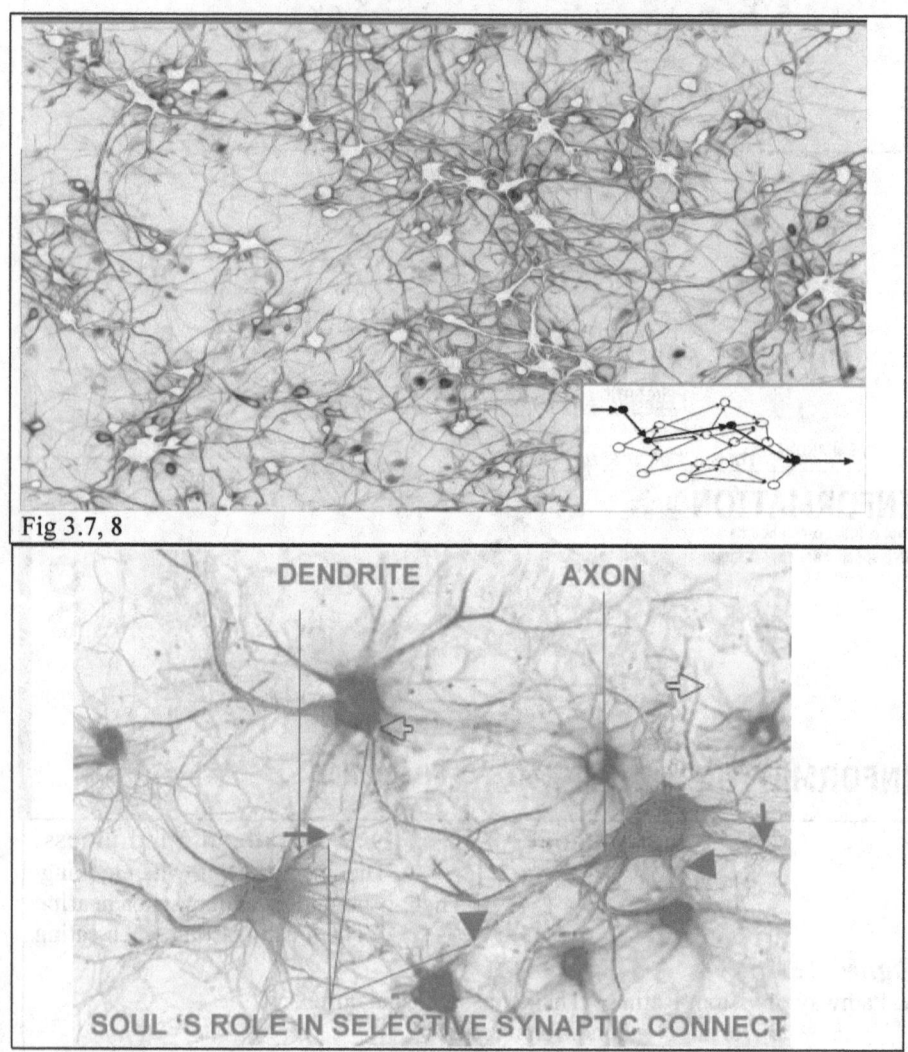

Fig 3.7, 8

Figure 1.9
Diagram showing the selective synaptic connection to establish the pathway of neuronal conduction. Given the multiple choices for selection of connection, the choice is specific for each information

Chapter IV: Categories of MEMORY

I. The Working Memory (WM) (Fig 4.1)

WM is the capacity of the brain to collect all information from the MM store of different types of the current life in a readily accessible form available for comprehension, planning, reasoning, presentation in speech, and problem-solving. Atkinson, in 1968, proposed the "Multi-Store Model" composed of multiple modules. WM mainly involves short-term MM

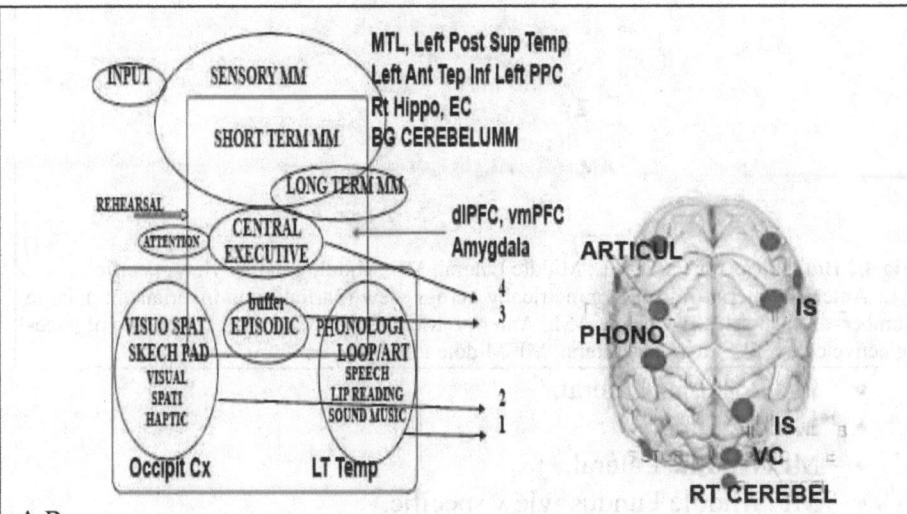

A B

Fig 4.1A: Diagram showing different types of MM connected to the Executive center (4), Phonologic Loop (3) and AudioVisual (2). Episodic Buffer (3) is replaced by attention.
B:. Different areas of cortex activated in Working MM: Parietal, Temporal and Occipital.
AR, articulatory rehearsal; IS, inner scribe (spatial rehearsal); PS, phonological store; VC, visual cache (Braddeley 2000, 2003)

PFC plays an essential role in the WM due to the dlPFC connecting with IPS forming a network in managing, controlling and coordinating different centers:
Visuospatial Sketchpad=inner eye: cortices parietooccipital RT and LT, RT more important than LT
Phonological loop/Art and Speech, Lip reading, Speech
(LT Temporal cortex) (required duration: 1-2 seconds
 Episodic Buffer connecting Phological loop with episodic MM centers and remote MM, sensory MM, and recent information-input

II. MM of Space, Navigation map and Smell
In 1960 O'Keefe incidentally identified the map of navigation in CA1 (where the Place cells are located) of the HIPPO while inserting a needle in the thalamus for his study of the Thalamus. He was awarded Nobel Prize)

III, MM of Face with Face cell Area in the Temporal lobe (Fig 4.2)

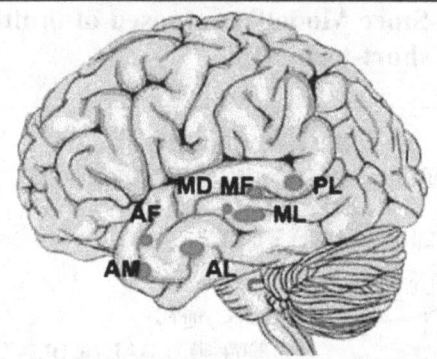

Fig 4.2 Brain areas for Face: ML: Middle Lateral, MF: Middle fundus=view specific AL: Anterior Lateral, mirror symmetrically across view (partial view invariance): a large number of face-selective cells. AM: Anterior Medial=full view: large number of face-selective cells, PL: Posterior Lateral, MF Middle Fundus

- PL: Posterior Lateral,
- Middle Fundus
- ML: Middle Lateral,
- MF: Middle Fundus=view specific,
- AL: Anterior Lateral= mirror symmetrically across view (partial view invariance): a large number of face-selective cells,
- AM: Anterior Medial=full view: a large number of face-selective cells

IV. Severely deficient autobiographical Memory (SDAM) in healthy adults: A new mnemonic syndrome).
With regular CS and standard semantic MM but reduced episodic MM regarding time and space

V. False episodic Memory
Plays an important role in a witness inn legal court cases
- Verbatim MM having, explicit MM. Located in vmPFC, and Gist MM, faint traces because the MM traces semantic forms of the objects or events. Verbatim MM is lost in aging, but Gist MM is in MTL. The combination of explicit and False MM renders the False MM more detailed

- Explicit MM tends to recognize False MM as false, while semantic MM is often influenced by median hypnosis
- Gist MM increases with age, therefore False MM often encountered in the young age group
- False MM has often seen in **deficient autobiographical Memory**
- False MM is long-lasting

VI. Hyperthymnesia= Highly Superior Autobiographical Memory
(Photographic Memory= Persistent Eidetic Memory),
There are about 50 reported cases of Hyperthymnesia
Characterized by the MM as being represented by video recording all sensory inputs (including audiovisual). The recording is automatic represents cases of Ananda remembering all Buddha's teachings when he attended
People with Hyperthymesia are autistic
The MM is not always perfect but more superior in quality to normal persons and can be influenced by media. The Iq is normal
fMRI in cases of **Hyperthymesia** showed hyperactivity of areas related to MM : left **HIPPO**, Premotor, PFC, Retrosplenial, Cingulate Cortex (Brandt 2018) , Precuneus and large Temporal lobe and Caudate nucleus. Connection between PFC and HIPPO is also enhanced. In one study, the Forntostriatal connection is suspected to be the cause of the problem (Santangelo 2018).

Eidetic Memory (Fig4.3)
Is hyperthymesia in children
VII. Recognition Memory
Represent the episodic MM in the circumstances for immediate and fast MM recovery:
The feeling of knowing (semantic MM) is fast but lacks the accuracy
Neurobiologically: connections are between the Mediodorsal of the Thalamus, Entorhinal, and Perirhinal cortices, vmPFC, and Cingulate Cortex Mammillary body.
  There may be involvement in two mechanisms: retrieval of MM (ventral pathway is involved) and feeling of knowing (vmPFC –Nre- HIPPO pathway is engaged)

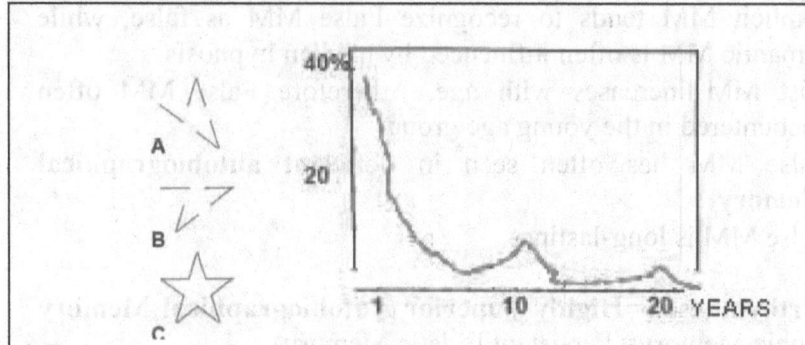

A B
Fig 4.3 Left A: combining A+B →C is easy in children with Eidetic MM
(Gray 1976, 1975, Haber1979,1969)
RECOGNITION TEST
Right B: Cases of Eidetic MM decrease with age

VII. Reference MM

Is related to Place MM in the experiment of mice remembering the path to find the foods
Neurobiologically, there is involvement of implicit MM (temporal cortex) and explicit MM (vmPFC-HIPPO

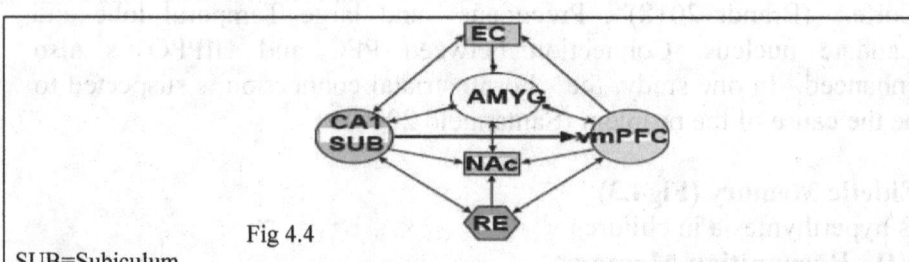

Fig 4.4
SUB=Subiculum
Connections HIPPO with CA1/SUBiculum and MPFC are direct or indirect through Nucleus Reuniens /RE, N. Accumbens/ NAc, Amyg , through Entorhinal cortex

Chapter V: COGNITION or CONSCIOUSNESS

Summary

Cognition or Consciousness (CS), like Memory, is very close to human life and is continuously used in the wakefulness and dream state. However, it has been considered part of the metaphysical aspect of life because the CS has yet to be understood regarding the structure and its mechanism of formation from the information. Therefore, the investigation of the transformation of the data into CS is considered a critical step in understanding this vital part of life. Francis Crick, who won the Nobel Prize for revealing the structure of DNA, turned his research interest to Neuroscience and posed the thorny problem of the minimum mechanism of formation of Neural Correlate of Consciousness (NCC). This manuscript aims to review the understanding of the NCC, and the more fundamental question is how living beings recognize and understand the environment.

Results

All living organisms can be grossly divided into two parts. The body is the physical part that follows the cycle of birth, development, aging, and death. The Mind or the Soul is the metaphysical part created from the Ultimate Omniescience when bad karma is added. As a result the metaphysical part will eventually cease to exist when it can get rid of bad karma to become its original state of Ultimate Omniescience. This metaphysical part has caused many questions, discussions, and hypotheses among the Western, Middle-Eastern, and Eastern philosophers about the characters and their existence. So far, the identification of the Soul has failed even with sophistical instruments like EEG, fMRI, single cell stimulation technique, or even with the involvement of current advanced knowledge in the quantum field. Nevertheless, Neuroscience has shed some light on the question of NCC:

1. CS is all information coming to the Cortex with a component of attention. Information without attention is discarded or registered in the Subconsciousness or Unconsciousness

2. CS is the summation of the integration of information at different zones of the Cortex. Example: Visual inputs to Visual Cortex V1 are sent back to the cortices in two streams toward the anterior parts of the brain.. The dorsal stream that goes to the motor cortex and the Frontal Eye Field/ FEF is responsible for reflexes controlling the movement of the eyes, head, and body i.e, closing eyes when a fly comes close to the eyes before the CS is involved. The Ventral stream, which goes to the

Inferior Temporal Lobe, generates the CS, including emotion. The CS is formed by the ACC's participation, detecting similarities/ dissimilarities with the ICS, primarily the Default Mode (PCC, MTL, vmPFC...), semantic MM, and sensory cortices storing sensory MM. The ICS plays the role of the model of referral CS.

The two streams proceed in parallel fashion, with the dorsal stream slightly faster than the ventral stream. The evidence for the existence of the two streams theory is the unconscious reflexes, the hemineglect syndrome, consistent with the stream of information carried by the Dorsal stream, which is fast, objective, short-lived and subserving the purpose of reflex. The other stream, the Ventral stream, is slow, subjective, longer lasting, and covers large areas of the Cortex subserving the purpose of CS.

3. Referral model of CS or Inner CS (ICS): The site of the brain for the ICS is the cortices of the Default Mode Network (DMN) + Medial Temporal Lobe (MTL) Basal Ganglia (BG), sensory cortices and the Cerebellum. These areas are essentially regions of storage of semantic MM. The previously stored MM are used as referral models or templates for the brain to compare with the new information. Since the brain is a box of prediction, initial information activates the Cortex (related to the new data) to send a request for more information based on the Bayesian effect (Bayesian effect: the propriety of calculation of the probability of occurrence of a recent phenomenon based on previous experiences). The second information phase is sent up to the sensory Cortex, then sent to the ICS, for ACC to compare the two types of information. Information similar to that in the ICS is labeled as such. Novel information adds to ICS as an update.

As described in the previous chapter of MM, the MM is stored in the ICS, which is the storage of the model of CS. CS is the critical component of the Soul/ Mind. After the ensoulment, the store MM in the Soul is pasted in the ICS of the newborn's brain. Since then, the new pieces of MM (equivalent to the first parts of CS) have been updated in the ICS of the baby. As a result, the ICS contains all information/ MM of the newborn person and serves as a personal encyclopedia.

Furthermore, there is another step in forming the CS: The Soul and DMN share the same brain areas. The new information is also shared with the Soul. The Soul is always connected to Buddhahood or True or Holy Spirit/ Ultimate Omniscience. Since Buddhahood or True or Holy Spirit is the highest level of omniscience, the CS only forms from this ultimate level. *It is a grave error to say that CS is generated from the brain or the ICS.*

4. In addition to the above processes, for the information to be made into CS, other involving factors are the Thalamic Reticular network (playing the role of purifying similar to the noise-canceling system in the headphone), Thalamus, Hippo for receiving the information, Basal Forebrain, the Ascending Reticular system and other center controlling awakening

In sleep, there is a decrease in the connection of different regions of the brain. In NREM sleep, there is no CS, but in REM sleep, there is Consciousness, as in cases of dreams, because the level of Acetylcholine is still high in the brain in NREM sleep 5.

I. GENERALITIES

A widely accepted concept of all times in Human history, such as that of Philosopher Rene Descartes, is the distinction between the Mind and Body. According to Oriental philosophy, the whole reality, conventionally considered the Ultimate Pole, can be considered an examination object. The division/analysis in the process of examination can be performed in two manners:

a) Egoless principle: the role of the observer is ignored, and the observation is objective. In this division, the initial step creates two new parts. Since the division can not be equal or complete: Call the two parts Emptiness or Yin (Negative) and Form or Yang (Positive). The further division will lead to 4 domains.

Since the division can not be perfect or complete, the Negative contains a small area of Positive, and the Positive a small area of Negative: Therefore, the Easterners design an ideal way to represent the above four dissimilar parts as Tai Yin or Greater Yin/Negative, Shao Yang or Lesser Yang/ Positive, Tai Yang or greater Yang/ Positive and Shao Yin or Lesser Yin /Negative.

Each part, Yin or Yang, is represented by a line also called an oracle. The broken line represents Yin and the continuous line represents Yang.

Further division of the Ultimate makes it into four unequal /dissimilar parts.

To represent the nature, the four parts are represented by pairing two lines to form four different diagrams that may represent different attributes in nature, such as four seasons in one year. In each diagram, the inferior line represents the basic nature of the diagram, the upper, the change or the manifestation.

b) Egocentric principle: the above division or observation/ discussion can not be logical or scientific if the role of the observer is included in the process of division/ observation. As a result, another line/oracle is added to the diagram to establish the trigram. Since trigrams are composed of any combination of Yin and Yang. Each trigram can be considered a basic component in Nature. There are in toto $2^3=8$ trigrams that can be attributed to different sets of 8 phenomena or entities such as Heaven, Sun, Lake Water, Fire, Mountain, Thunder, and Earth or different organs or parts of the body, different types of Consciousness

Because in nature, there is always an interaction between different components, a design was developed to pair two different trigrams to create hexagrams to represent the circumstances of interaction in nature, such as the interaction between each individual in a society. This concept of interaction seen in I Ching to describe the activities or changes in nature was also mentioned in the Buddhist sutra when Buddha told how nature is created initially. The above pairing represents all activities in nature in general between individuals in human society. The above pairing of two different trigrams creates $8^2 = 64$ combinations. In antiquity, the use of the 64 hexagrams for prediction in farming, fishing, and human behavior...is documented. In this prediction, the attribution of each line in the hexagram occurring at different times, seasons, and circumstances represents multiple eventualities in life.

An alternative way to represent nature is by successively dividing the ultimate pole in two and then into four parts. The four-component is then added with the fifth component, representing the observer, into the set of five essential elements. Each of the five elements, represents one of the following: metal, wood, water, fire, earth, or five primary colors, ...The set of four elements does not represent any forms in the earthly world but represents any set of four parts in the egoless world.

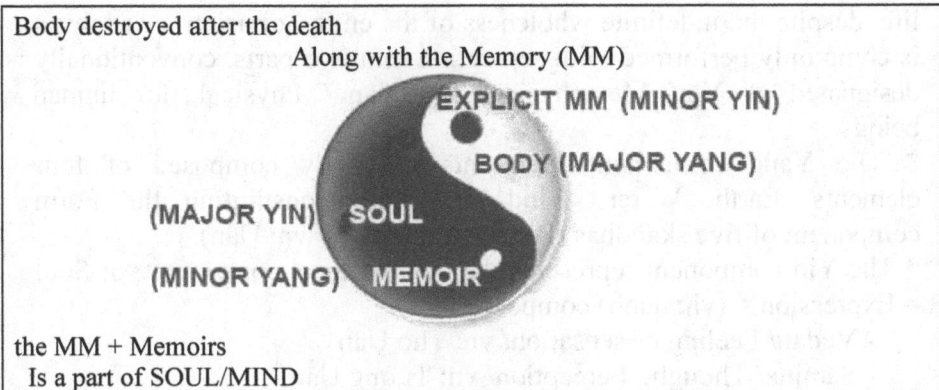

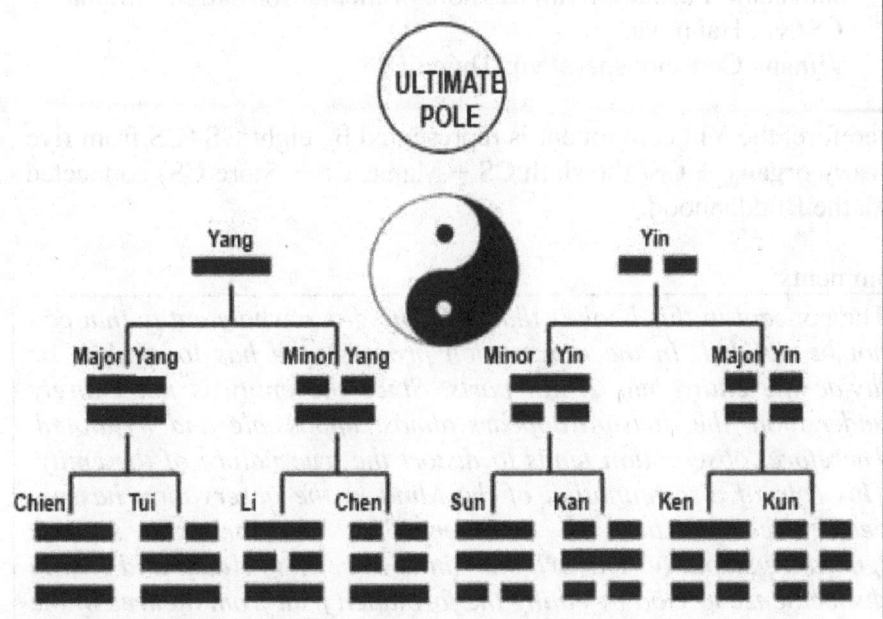

Fig 41B: Octagram having eight trigrams. Each trigram has three oracles: of either Negative (- -) or Positive (---).

Chien (Càn) = Creative = General CS
Kan = (Khảm) Abysmal/ Water = Gustatory CS
Ken (Cấn) = Still/ mountain = Tactile CS
Chen (Chấn) = Arousing = Olfactory CS
Sun (Tốn) = Gentle = Manas CS
Li (Ly) = Clinging = Visual CS
Kun (Khôn) = Receptive = Store CS
Tui (Đoài) = Joyous = Auditory CS

II. The Metaphysical Realm

Along with the above concept, Ultimate Pole is an indefinite composition called Tao. Tao can not be expressed in words. In worldly life, despite the indefinite wholeness of the entity, division or analysis is commonly performed with separation into two parts, conventionally designated as Yin/ Metaphysical and Yang/ Physical, for human beings.

* The Yang component represents the body composed of four elements, Earth, Water, Wind, and Fire, constituting the Form component of five skandhas (Panca/ Aggregates/ vn: Uẩn).
* The Yin component represents the metaphysical component=or Soul = Expression = (vn: danh) composed of:
    - Vedan/ Feeling or sensation/ vn: Thọ Uẩn
    - Samjna/ Thought, Perception/ vn: Tưởng Uẩn
    - Samskara/ Formation, impression, or mental formation = Manas CS/ vn: Hành uẩn
    - Vijnana Consciousness/ vn: Thức uẩn

Therefore, the Yin component is represented by eight CS (CS from five sensory organs + CS/ the sixth CS + Manas CS + Store CS) connected with the Buddhahood.

Comments:

> The concept in this book is that everything is a whole entity that can not be divided. In the observation process, one has to analyze or divide the entity into 2, 4... parts. Since the entity is not entirely understood, the division appears almost impossible and irrational. Therefore, observation tends to distort the true nature of the entity. This role of discrimination of the Mind in the observation has not been accepted and is considered as the original sin or ignorance/stupidity. The original sin is to refer to Adam and Eva in disobedience to God by eating the forbidden fruit from the tree of the knowledge of good and evil. After the eating, Adam and Eva start having discriminative Minds distinguishing the difference between good versus evil, man versus woman... the acquisition of a discriminating Mind is the primordial sin, leading to the subsequent sins as seen nowadays. In Buddhism, the non-conceptual Mind is equivalent to the non-discriminative Mind= no thinking, no perceiving, and no imagination= Mind of equality/ Mind as Bhutatahata/ Buddha nature/ peaceful Mind. Therefore, the product of the observation is only correct in a conditional/ conventional and limited paradigm and always deviated, distorted, and deluded in the entity's true nature. Lao Tzu said Tao cannot be expressed in words. With this concept, the separation of the body and Soul is relative and conventional in the understanding that division is impossible. In the body, there is always a component of the Soul that is represented by

> *the synaptic connection for the MM. In the Soul, part of the body is represented by the MM in the form of physical neuronal connections for autobiographic and storagw of MM. that enable the process of imprinting or pasting MM into the newborn's brain.*

A. Metaphysical Realm, easily Recognizable: CS related to Five Senses, General CS and Manas-CS

CS=Cognition=Knowledge= the Sixth CS
The first to 5th CS is CS related to the five Sensory organs. The seventh CS/ thinking is part of CS having the management role and the capability to use the information in all CS, including store CS if possible, for the transformation, repression, or Creation of new CS. In Śūraṅgama Sūtra/ vn: kinh Lăng Nghiêm, Buddha pointed out that the Nature of Hearing (or seeing, smelling...) does not originate either from outside Dharma (***Dharma is the universal truth taught by the Buddha, and is therefore all existing physical and metaphysical entities including cosmic law after the Creation of the Universe***) or from the ear/ Brain/ CS. This nature of hearing is permanent and not associated with birth or death because it is from Buddha's nature. That is why one can hear without an ear (or see without eyes... *because sensory organs are impermanent and the nature of hearing , seeing... is permanent*) when one can clean up the obstructing barrier/ filter of the CS in the Brain. The obstructing barrier/ filter is the ICS, the veil of ignorance (Avidya)/ delusion/stupidity/ unenlightenment/ False Ego.

B. Metaphysical Realm, Recognizable with difficulty
Composed of
  Sub-CS and UNCS
  Store CS Alaya (the eighth of the eight vijnanas/ vn: Tạng thức A lại da thức)

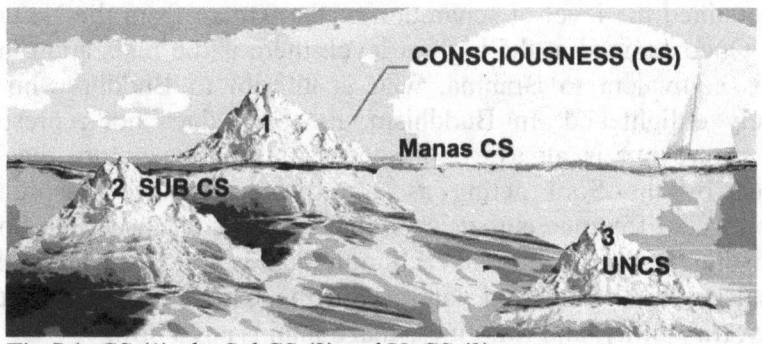

Fig 5.1 CS (1) the SubCS (2) and UnCS (3)

The distinction between Sub-CS and Un-CS depends on the depth of the information buried in the information store. Store CS is more profound than the Un-CS and belongs to the imprinting of the MM of the previous lives into the current brain/ Soul. The depth of these types of CS also reflects in the difficulty with which the MM is retrieved. In addition, in most instances, these types of CS cannot be voluntarily recovered but spontaneously retrieved in certain circumstances like intuition, dreaming, and hypnosis. Furthermore, they can play as cause or Karma in activities in life as impulses in daily Non-free will activities. They can play an essential role in the physiopathologenesis of many psychosomatic diseases ranging from anxiety, depression, insomnia, gastroduodenal ulcers, ulcerative colitis, lung, thyroid skin diseases...panic phobia, Freudian slip of tongue, and error in intuition. They also contribute positively to other activities like intuition, fast but correct decision cognition, and many talents in young persons.

C. Soul is different from Holy Spirit/ Buddhahood/ Buddha nature

from Atman, conceived by externalists like in Hinduism. Atman in Hinduism has the same features to those in Soul. The main difference is that Atman is permanent. This is the primary and fundamental difference between Buddhism and Hinduism. Because Atman represents one's self as opposed to Mahatman represents Brahma, the creator. Of six branches of Indian philosophy, Yoga branch of Sankhya philosophy is most devoted to Meditation. Sankhya's philosophy is dualistic with the concept of Self/ Atman, associated with CS /intellectual capability, and Nature /Prakriti associated with a motile capability. A combination of Atman /Self and Prakriti / Nature forms an individual with intellectual and motile ability. In this system, the concept of Brahma (chief of Hindu Gods composed of Bramanical Trimurti: Brahma, Vishnu, and Shiva=Devas, with Brahma/ vn: Phạm Thiên) is only used in the initial step of Meditation when the meditator has not attained the level of separation of the Atman from the Nature/ Prakriti. Once attained at the highest level, there is the realization that Atman is equivalent to Brahma, who is inferior to Buddha who is completely enlightened. In Buddhism, the Soul does not represent Buddha nature but is always connected with Buddha nature that is obstructed by the Soul acting as a veil of ignorance/ stupidity/ inversion. The difference is that when the Karma is cleansed, the Soul stops existing, Buddha Nature in each individual is revealed, and let the individual enjoys the Buddha nature with the Ultimate Omniscience, eternality, tranquility, and Emptiness that contains any Form.

Therefore, the teaching of Buddha had created heated discussions among Yoga and Buddhism practitioners and opposition from arrogant Brahma

In Summary, Soul/ vn: Hồn exists in two entities. In Living beings, all types of CS (eight) represent the Soul; after death, the Soul is only represented by Alaya/ Store CS/ Karma power/ vn: Nghiệp Lực. The Soul is different from Atman and Holy Spirit due to the impermanence and the presence of Karma in its content. The Soul belongs to the metaphysical realm and last much longer than the body having short existence

Soul and Memorial/ vn: Hồn và Vía)

The Easterners, as well as the Westerners, distinguish the Soul that belongs to the Metaphysical Realm distinct from the body; after death, with the disintegration of the physical body, the Soul is disconnected from the body. As discussed in the previous paragraph, the separation of the Soul and body is never complete. The remaining part of the Soul after its leaving can be conceived as traces of MM among the relatives of the dead. This part represents the Memorial Souvenirs as seen in the physical world as represented by the dark dot in the paradigm of the Ultimate Pole.

D. Concept of Multiple Souls in one body

1. The current common sense is that each individual only has one Soul. When treating patients with double or multiple personalities, Freud always considered the second identity in the patient as representing part of CS repressed. The part of the CS being repressed is frequently caused by physical and emotional trauma. The process of amnesia disconnects the two parts of CS. The amnesia is hypothesized to explain the disconnection of CS that persists after the trauma and MM that is temporarily abolished. However, the implicit MM of the two separate IDs is not disconnected, as evidenced by the two ID using a common skill like dressing, eating, talking, walking ....

On the other hand, Pierre Janet proposed that humans could have multiple CS.

- First Janet's law: Each CS is responsible for the behavior and conduct of each ID and may share some common set of MM, but there is no disconnection of MM (explicit MM) or amnesia lasting from minutes to years.

- Second law: Multiple CS is created, not by division or splitting. This proposal is a surprise to many people at that time.

- The multiple CS can co-exist at the same time. One CS acting as automatism.

The second CS can completely replace the first CS, consistent with a typical case of double personality.

- The two IDs can alternatively take control of the body. Despite the proposal of multiple CS, Janet did not go further in his hypothesis regarding the mechanism of the existence of multiple CS

2. Proposal of Cases with Multiple Souls in One Body

The concept of the Ultimate Pole is widely accepted in Eastern philosophy. In the succession of the development of nature to create this world from the Ultimate Pole, the concept of Duality or Multiple branching is easily verifiable in its ordinary sense. As a result, in this world of duality, the presence of cases with multiple Souls in one body should be seriously considered. Since the Soul is composed of all types of CS, Janet's cases of multiple CS in Multiple personalities likely represent the typical example of multiple Soul in One body. As suggested by Freud, only one Soul plays a dominant role in commanding the body, whereas other Souls are repressed as Sub-CS or Un-CS. These Sub-CS and Un-CS can be manifested when the dominant Soul becomes traumatized

An example of duality can be seen in the fertilization process: The egg (ovum from the ovary) only contains 23 chromosomes (n=23 in humans) and is usually fertilized by only spermatozoa, which also has 23 chromosomes. The resulting fertilized ovum comprises 23 pairs of chromosomes (=46 chromosomes). After being fertilized by one spermatozoon, the ovum immediately forms a thick layer of material called zona pellucida that impedes the entry of other spermatozoa into the ovum.

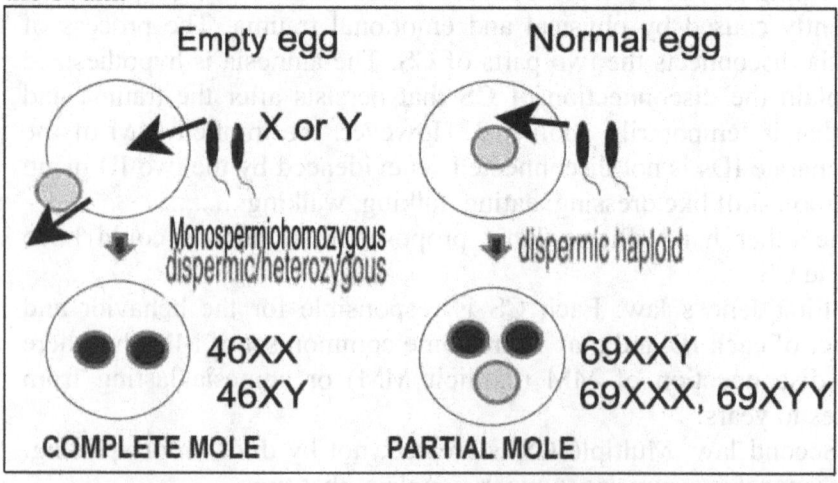

Fig5.2 Fertilization may occur by two spermatozoa, each has **n** chromosomes. When two spermatozoa enter into an ovum without a nucleus (the nucleus was eventually expelled by two spermatozoa, a complete mole will develop and contains cells having 2**n** chromosomes with two Y chromosomes. When two spermatozoa enter into an ovum without a nucleus with **n** chromosomes, a partial mole will develop and contains cells having 3**n** chromosomes (trisomy).

In some cases, the ovum can be fertilized simultaneously by two spermatozoa. The zona pellucidae is not efficient in its role. There are three eventualities

- If the entry of two spermatozoa occurs after the division of the nucleus of the ovum, each spermatozoon fertilizes each nucleus. As a result, this is the case of twin pregnancy is non-identical twin

- If the entry of two spermatozoa occurs before the division of the nucleus of the ovum:

· There is the formation of a nucleus with 3n= triploidy with the following development of the partial mole

· There is the expulsion of the nucleus of the ovum with the formation of a nucleus with 2n coming from spermatozoa. The following development is the molar pregnancy

III. CS-related Terminology
IV. CHARACTERISTICS OF THE CS
- **Inner CS :**
- CS/ Cognition is defined

in Merriam-Webster as "the quality or state of being understandable, especially of something within oneself, conscious of an external object, state, or fact of the inner and outer existence
- Ultimate Omniscience/ Prakriti-buddhi, Original Bodhi / vn: Bản Giác /Absolute Emptiness, Absolute Void/ Original Mind
- A wakefulness (vn: Tỉnh) is a recurring brain state of Consciousness with behavioral responses in responses to Inner or external stimuli
- Mindfulness knowledge results from focusing the Mind on a restricted area of information (i.e: the rhythm of respiration, the third eye, the sound, tip of nose....)
- AWARENESS: quality of acquisition of the inner or external information based little on the attention, but mainly on the original Mind that is still partially obliterated by the veil of ignorance/ the Consciousness, somewhat close to its true nature, a "Thus Come One" / vn: như thị.

- Knowledge: composed of Understanding (/Consciousness / Mindfulness) and Awareness.

$$\text{Ego }(0\text{-}1) = \frac{\text{Consciousness}}{\text{Knowledge}} = \frac{\text{Consciousness}}{\text{Consciousness+Awareness}} = \frac{1}{1 + \frac{\text{Awareness}}{\text{Consciousness}}}$$

Consciousness increases Ego and obliterates Awareness and Ultimate Omnisciene. Awareness decreases the ego. The Ego is always less than one.

- Recognition: being knowledgeable Concern: knowledge with anxiety.
- Apprehension: Perception with Consciousness/ **vn:Tri Giác**.
- Realization: Awareness applied in reality in its fullness and in-depth.
- Discernment Acquisition of CS in an obscure subject.
- Regard attention with wishes.
- Heed Heedfulness attention.
- Subconsciousness, Unconsciousness: Buried CS with various difficulties in voluntary retrieval, but with possible spontaneous retrieval in a dream or intuition.
- Subliminal self Submerged Mind Under sensitivity.
- Psyche : Mind/ Spirit/ Soul/ Essence.
- Remembrance, recollection, Flashback memorization.
- Thought, thinking.
- Recollection/ Recapture/ Recall reminiscence.
- Dead eye/ Mind's eye/ Camera eye: different status of seeing
- Alertness: status of active attention with high sensitivity to outside stimuli.
- Intuition CS resulting from spontaneous retrieval from different levels of CS.
- Metacognition: CS review of CS in the past.
- Holy Spirit/ Buddhahood.
- Wisdom: high quality of experienced CS resulting in appropriate judgment/reaction
- Intelligence Ability to acquire and apply CS and skills.
- Jnana and Prajna wisdom (Trí Huệ).
- Prajna-paramita / vn: Trí Huệ Bát nhã/ Wisdom-paramita. Prajan Transcendental/ supreme /sublime/ ultimate/ real- unimaginable/ Infinite knowledge/ universal (six paramitas).
- Knowledge of Tathagata. CS of Buddha. Ultimate Omniscience.

- Jnana/ v: Trí tuệ/ wisdom.
- Samaya wisdom—The characteristic of a Buddha's or Bodhisattva's wisdom. Wisdom attained in Samadhi.
- Wisdom paramita WISDOM: Knowledge with experience, good judgment and behavior "listen to his words of wisdom."
- Parajnaparamita Tri Kiến Ba La Mật.

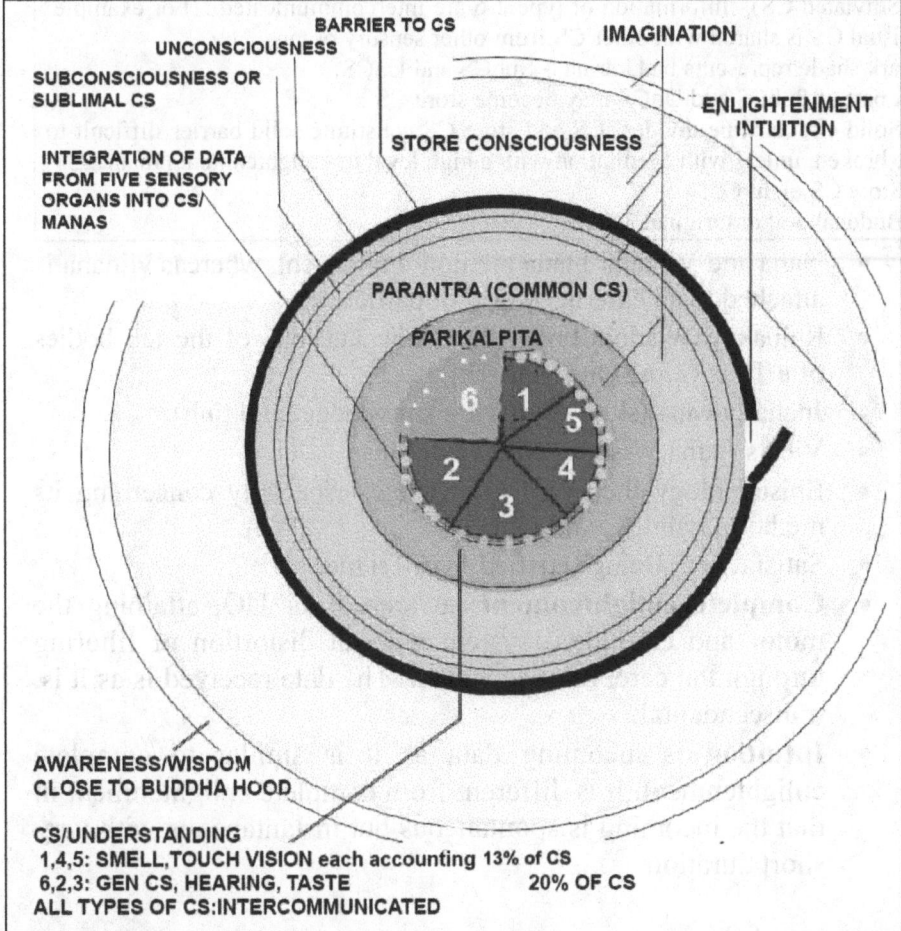

BUDDHAHOOD ALL OVER THAT INCLUDES LARGE THINGS AND DOES NOT EXCLUDES SMALL THINGS

Fig 5.3 DIAGRAM OF THE CONSCIOUSNESS AND RELATED ENTITIES
Original Mind/ Buddhahood obliterated by Veil of IGNORANCE.
Without the Brain/ Veil of IGNORANCE the Original Mind/ Ultimate Omniscience are overwhelming in the Universe
Information of five sensory organs and Consciousness, SubConsciousness UnConsciousness Store Consciousness: From inside to outside. The center: the object displaying information to sensory organs
- 1-6 represent 6 types of CS

According to Surangama sutra, auditory and gustative CS 1200 merit points each calculated as follows: 3 (time: present, past, future) X 4 directions X 10 directions (up,down+ 8 directions...); 2, 5, 6 are limited with only 800 merit points).
- Thinking the seventh CS
- Inner CS:
Composed of Parikalpita (wrong/ distorted CS and {Paratantra (common CSdiviated CS). Information of type 2-6 are intercommunicated. For example, Visual CS is shared with other CS from other sensory organs
Dark shade represents bad karma, - SubCS and UnCS:
A part of SubCS and UnCS may become store CS
- Solid circular line divides CS and store CS constitute solid barrier difficult to be broken, unless with Meditation with a high level of enlightenment (Jhana2-3)
- Store CS eight CS.
- Buddhahood or Original Mind.

- Nana and Vijnana Jnana are non-attachment, whereas vijnana is attached to an external world of particulars.
- RajnakayaWisdom-body, the Tathagata, one of the ten bodies of a Thus Come One (nhu Thị).
- Jnana-pavana (skt)—Purifying knowledge. (Trí tịnh).
- Views Jnana Wisdom-Knowledge.
- Epistemology theory of knowledge, especially concerning its methods, validity, and scope.
- Satisfaction. Being satisfied / vn: Tri túc.
- **Complete enlightenment**: awareness or UO, attaining the motor and emotional system without distortion or filtering through the cerebral mechanism. The data received is as it is, transcendental.
- **Intuition** is incoming data as it is similar to complete enlightenment. It is different from complete enlightenment in that the incoming is spontaneous but instantaneous with very short duration

Antonyms of CS
· ignorance, · disregard, · neglect, · thoughtlessness, · stupidity
· carelessness, · inattention, · negligence, · senselessness, unconsciousness

IV. CHARACTERISTICS OF THE CS
Cognition, CS, Understanding, Awareness (the extent of CS), Mindfulness (the depth of CS), Awakefulness, Attention, Sleep, Mindfulness Meditation, Hypnosis, Dream, Sleepwalking and Intuition are thought to be the work of the brain. The basis of the work is the CS.

CS is the state of Awareness about the surrounding environment with a possible choice of response. CS can be

categorized into: innate, instinctive Acquired following learning, experience and investigation. Usually, CS increased with the effort of learning or exploration. Therefore, the process of CS involves receiving the information from the environment and the body, storing as MM, using the information for reaction and for thinking. Thanks to the rapid progress in Neuroscience, Neuroscientists are now able to unlock to some extent the mystery of Neural Correlates of CS (Crick 1998).

- **Inner CS :**

Reflexes of the newborn baby as crying, sucking. These types of reflexes like crying, sucking... are similar to the built-in features of operation in smartphone. Similarly, eating walking ...are the type of skill that human being automatically develop. Teaching does not influence very much on the development. During fetal development, the neural connections for this style of skills have been imprinted from the Soul into the embryonic brain.

Note: The development of a number of activities occurring in the early period of life can be considered fundamental, basic for survival, at the level of unconsciousness. Example:

i. Jerk reflexes of the tendons such as the knee, ankle reflexes are simple reflexes involving arc consisting of sensory neurons, and motor neurons.

ii. More complex reflexes like the postural reflex, consisting of the simple arc of reflex as above and the involvement of interneurons in the spinal cord and the ascending and descending neurons from the reticular system in the Medulla Pons and Midbrain. The operation of the above processes is unconscious or subconscious without the involvement of attention. Similarly, sleep and activities of viscera are not mainly controlled by the CS

B. Acquired CS
Example: Conditioned reflex of Pavlov. This type of reflex consists of learning and practicing with the involvement of:
 -Hippocampus for storing the MM for initiation of the reflex.
- vmPFC for the extinction of reflex

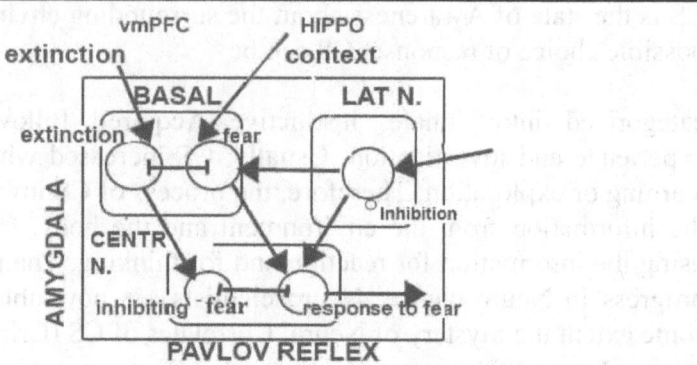

Fig 5.4 Upon receiving the information in the Lateral nucleus, the Basal and Central nucleus are activated with moderation from vmPFC and HIPPO. The Basal and Central nucleus respectively control two centers in the Central Nucleus for fear activation and fear (input from HIPPO) extinction (input from vmPFC). The final responses are sent out from the Central nuclei

- Amygdala containing Lateral Nucleus for receiving sensory input
- Basal Nucleus: Upon the arrival of the sensory input, modulation of the response by the control from Hippo regarding the memory of previously and recently relayed events and remote memory associated with behavioral, social, and conscious experiences from vmPFC.. In the Central and Basal nuclei there is an interplay between fear and extinction of fear (Fear input from HIPPO and Fear extinction input from vmPFC): The successful key in learning is the balance of fear and extinction of fear with reward.
- Central N: output for response

While the innate CS is primarily for simple and vital functions for the body, the conditioned CS is necessary for high-level functions such as learning, playing for development, and adaptation in the surrounding environment for the evolution of species such as learning, fighting to manufacture creating, and imagination. For example, thinking is also a conditioned reflex. Store MM experiences influence this kind of reflex.

One can compare the two above types of CS to the various operation of a popular smartphone: Some processes are built-in/ preloaded by the manufacturer, others are of a high level of function like the game, working program for business are acquired and must be installed by the owner.

V. NEURAL CORRELATE OF CONSCIOUSNESS (CS)

CS is very close to all humans. We use it every day during awakening and even during dreams in sleep. However, we only know a little about it. But we are very sure that it is present in all animals. Investigation and understanding of the mechanism of formation of the CS is of critical importance in neuroscience and understanding of the mechanism of mindfulness Meditation.

A) Principles of the investigation of CS in Neuroscience with the progress in understanding basic principles of the Neurosciences about Memory, and clinical investigation of patients with perturbation of the CS, CS has become gradually attributed to specific brain regions. In addition to studies on laboratory animals, improving imaging techniques, especially with functional MRI and other supplementary techniques, has allowed investigators to look at the functional changes in different brain areas associated with different CS states.

When dealing with the universal force of attraction acting between all matters, we designate gravity as this common force. Scientists designed conventional devices to compare weight, gravity, and force. In comparison to light, electromagnetic energy is made up of photons; we do not understand the structure of gravity. Similarly, the essential nature of the constitutional particles of sound remains elusive to scientists, for the Soul is mainly composed of Memories, CS, with associated different manifestations through emotion, behaviors, and interaction with the environment. We use components of CS every day, and therefore CS is very close to life. CS is the capability of learning, understanding, thinking, imagination, and creativity to adapt to nature. Other related well-documented phenomena are mediumship, Memories of previous lives, and sleepwalking, which are shreds of evidence of the existence of a metaphysical world that isn't easy to dismiss. We are often misled by the conception regarding the relationship of the Soul with the brain and the role of the Soul in extrasensory phenomenons. As a result, Souls appear to the scientists not accessible to the investigation. According to Darwinism, the acquisition of CS is followed by changes, including the physical and metaphysical aspects of the individuals and, to some extent, the environment.

Since antiquity, the nature of the CS has posed philosophers, psychologists, and scientists with curiosities and problems in investigations of human and animal behaviors in normal status, sickness/ injuries, and experimental laboratory conditions. With the advances of science, new techniques of inquiry including imaging techniques of visualization of changes in the brain in various conditions of CS have remarkably shed light on the understanding of CS. In the 1980s, Francis Crick, a Laureate of the Nobel Prize for his and James Watson's discovery of the structure of DNA had turned his interest in

Neuroscience, proposed a question of Neural Correlate of CS in Neuroscience to identify the minimal neuronal mechanisms jointly sufficient for any one specific conscious precept". (Crick 1990).

An example: Research is performed to investigate the lifestyle of an ethnic group of immigrants. If the researcher only contacts a limited number of ethnic immigrants, the information is insignificant. If the information shared by people in contact is significant in content, and if the researcher has experience with similar groups of ethnical immigrants to compare the results, then the end results are of high and significant value. The last step of comparison of the results makes the research more significant in revealing the characteristics of this group of ethnic immigrants.

B. Theories of Formation of CS
1. Global Workspace Theory of Consciousness, (GWS) of Bernard J Baars. (Baars 2005) The theory originates from the analogy between the brain and screen for receiving the projection of images (Newell 1994), CS is the combination of information composed of perception, emotion, and learning related together in the brain. Because of the composition of multiple pieces of information simultaneously received, only those associated with attention are registered as in cases of "the winner takes all". Information without attention is only registered in the Sub or Un-CS

The phenomenon is similar to actors in the stages. Only those with focus, usually 1-4 actors, with focus attention or under bright light are mentioned by the audience.

fMRI or Positron emission Tomography scans demonstrated that the FrontoParietal and Medial Temporal Lobe are activated in the CS processes. In visual CS, areas of the face, light, color, contrast, motion, retinal size, and location,... in the Ventral stream in the inferior Temporal Lobe are activated. Small lesions in these regions only selectively affect the selective CS without affecting the other types of CS. Of interest is that the lesion of the Hippo can influence a large extent of CS because of HIPPO's role in retrieving MM. For the emotional component in the CS, neurons in the Amygdala may be activated. However, the Amygdala does not play a role in MM retrieval (Moscovitch, 2001). In contrast to states of Un-CS and locked-in syndrome, persistent vegetative state, sleep, and general anesthesia are associated with varying degrees of decreased activities in the regions of interest in CS.

2. Consciousness as Integrated Information Theory = IIT
CS integrates information from related cortices and labels the CS by comparison with the reference ICS.

The theory developed after the Global Workspace Theory of Consciousness (GWS) of Bernard J Baars. Tononi characterized the summation of cortical activities as Perturbational Complexity Index (PCI) (Tononi et al., 2016) and used the index to access the cortical changes. Since each cortex area has respective and specific functionality, combining different activities is essential for forming the CS. Example: The Hippo and inferior Temporal Lobe are necessary to retrieve face MM (from face cells), shapes, colors, areas of TE, and TEO in the Ventral Visual stream. The other sites like hMT, IPS, TPJ, and Premotor FEF in the dorsal stream are required in a dorsal stream for motor reflexes to visual stimuli.

In the CS examined with imaging techniques, the cortices are activated in extended areas compared to other states like sleep, anesthesia or dreaming. The FrontoParietal and Temporal cortices consistently display enhanced images. It is worth noting that the PF C predominantly plays important behavioral, social, and managerial roles that control the entire activities of the brain.

In deep sleep of stages NREM 3 and 4, there is no CS: Slow Wave (Sleep) SWS with the following changes:
-Low activity in the FrontoParietal cortex,
-No connectivity with other cortices,
-Limited Thalamocortical connections,
-Limited activity in the sensory cortex when the sense organ receiving stimuli

3. A unified 3D- default space Consciousness
Model Combining neurological and physiological processes that underlie conscious experience In 2015, Jerath proposed a unified 3D- default space Consciousness model that underlines the aspect of previous experience. New information on the general CS type and specific sense organ type is referred to the previous CS stored in the so-called "D Default Space Consciousness" to fill the space and create a seamless CS. The theory is somewhat similar to the concept of ICS emphasized in this book based on the teaching of Buddha to the Public and Ananda in Surangama Sutra. (James 1958, 1900, Holt 1890, Kelly 2007, Meyer 1903,)

Furthermore, even in 1890 James suggested that the brain is not the relay station for transmission to the transcendental sphere. According to James, each person has their own threshold of CS, low or high. Recently. Pin van Lommel, a cardiologist, turned his interest to the study of near-death experience (NDE). To explain the phenomenon of Out-of-body experience (OBE), Lommel suggested the theory of "Unified field of Consciousness: in which the concept of "Non-Locality of CS" is the attempt to explain that CS is not limited to the brain but extends beyond the being connected to the endless or non-local/ higher/ supreme or cosmic, divine/ boundless, transpersonal ultimate, unitary, eternal CS. These terms refer to a powerful source of knowledge. The nature of the source is clarified in Surangama as the ultimate source of CS in Buddhism, known as the original Mind or Buddhahood. In summary, the Unified 3D default space Consciousness model proposed by Jerath only represents the unified concept. But only limited to the brain. The unified theory of James and Pin van Lommel extends beyond the brain, with a connection to the Ultimate CS. The former theory likely ignores the supernatural phenomenon of CS; the latter includes phenomena like NDE, and OBE. However, it falls short of mentioning the Original Mind consisting of the Emptiness that has the potential to generate any Form of this world triggered by arising Transient/ Subtle Delusional/ Erroneous / False thoughts (Vitathavitakka =Vọng Niệm)

4. High Order Perception Theory = HOT and Inner Sense Theory. The HOT was conceived by Aristotle, Descartes (1641), Locke (1711) Kant (1787). The CS is connected to the High order of CS through a connection called Transitivity Principle(Carruthers 2005; Rosenthal 2006; Lau and Rosenthal 2011, Fauchon 2019, Rosental 2005, Lau 2010, Weiskrantz, L. 1997, Lycan 1996, Diene 2008, Carruthers 2000, Pasquali 2010, Gennaro 201, Dretske 1995, Block 2007, 2009,).

The Inner Sense Theory is similar to the concept of Inner CS emphasized in this book. The theory is fervently supported by DM Amstrong (Armstrong 1980) and Willian G. Lycan (Lycan 1996).

The above two theories emphasize the role of a high order of CS as the ultimate order forming the CS. As a result, the five sensory organs or the brain only play the role of relay center for transmission for the formation of the CS.

5. Other Supplementary Theories of CS

The two theories of CS: GWS and IIT, are two mainstream theories but insufficient to clarify exceptional cases of CS as seen in the following circumstances

a) CS in Dreaming. Role Acetylcholine

NREM 3-4 sleep has no CS and no dreams. In the REM stage (stage of sleep 5), there is no increase of nor-epinephrine in the sleep, but there is an increase in acetylcholine from the Forebrain and the brain stem LDT. Due to the role of CS in dreams, acetylcholine is accounted in paradoxical sleep (sleep but with CS). Furthermore, REM sleep is associated with gamma waves seen in the wakeful state.. Acetylcholine is responsible for wakefulness that the ARAS system, TRN, and the Thalamus, with an abundance of thalamocortical connections, usually control. Acetylcholine is the important neurotransmitter along with Glutamate in the link of the BF with the entire cortex necessary for wakefulness and the attention

b) CS without Cortex (Bjorn Merker)

i. Hydranenephaly, Merker is a neuroscientist who studied psychology and neuroscience at MIT, USA. He was interested in work in Midbrain. A study of CS in cases of hydrocephalus showed the evidence of CS: Turning the head to lights or objects, understanding some word said to them, emotionlike cry, smiling, fear... ranging from distress, contentment, pleasure, and joy with an emotional expression like laughing giggling crying. The minimal CS is thought to be attributed to the Midbrain structures like colliculi Red nucleus hypothalamus, the Substantia Nigra, and the Zona Incerta rather than to remnants of the cerebral cortex (Merker 2007, Aleman 2014, Marin-Padilla 1997, Takada 1989, Baars 2003, Sporns 2000)]).

Cases of patients with hydrocephalus with paper-thin cortex **but an IQ average or above the normal**. The cases illustrate that semantic MM and related CS are not only stored in the brain but also stored in the Soul. The Soul uses the brain (particularly DMN) to connect with emotional and motor centers for behavioral, verbal expression, and motor function.
https://www.sciencedirect.com/science/article/pii/
B9780444529770500103/imag

ii. **Sprague effects**: The Effects were seen in cats with unilateral resection of the Visual cortex. The cat is blind in the contralateral side. However, with resection of the superior colliculus on the other side, the cat regained vision for orientation. Similar Sprague effects are also seen in humans. The physiopathogenesis is the abolition of the inhibitory mechanism exerted by the contralateral colliculus on eye movements (Krauzli 2013, Weddell 2007, Jiang 2009, Lomber 2007) (H4.7). The inhibition of the contralateral SC is necessary to ease the function of the ipsilateral SC in the control of eye movement

iii. In insects: In insects: the centers of CS are neural ganglia. Therefore, CS is somewhat preserved even after the removal of their head.

iv. In craniotomy for epilepsy with local anesthesia, Penfield and Jasper noticed that removing part of the Frontal, Parietal, and Posterior cortex of one side, even complete one hemisphere, did not cause the loss of wakefulness. Of course, the removal of both sides causes the loss of CS.

In temporal lobe epilepsy, EEG c shows evidence of electrical waves originating from the brain stem interfering with the ARAS (that usually activates the cortex) with subsequent loss of CS. The phenomenon is consistent with the function of the brain stem (ARAS) in maintaining wakefulness

(Penfield 1958, Selimbeyoglu 2010 Englo, 2010) (Fig 5.5)
Zhong, Y.M.,Rockland,K.S.,2003.Inferior parietal lobule projections to anterior inferotemporalcortex (areaTE) in macaque monkey.Cereb.Cortex13.

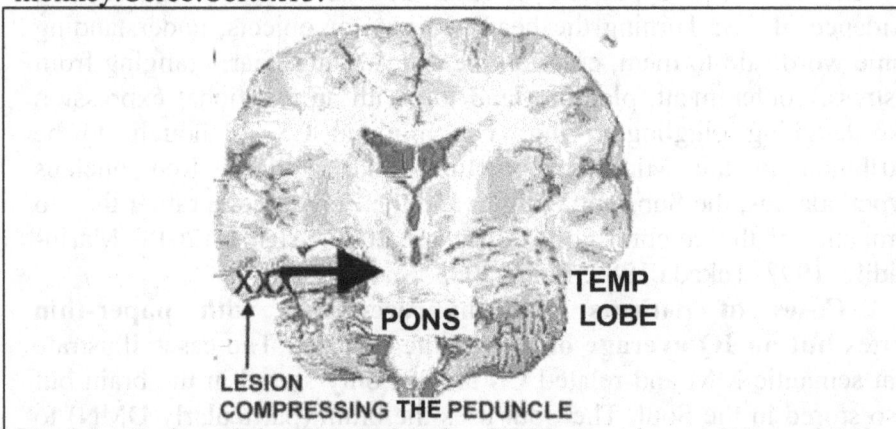

Fig 5.5. Lesion of the Temporal Lobe (XXX) can trigger Temporal Lobe Epilepsy/TLE. TLE may inhibit the brain stem, Thalamus and Ascending reticular Sytem and cause the loss of Wakefulness.

v. Artifical Inteligence/ AI

It is evident that in AI, parts of the cortex are represented by interconnections of different electronic components. Ultimately, these interconnections are the most ideally conceived product of human CS that does not deviate or is minimally obstructed by chance, emotion or other environmental factors.

6. Other new philosophical concept of CS

a) Minimal Unified Model (MUM) (Wiese 2020). In 2019 Kanai and 2020 Wiese proposed a theory in which the incoming information generates a model of CS considered a Minimal Unifying Model (MUM) that entails the minimum requirement for the determinative generation of CS based on recent and existing theories of CS. The minimum requires the input information consisting of intention, imagination, planning, shortened MM, attention, curiosity, and creativity. Therefore, the model requires continuous updating and refinement to generate the internal model, which endows the organism, the non-reflexive, flexible, and intelligent behavior. This *"information generation" (coined by the authors)* is consistent with the top-down process, similar to the Creation process of ICS.

b) Relativistic Theory of Consciousness
Lahav, Nir and Neemeh Zachariah from the Department of Physic and Philosophy of the University of Memphis, TN, USA proposed
this theory. In this theory, in the examination of the CS of Alice, who is in a fixed position, and Bob in a car moving toward Alice. For Bob, Alice is moving toward him. The authors proposed a mathematical equation for the expression of this relative phenomenon. The purpose is to point out that CS is a relative phenomenon. CS, as conceived, varies according to the context, and its generation depends on the prior information obtained by each individual. The preceding information is equivalent to the ICS

c) Consciousness as a Memory system (Budson AE 2022)
CS is only generated about 500ms after the perception of the information input, while a reflex reactive to the visual input is much faster. Therefore certain reflexes are unconscious such as closing eyes when a fly flies toward the eye. In other cases, reflexes, such as hitting a tennis ball, come late after the CS is generated. In latter cases, the input information is referred to preexisting MM of a different type, such as episodic semantic and implicit MM, rearranged with flexible combination to form CS with subsequent production of various functions that may or may not be relevant to the recent information. The authors called the relevant MM a *"Conscious MM system"*

In summary, in 1990's, Francis Crick posed the question of "the neuronal correlate of Consciousness," defined as the minimal neuronal mechanisms jointly sufficient for any one specific conscious percept. Numerous studies have successfully revealed the neural mechanism of the formation of CS involving

-The reception of the exteroceptive (external, environment), interoceptive (internal organs), and proprioceptive (sense of action, location of the body) information
- The integration of the information into the cortices
= The requirement of attention for the information, status of wakefulness

Lately, the role of the inner CS that constitute a model of reference for new information has neen recognized. This model is termed High Order Perception or Inner Sense
(Rosenthal 1986, 1993, 2005 Dretske 199,) Seager 1994, Byrne 1997, Jerath 2018, Wiese 2020), and Conscious Memory system (Budson 2022). In the concept of the inner model of CS, the Memory of explicit and implicit plays a critical role in generating the inner model of CS, the the main component of CS.

Although CS belongs to the metaphysical realm, it interfaces the physical and metaphysical worlds. Therefore, it is reasonable to conceive that CS may be related to some aspects of the physical structures. In Paragraph of the Soul, it is proposed that neutrino may play the role of an elementary particle of the CS

VI. Theory of Dorsal and Ventral streams
A. Dorsal and Ventral Pathways of Conduction (Fig 5.6,7,8,9)

The concept of Dorsal and Ventral streams is consistent with the concept of duality in this world. The brain is created to manage the mobility of organs and voluntary muscles. On the other hand, the CS is ultimately generated by the Original Mind, which is the Buddhahood. Inputs from the sense organs have to be relayed to the brain to elicit the movement of parts of the body, known as conditioned or unconditioned reflexes. As a result, the input/ information is filtered and distorted by the brain before reaching the Ultimate (original) Mind/ Buddha.

1. (Top Down) The Dorsal Stream is related to the Motor-Premotor cortices

The pathway originates from the visual cortex to the Superior Parietal Lobe (SPL) to FEF for control of the eye field movement. (IPS is associated with perceptual-motor coordination of eyes for reading and hands for procedural movement. FEF is related to the direction of the eyes)

a) **Commonly, this pathway is for reflex movements.** For example, a foreign thing flies toward the eyes, inciting the movement of eyelids, head, or arms to protect the eyes.

Because attention is not yet involved, and the pathway only involves the motor cortex, there is no CS.

    b) **In attention**, such as in learning
- the pathway controls the motor centers that focus attention on the head, eyes, ears, extremities, and CS. Due to the extended attention time, the Ventral Stream is involved in the attention. This will create a CS that is rich in detail.
- attention to the target; for example, in learning or in case of a predatory animal looking at prey: this is subjective mindfulness attention that depends very much on Consciousness obtained from previous knowledge and experience and not much on Awareness.
- in case of attention to large areas, the Awareness is involved and may provide new data. Therefore, the subject can change the focus of attention

    The shifting of the attention focus consists of two steps.
      . Leaving the current focus
      . Redirecting the attention to the new focus controlled by the Dorsal pathway (the SPL

    c) **Volitional navigation**
    The navigation is a complex process requiring the map of navigation in the HIPPO, **ParaHIPPO Place Area/PPA** (landmark of the map: for example a very tall building...) **and RSC** (likely for reference frames for comparison between new frames and stored frames.

    d) **Spatial cognition**
while the direction of Right-Left and Superior-Inferior is provided by the Retinotopic map (two dimensions); the depth that is crucial for the understanding of space is provided by neurons in MT area (binocular disparity and motion parallax [near objects are displaced on the retina more than far away objects].

    e) **In Meditation,**
this is the pathway of mindfulness. Contrast to b) the focus of attention is narrow, simple, but extended with the duration of Meditation. This will result in the participation of Ventral Stream, which provides an additional choice of information to the IPS. The IPS can discard the ventral information or switch the attention to a new focus from the Ventral pathway. In the latter case, a wandering Mind occurs in Meditation.

2. (Bottom Up) The Ventral Stream :

Is a critical system but often needs to be more appreciated.

It is a system based on Awareness, closer to Buddhahood than it counterpart, the CS. The pathway is responsible for an overview of the entire environment and quality close to its true nature.

Primary sensory cortex through the cortex of the Temporal Lobe for CS formation, usually after the occurrence of reflex movements', usually 100-300 msec.

This pathway originates from the primary sensorial cortex (visual, auditory, tactile...) and goes forward to secondary sensory cortices of the respective sense. The pathway involves TPJ, the supramarginal Gyrus, and the Ventral Prefrontal cortex. TPJ is connected with the thalamus, the limbic system, and the visual, auditory, and somatosensory systems. The TPJ and particularly the IPS also integrates information from both the external environment as well as from within the body.

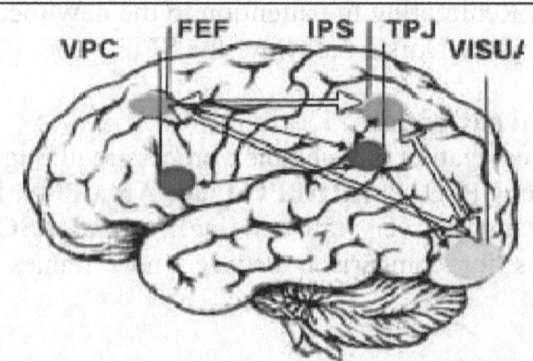

Fig 7.2 Pathways in the Attention: FEF-TPJ IPS VPC.
Open arrows: Dorsal pathway. Thin solid arrows: Ventral pathway
FEF: Frontal Eye Field, VPC: Ventral Prefrontal cortex, IPS IntraParietal Sulcus, TPJ TemporoParietal Junction

3. The IPS is the site of integration of the two pathways,

Dorsal and Ventral. *As a result, IPS is likely the area where the selection of the new focus of attention is made and is the main culprit of the Mind wandering in Meditation. In Meditation on the Sound, according to some meditators, the inner sound current soothens the brain, likely inhibits this region, and is very helpful in calming the Mind.* Other areas like TE (connected to IPL, Prefrontal and TEO are connected to IPS. In turn, IPS and Parietal Lobe are connected to

Middle Temporal Gyrus/ Fusiform Gyri and Supramarginal G. In addition to sensory input, preattention, and intention from the Salient

network, the ventral pathway controls the dorsal pathway. This pathway of Awareness plays the role of adjusting and accommodating attention appropriately via the IPS.
   a) Commonly this pathway lasts as long as the attention
   b) Because much attention was used in the Dorsal stream, the areas of attention are much larger than in 1b), the acquisition of data is broad but superficial. In learning, the attention is overwhelming, and the data acquisition is entirely subjective ("we see things as we are")
   c) In Meditation, this is the pathway of Awareness, attaining omniscience/ enlightenment. The attention gradually decreases its importance in acquiring data; therefore, the data acquisition is slowly objective (we see a thing as it is). (please see Figure 1.

In the formation of CS, the two streams, Dorsal and Ventral, must control each other, particularly the Ventral stream that feeds the Dorsal stream at IPS.

4. Different Sensory Pathways
a) Visual MM

- Dorsal stream- *CS/ Mindfulness) V1→hMT→PPC→motor cortex→Frontal Eye Field:* for eye movement (HOW?) for defense reflex, hMT: The extrastriate area MT [V5 (hMT+) in humans] located in the "dorsal pathway" of the primate brain is specialized in the processing of visuospatial coding of movement direction for visual motion information , i.e., in the control of visually guided hand movements

- Ventral pathway: an overview of the environment *(Awareness)*, involving large areas of the temporal lobe for the acquisition of CS:

   - Successively from V1 → V4 → Medial Temporal Lobe (MTL), TEO,TE (three dimension shape), AMYGD/fear, N. Accubens/joy: for emotion in vision (WHAT?); (TE, TEO: particular areas for visual CS about to the shape) (note: TE in connected with IPL)

   → *Pulvinar/ Superior Colliculus for location (WHERE).*

   → *Finally the information transferred to Inner CS (Carruthers 2005; Rosenthal 2006; Lau and Rosenthal 2011)*

   - The images are kept in the Visual cortex V1 for a short time, under 30 secs. Studies on Visual MM showed that the MM is stored in vmPFC and IF junction (Pars triangularis=BA44). In children, the lateral visual and Fusiform gyrus are related to faces, objects, shapes, and colors.

   - Furthermore, the HIPPO is related to the geographical map and location.

- Fusiform gyrus
-Lesions of the Occipito Temporal lobe can cause Agnosia, most commonly Visual Agnosia. But Agnosia can affect auditory tactile senses. Visual Agnosia comprises two types: apperceptive (perception/preliminary, such as the object's location) and associative (meaning). Agnosia can be specific to certain visual classes like: prosopagnosic (the most common class with an inability to recognize face due to damage of the fusiform face area/FFA).

Other types of Agnosia are Achromatopsia (color blindness, pure Alexia *Word Blindness), blindness to spatial layout, and gestures. Body language: The patients are otherwise neurologically normal.

- Nucleus Accumbens/NAc, AMYGD, PFC, Substantia innominata of Meynert (related to MM Amygdala), auditive, gustative and touch cortices..., vmPFC, TPJ, Supramarginal (sounds, voice, and empathy).

b) **Auditory pathway**. (clear arrows) https://human-Memory.net/somatosensory-cortex/Auditory pathway: Dorsal and Ventral.

Dorsal stream A1, A2 connected to IPS/ Intra Parietal Sulcus corresponding to the head top (Rauschecker2011) → Promotor →DLPFC, cortex for phonation, attention, vlPFC. In Meditation, the inner sound appears to arise from the head top. From IPS, the auditory stream reaches the motor cortex, which accounts for the transitional rhythmic movement of the head Ventral stream → vlPFC (BA45/47)→dlPFC , AG (angular gyrus), PFC Amydala (Fear) and Nac (reward)

c) **Touch MM**: (thin arrows) Dorsal (PFC) and Ventral (Amygdala/ vn: Hạnh Nhân) (Fig 5.8,11)
(Gardner2008.
*h*ttps://www.researchgate.net/publication/285222357_Dorsal_and_Ventral_Streams_in_the_Sense_of_Touch , Camlier 2012)
Sensory cortex S1 →S2
- Dorsal stream (Touch/ Haptic Memory): TPJ and superior Parietal, texture: Parietal operculum)
- Ventral stream from S1 → S2 Parietal opercula → PCC. → Broca area 47/45=Pars Triangularis +Parietal operculum +Nucleus accumbens/ Nac (reward), Amyg/ Fear. (Kostopoulos 2007).

PULV: Pulvinar of Thalamus for Vision, TRN: Thalamic Reticular Network, Nucleus Accumbens/ NAc: Nucleus Accumbens reward,

joy, AMYD: Amygdala/ Fear, SC: Superior Colliculus relay center for visual pathway.
MTL: Medial Temporal Lobe, V1,2,3... Visual cortex, SI: Substantia Innominate learning/ ACC: Anterior Cingular Cortex/ error detection, OFC, PAG/ emotion
FEF: Frontal Eye Field: eye movement, dlPFC: General management, TE/TEO: special areas for vision.

hMT of the extrastriate area MT [V5 (hMT+) in humans] located in the "dorsal pathway" of the primate brain is specialized in the processing of visuospatial coding of movement direction fö visual motion information, , i.e., in the control of visually guided hand movements.
S1,2 Cortex to sensory cortex level 1, 2.
40: Broadman area BA40 Supramarginal gyrus of Parietal cortex, Inferior Posterior Parietal cortex,.language, reading
47, BA47 Orbital Area Music Language
BA44,45: phonation

5. Diagrams
a) **Visual pathway: Fig 5.6**
The ventral pathway: for Consciousness (emotion, face, color, form)

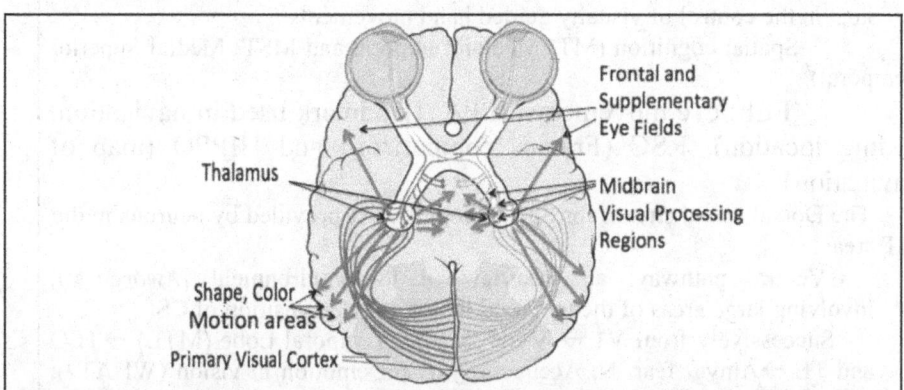

Fig 5.6 Visual pathway connected to Superior Colliculi and Lateral Geniculate Body the to TPJ (colors and forms), motor cortices,

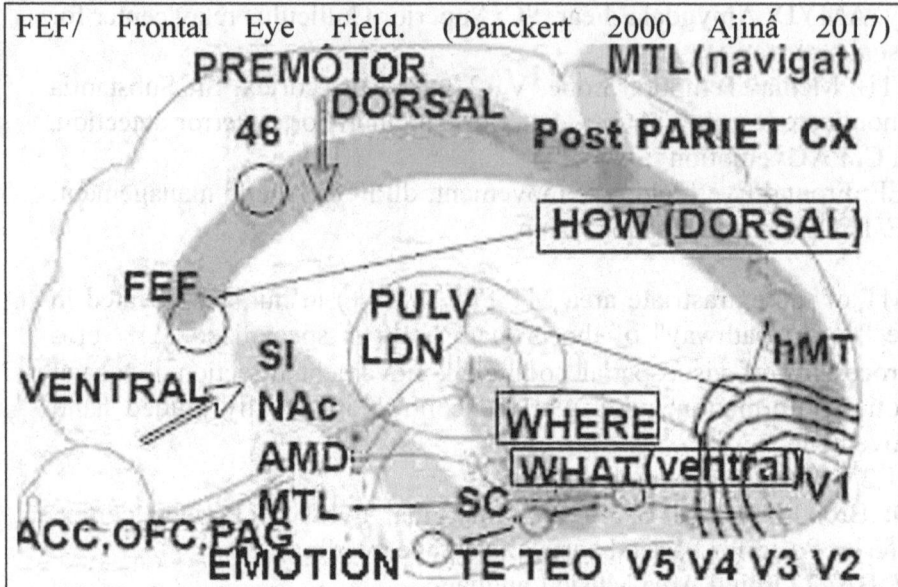

- V1: Retinotopic map of space. This is not enough for CS because CS needs to know about the relationship with areas not seen at the same moment
- Dorsal stream- *CS/ Mindfulness) V1>hMT>PPC>motor cortex>Frontal Eye Field:* for eye movement (HOW?) for defense reflex. hMT: The extrastriate area MT [V5 (hMT+) in humans] located in the "dorsal pathway" of the primate brain is specialized in the processing of visuospatial coding of movement direction for visual motion information, , i.e., in the control of visually guided hand movements

Spatial cognition (MT/ Medial Temporal and MST/ Medial Superior Temporal)

FEF: eye movements, PPA (landmark used in navigation, coding location), RSC (Frames comparison) and HIPPO (map of navigation)

The Dorsal visual stream for Spatial cognition is provided by neurons in the MT area

- Ventral pathway: an overview of the environment *(Awareness)*, involving large areas of the temporal lobe for the acquisition of CS:
  - Successively from V1 → V4 → Medial Temporal Lobe (MTL) →TEO and TE →Amyg/ fear, N. Accubens/ joy: for emotion in vision (WHAT?); (TE, TEO: special area for visual CS)

  → face cell area----color, form, emotion (cognition)
  - → Pulvinar/ Superior Colliculus for location (WHERE). → >> Finally the information transferred to Inner CS (Carruthers 2005; Rosenthal 2006; Lau and Rosenthal 2011)

Inattentional blindness/ Blightsight/ hemispace neglect is observed when the ventral visual stream is not probably working due to lack of attention or a temporal lobe lesion. In this syndrome, the dorsal stream is intact. In the Hemi Neglect syndrome, the lesion in the

Right temporal causes the loss of visual CS from the LEFT side input with loss of vision on the Left side. The dorsal stream is for reflex reaction to L hemispace vision.

It is important to note that all information, regardless of its source (from eyes, ears, skin etc.), is specific information-related CS available to other types.

b) Auditory Pathway Fig 5.7,8

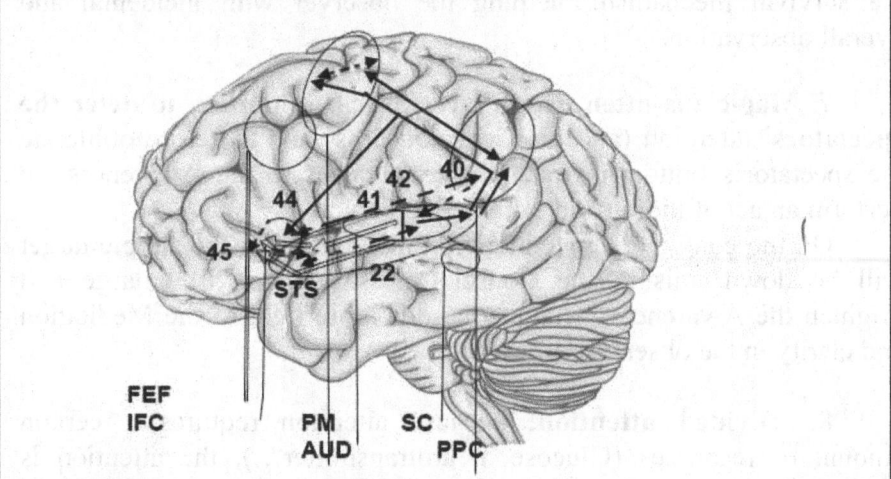

Auditory pathway. (clear arrows) https://human-Memory.net/somatosensory-cortex/Auditory pathway: Dorsal and Ventral.

Fig 5.7 Dorsal stream A1, A2 connected to IPS/ Intra Parietal Sulcus corresponding to the head top (Rauschecker2011) > Promotor >dlPFC, cortex for phonation, attention, vlPFC. In Meditation, the inner sound appears to arise from the head top. From IPS, the auditory stream reaches the motor cortex, which accounts for the transitional rhythmic movement of the head

Ventral stream → vlPFC (BA45/47)→dlPFC, AG/ angular gyrus, PFC Amydala (Fear), and Nac (reward)

c) Touch Pathway and Proprioceptive. Fig 5.8

6. Two pathways in the Sensory System of Attention in the Meditation:

As described previously in this chapter

The visual, auditory, and touch sensory systems are all composed of top-down and bottom-up streams.

a) The top-down pathway is more related to CS (because CS is only created when mindfulness is involved). Initiated about 100ms, but lasting CS is made in the brain and consists of Memories/ information of the present and even of previous lives. It is usually considered as the veil of ignorance that obliterates the Buddhahood and Awareness.

b) *The Bottom–up system* is more related to Awareness, less influenced by experiences, reflecting more closely to the Original Mind/ Buddhahood. This pathway comprises of the fusiform gyrus, inferior Frontal Lobe, and TPJ. The pathway is responsible for the overall view of the background and providing information on the environment around the target. This pathway is objective, exogenous attention, creating CS about 500 ms after the dorsal top-down pathway. This is the survival mechanism, helping the observer with incidental and overall observation.

7. **Magicians often use this reciprocal inhibition to deter the spectators**' attention (and the Consciousness) and therefore obliterate the spectator's bottom-up pathway (equivalent to the Awareness) to perform an act of hiding and exchanging objects.

On the contrary, in Meditation, the attention to the simple target will be down, raising the Consciousness, which will enlarge and brighten the Awareness. This result adds more light to the Meditation and clarity in the observation.

8,. **Divided attention**: because attention requires a certain amount of resources (Glucose, neurotransmitter…), the attention is limited

    a) involving two or multiple targets of different sensory cortices like Visual and hearing: easy, even natural to perform: one can see and hear at the same time

    b) involving the same type of sensory cortex: difficult to perform and requires training.

The control center is the dlPFC.

- **COMBINED DIAGRAM OF THREE TYPES OF CONSCIOUSNESS/MM AND RELATED ENTITIES (Fig 5.9)**

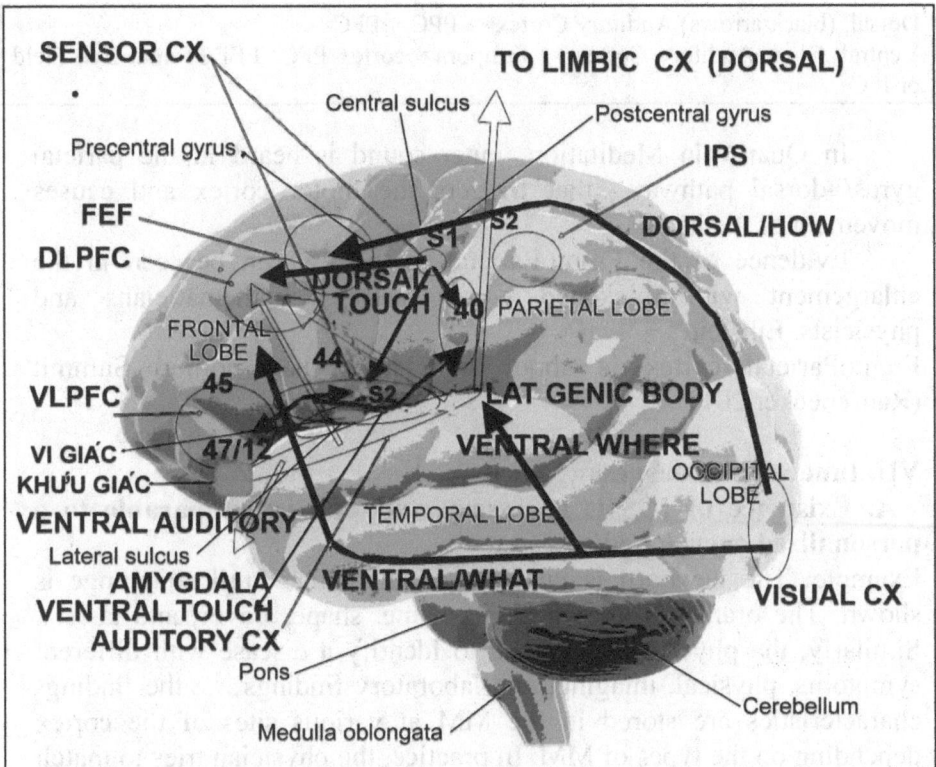

Fig 5.8 Diagram of pathway for **Vision** (bold black), audition (clear) and Touch (thin black)
PCC/ posterior Cingulate Cortex (Autobiographic MM), IPS/ IntraParietal Sulcus (Parietal-Temporal-Occipital), PPC/ Posterior Parietal Cortex, FEF/ Frontal Eye Field= eye movements) AMYG/ Amygdala (Fear), Inferior Temporal Lobe: emotion and shape in vision, SC Superior Colliculus for the control of the eye. AMYG or Nucleus Accumbens/NAc then to PCC, ACC to identify the emotion
Note both ventral and dorsal pathways are connected to PCC (Katsuki 2014,Mohanty 2013**Touch** (thin black) **Touch MM**: (thin arrows) Dorsal (PFC) and Ventral(Amygdala/ vn: Hạnh Nhân)
(Gardner2008.
*h*ttps://www.researchgate.net/publication/285222357_Dorsal_and_Ventral_Streams _in_the_Sense_of_Touch , Camlier 2012)
Sensory cortex S1 →S2

Ventral and Dorsal Visual Pathways
Dorsal stream (Touch/ Haptic Memory): TPJ and superior Parietal, texture: Parietal operculum)
Ventral stream fron S1 → S2 Parietal opercula → PCC. → Broca area 47/45=Pars Triangularis +Parietal operculum +Nucleus Accumbens/ Nac (award) Amyg (Fear.) (Kostopoulos 2007).
Dorsal and Ventral Auditory Pathways:

> Dorsal, (black arrows) Auditory Cortex>> PPC>>PFC
> Ventral A1 or Auditory Cortex >>Tempora>>cortex PFC   FEF/Frontal Eye Field or IFC

In Quan Yin Meditation, inner sound is heard in the parietal gyrus/ dorsal pathway that triggers the motor cortex and causes movement of the head.

Evidence of the FrontoParietal in the CS can be seen in the enlargement with thickened cortex among mathematicians and physicists. Einstein's
FrontoParietal cortex is about 15% larger than normal Summit (Rauschecker 2011

VII. Inner Consciousness

A. Existence of ICS is necessary and ICS is comparable to a personalized encyclopedia

Example: The newborn is unaware of an orange until an orange is shown. The orange is known by its name, shape, flavor, and taste.... Similarly, the physician is trained to identify a disease with different symptoms, physical, imaging, and laboratory findings, ... the findings characteristics are stored in the MM at various sites of the cortex depending on the types of MM. In practice, the physician tries to match the current characteristic features with those in his ICS and in books to arrive at the diagnosis. Before showing the orange or the patient to the baby or medical student, all the information has no significance and is discarded or stored in the Un-CS. (Adler-Neal AL 2017, Bansal A, 2016)

B. Types of ICS

Therefore, the final and essential step in forming a CS is referencing the information input to the ICS for comparison. Information similar to those in ICS is labeled as such. New information not existing in the ICS is added to the ICS. In this final step of the formation of CS, there is the involvement of :

- MM stored in ICS, the sensory cortex and the Soul. As previously discussed, the Soul not only shares the MM with ICS but also facilitates the retrieval of MM.

- Detection of the difference between the information in the input and the ICS by the dACC and Zona Incerta

- Connection of the Soul to the Original/ Fundamental Mind/ Buddhahood/ Holy Spirit. The significance of this process has been discussed in the previous paragraph.

According to the Vijnaptimatrata-Trimsika and the Studies of Consciousness, CS can be categorized into three types of knowledge based on the ICS.

1. Parikalpita/ vn: Biến kế sở chướng is wrong discrimination of judgment due to the fabrication from a biased individual Mind and illusional imagination. The views are considered discriminative by the public.

If the ICS is poor in quality, the resulting CS is wrong and biased (in the above example, MM and reference book are inferior in quality, containing inaccurate, insufficient information: the diagnosis is wrong). The encyclopedia is inaccurate, biased, and distorted, leading to the wrong, biased, distorted CS. For example, the yellow color is indicative of the color of death. Similarly, suppose the MM stored in the brain is incorrect. In that case, the Consciousness is biased or based on bad judgment or an imaginative construction of the view that regards the misconception of things as accurate.
Consistent with incorrect, distorted encyclopedia

2. Paratantra (Relative knowledge/ vn: Y tha Sở Tánh):
Suppose the MM and reference books are close to objective but need to be revised in quality and reflect the up to the level of true nature of reality. This is the knowledge commonly known as the common sense of the public. The encyclopedia is consistent common knowledge widely known as the public's common sense.

The two above types of CS are consistent with the concept that "we see things as we are." The view which sees Things as derived. That is not the view which sees things in their true nature.

Literally, "depending on the common sense of the public" is knowledge based on some facts commonly sensed to correspond with the fundamental nature of existence. The characteristic feature of this knowledge is formulated from generation to generation of human beings in the Creation, which is an illusional/ upside down/ inversion world. This world commonly displays illusional views.
- There are seven illusional/ evil views (vn: thất Điên/ skt Viparyaya: Wrong/ Evil / illusional/ misleading views on
1) Permanence   2) Impermanence,
3) Egoism,   4) Non-Egoism,
5) Emptiness   6) Purity and Impurity,
7) Worldly happiness and unhappiness

- There are eight inversions or upside down/ Heretic views/ vn: Bát đảo: wrong views on
1) Permanence   2) Impermanence
2) Non-Egoism   3) Egoism
3) Purity/Tịnh in Nirvana 4) No Purity in Nirvana
4) Pleasure in Nirvana   8) No Pleasure in Nirvana
Consistent with encyclopedia of common sense but still incorrect

3. Parinishpanna

Perfectly-attained knowledge:
Right/ Perfect Knowledge and Suchness/ vn:Viên Thành thật (Tathata) (Samyagjnana) of the five Dharmas, consistent with the concept "we see things as they are." It is available as one reaches the state of indestructible Buddhahood, self-realization beyond Names and Appearances, and all forms of discrimination or judgment.

The thing is perceived as we are. This type of CS of Suchness/ Tathata is only obtained when one reaches the state of Buddhahood, beyond the level of Formlessness, without a Mind of discrimination, and where there is no death, birth or destruction. This suchness is not perceived through the six senses or cerebral mechanism of integrating the information for CS but is perceived directly at the Buddhahood/ Parinishpanna level. The six senses, brain, and veil of ignorance (Parikalpita+ Paratantra) distort the information coming from the five senses (However, the brain and the body are necessary for the spirit to communicate with the physical world through the motility-like mental expression, talking, and movement of all parts of the body including viscera.
The encyclopedia is correct.

VIII) ACC is the anterior part of the Cingulate Cortex

a) ACC has connections (Fig 5.12) with PCC (autobiographic MM), MTL (Semantic MM), mPFC (social interaction), OPFC (high-level social interaction, Mental value), Amygdala (anger/ fear), Zona Incerta (detection of error). Since the brain is comparable to a box of prediction: when there is a piece of preliminary information (such as a sound from the door), the brain sets up a mechanism to predict the type of information (that could be the wind, somebody knocking at the door, animal hitting the door). After checking at the door, it turns out that the wind is causing the noise. ACC identifies the error by distinguishing the information's strength (intensity). The strength is strong or weak depending on the similarity with the predicted information. This world commonly displays illusional views. (Bush 2000, Posner 1998, Lrru 2004, Carter 1998 Holroyd 2004, Gehrin 1993, Carter 1998, Stern2010, Carter 1998, Van Veen 1998, Alexander 2017, Orr 2012.)

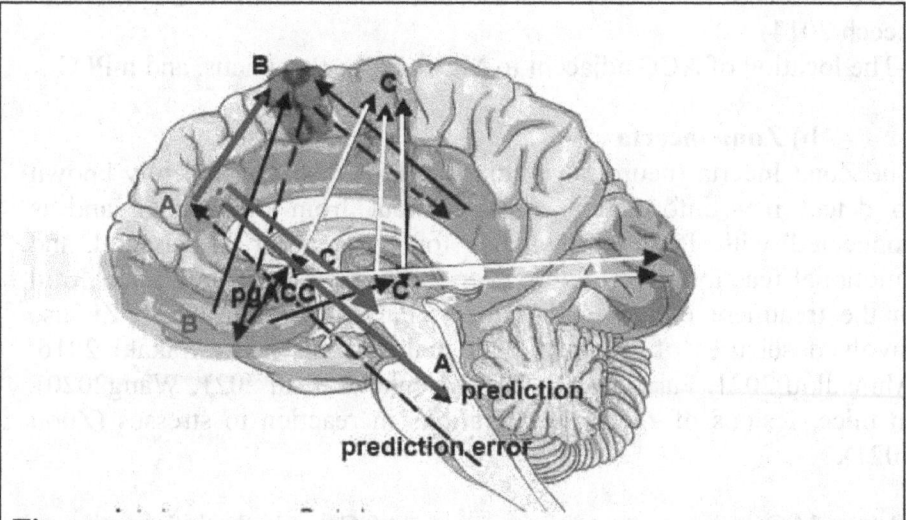

Fig 4.9 Three phases in the formation of the CS: interplay between ACC, Motor cortex (for action), and PFC (for moderation/ advice) with the participation of PCC and Sensory cortices (Vision, hearing, touch...) (for recognition of the reality)
A) Solid thick arrows: after receiving the information input, ACC sends the signal to the cortex. A, peripheral organs, and Hypothalamus for prediction and automatic emotional reaction
B) Discontinuous arrow to vmPFC, PCC for consultation, and solid thin arrow to the motor cortex for adjustment
C) pgACC: Clear arrow for the last prediction . Final information is sent to sensory cortices

ACC does not store the Memory very much but plays a role in detecting errors of the coming information compared to the ICS.

ACC can be a candidate for center detecting the similarities and differences between the coming information and the MM in ICS. Therefore, ACC is crucial in the labeling of the formation and in the construction of CS. (Paul E. Tibbetts. *Brain and Mind* 2 (3):323-341 (2001); C. Alan Anderson J Neurol Neurosurg Psychiatry. PMCID: PMC1083504
Frank R. Freemon PMID: 5158785; PE Tibbetts · DOI:10.1023/A:1014446623476)

The proposed theory is supported by the following:

-Connections of ACC with center involving in the formation of CS Thalamus, Hippo vmPFC ACC, ZI, and widespread cortices. PCC MCC (rarely involved by cerebrovascular accidents) are involved in the pathogenesis of Alzheimer's disease, attention, Autism ADHD (Schizo-Leech 2014)

-The location of ACC adjacent to MCC, PCC, Precuneus, and mPFC

b) Zona incerta

The Zona Incerta (neurotransmitter: GABA) has been recently known to detect new information, receive input from the dACC, and is connected with PAG, Amygdala, for an appropriate visceral and emotional reaction. In addition, deep brain stimulation of ZI is helpful in the treatment of essential tremors, Parkinson's disease. ZI also involved seizures of types of petit mal and absence (Arakaki 2016, Ahmadlou 2021, Farahbakhsh 2021, Venkataraman 2021, Wang2020). In mice, lesions of ZI cause alterations in reaction to stresses (Zhou 2021),

When dACC detects new information, dACC sends the information to ZI, which in turn activate PAG for appropriate emotional reaction expressed as joy (Enphakin), pain, or other emotional reaction (Ahmadlou 2021, Farahbakhsh 2021), and to Amygdala for fear, anger, with the participation of Hippo for retrieval of MM (Venkataraman 2021, Wang2020). In Mice, lesions of ZI cause the alteration of reaction to stress (Zhou 2021), likely related to detecting new information. Deep stimulation of ZI is helpful in the treatment of essential tremors, Parkinson's disease.

IX. Evidence of ICS

1. **ICS is the store of data preexisting** before receiving new information. With the principle of Original Mind/. Buddhahood of omnipresence and omnipotence, ICS also has the Original Mind but is obliterated by the veil of ignorance. The nature of the veil of ignorance is represented by the accumulation of the inputs from this world considered to be illusional, due to the distortion regarding the concept of ego versus egolessness, permanence versus impermanence..... The veil of ignorance does not block but distorts the input. The effect of the veil of ignorance makes the changes in which we do not see things as they are, but we see things as we are.

2. Mechanism of formation of ICS

For a long time, there has been evidence of a correlation between images in the retina and that in the V1 of the visual cortex that represents Mooney's low-resolution level. In 1957, Craig M Mooney introduced a face test (for children) to recognize the natural face by showing a series of faces drawn with only a two-tone method and no tone. The face will be recognized with a strong signal in fMRI of the V1 area if the face is known in advance. The significance is that the face MM is retrieved from the ICS to be compared with the current information; this will result in overlapping the recent and previous images and rendering the recent images identified with enhanced images detected with fMRI. Mental imagery is the brain's capacity to revisit the stored images in the ICS. In the process of imagery, the information pathway is opposite to that of the vision.

These findings were obtained in the experiment with fMRI and Multi-Voxel Pattern Classification (MVPC: special neuroimaging technique allowing the detection differences between conditions with high sensitivity): **the experimentee is shown a number of images and then contrast images in which the images are not pictured. The experimentee is asked to visualize the pictures in contrast images. The significance is that the viewed images are stored in the ICS and are retrieved in the imagery process to visualize the picture in the contrast images. fMRI and MVPC reveal the visualization of the picture of the contrast image represents the same pathway but in reverse direction.**

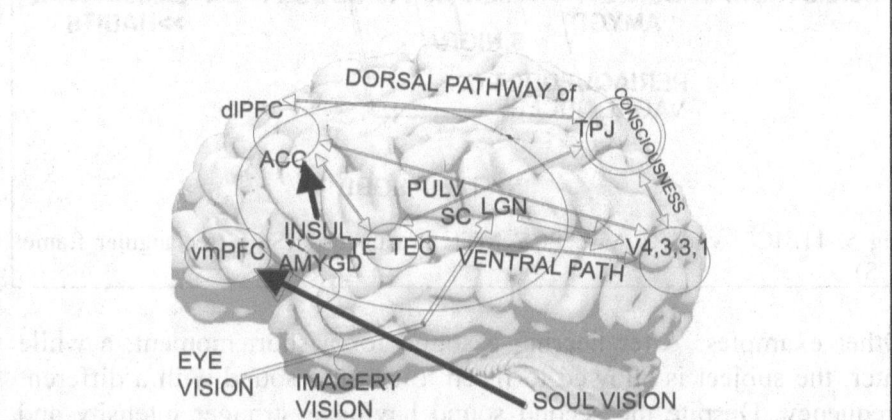

Fig 5.10 Three mechanisms of vision using the same pathway but with different directions.
 -**Eye vision**: → Lateral Geniculate body (LGN) →Occipital Visual cortex --: dorsal stream→PFC, and → Ventral stream → Temporal Lobe .
 - **Soul**: or vision without eyes →PFC → Occipital Visual cortex.
 - **Imagery** Salient network (Dorsal ACC –Insula) stimulating PFC/ dlPFC then using Soul pathway.

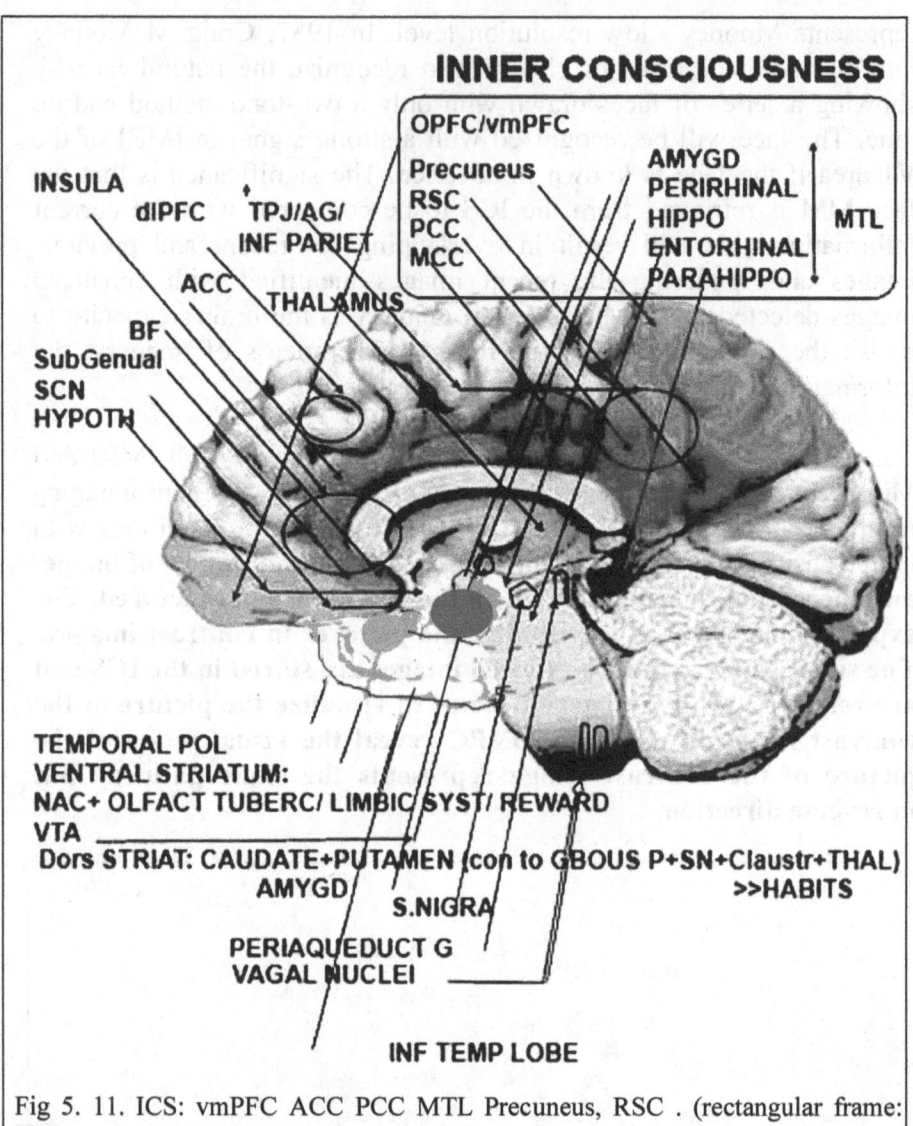

Fig 5. 11. ICS: vmPFC ACC PCC MTL Precuneus, RSC . (rectangular frame: ICS)

Other examples: After hearing a sound for a short moment, a while later, the subject is allowed to listen to another sound with a different frequency. Despite the second sound having a stronger intensity and lasting for a more extended period, the hearing of the second sound is influenced by decreased perception and partly superseded by the first sound.

Similarly, images or other features of the parents, first loved friends, teachers in primary schools, landscapes of native lands, and first music songs in childhood are difficult to forget. This MM represents CS in the

ICS used as referral CS for comparison, identification/ labeling, and transforming recent information into CS.

As discussed before, the ACC of the limbic system plays a role in detecting differences between the recent input information and MM stored in the ICS.
1. Updating the ICS Information or components of information similar to those in the ICS is labeled according to the component of ICS as such. As a result, these MM in ICS remained stable for almost the entire life of each individual. Information or components that could not be matched with any in the ICS is considered new information. Recent research on Zona Incerta, a small area of gray matter in the subthalamic region, revealed the role of detecting new information. The new MM will be added to the ICS as part of the updating process for ICS. Therefore, the ICS of each individual can be considered a personal encyclopedia for consultation when dealing with the problem of CS. Of course, ICS is not used in non-conditional reflexes such as tendino-muscular reflexes and startling reactions....

In ICS, mainly the semantic MM are copied into the Soul and carried over to the next life along with MM from innumerable previous lives. MM in the Soul imprinted into the newborn living organism becomes the Store MM/ skt Alaya/ vn: Stre CS likely in the RetroSplenial Cortex/ RSC. MM in Alaya is difficult to retrieve. The retrieval is possible when one attains the level of Meditation 3 or 4 (third or fourth Jhayana levels), and one can review the history of one's own life. In other states, like near-death experiences, some people with extrasensory power may have some abilities. The retrieval of MM of the current life is possible by the intermediate of MM traces in the Hippocampus. Retrieval MM of previous lives is likely possible by the intermediate of the soul.

3. Necessity of Original Mind/ Buddhda hood/ Holy Spirit for the formation of CS

As mentioned earlier in the Chapter on Memory, the information is integrated in the brain, compared with the ICS, and then ultimately connected with the Original Mind to become the CS. Failure of connection to the Original Mind results in the absence of CS/ Cognition, as seen in Cotard's syndrome. Dysfunction or Partial disconnection often results in an alteration of neurological functions or a focal neurological deficit known as **Conversion Disorder**.

4. THE BRAIN CONTRIBUTES TO THE CREATION OF INNER CONSCIOUSNESS or INNER MIND But Original Mind/ Buddhahood, but not the brain that creates the CS

The origin of the mind is a complex interplay of factors, not solely the Brain but also the Omniscience/Original Mind/Buddha Nature that contributes to its creation.

The Inner Consciousness area serves as a fascinating relay station, transferring integrated information to ensure it reaches the Buddha Nature.

In the Surangama Sutra, a pivotal moment unfolds when the Buddha, inquiring about the nature of the Mind, poses a profound question to Ananda. At this juncture, Ananda, having just achieved the fruit of Srotapanna, was yet to traverse seven more human lives to break free from the cycle of Birth and Death. The Buddha's query was simple yet profound: where does the Mind originate? Ananda's attempts to answer this question, repeated seven times with wrong answers, only underscored the complexity of the Issue:

    1) In the body: if in the body, why can't we see the six organs and five viscera

    (2) Outside the body: the mind and body are separate, why does the mind know, the eyes also know

    (3) Behind the eyes: why can't the mind see the eyes

    (4) The internal organs: internal organs are dark, the five senses outside are bright: the dark scene is not inside the body, the bright scene is not seen in the eyes

    (5) Thinking, (Thinking is related to at least two memories, for example, I saw Mr. A in Vietnam, the next day in Paris, thinking: Mr. A flew from Vietnam to Paris). The Buddha said that thinking has no self-nature, so it cannot be the Mind

    (6) In the middle of the object and the eye: back to the answer (2)

    (7) No attachment (no attachment): if there is no No attachment, there is nothing to discuss. If there is No attachment (like animals), it means there is Form. If there is thought, it cannot be the Mind)

Ultimately, the correct understanding of the Mind emerges: It is the Original Mind, the Buddha Nature, which appears after a thought is born. This Original Mind encompasses everything, with the large contained within and the small being included and beyond.

However, in the realm of sentient beings on Earth, the Original Mind is veiled by the Soul (Karma, or the Mind, the illusion of the previous life, the membrane of Ignorance). As a result, the Original Mind/Buddha Nature can only connect with the sentient beings through the intermediary of the Brain Soul, which only exists in the brain attached to the Inner CS and nruronal synapses all overthwee body. Any information that can reach the Omniscience has CS. The body's non-neuronal Soul, on the other hand, only generates strength, not CS.

Today, there are also many Buddhists who only recite the Sutras such as the Agamas who have many misconceptions about the issue of the Mind of the Five senses, typically as in the following passage

> *The Agama Sutra, "Whenever the issue of Perception is raised, the Buddha often cites the passage that says: "After the eye and forms come together, eye Perception arises." Not only the Abhidharma masters, but also the Madhyamika masters when referring to the appearance of Consciousness often cite this passage. However, the above passage does not mention the Original Mind. Another passage also says: "If the internal structures of the eye are not damaged, external forms enter the field of vision, if there is no proper attention, the corresponding Consciousness does not arise..."*

[The meaning of the passage is that, whenever and wherever there is a convergence of Attention, Senses and Objects, then and there Consciousness arises." This passage also does not mention the Original Mind

Attention, and Scenes are absolutely not enough to make Consciousness. Consciousness does not need only Attention, Senses and Scenes but also Buddha Nature/Original Mind.
William James, Principles of Psychology, Dover, New York, 1890:
If we could splice the nerves so that the excitation of the ear fed the brain centre conncerned with seeing, and vice versa, we would "hear the lightning and see the thunder"

It is known that no one is 20/20 blind when the visual cortex is damaged, because other parts of the Brain can replace the visual cortex (https://www.brainfacts.org/thinking-sensing-and-behaving/vision/2015/seeing-beyond-the-visual-córtex).
Brain specializes in Vision, accounting for 30% of the volume of Brain, showing that seeing is very important for living things. Blind people can receive stimuli from different sources from vocabulary, sound,

MM, suggestion, attention, hearing, transportation. It can be explained by adaptation for survival Darwinism. In Vietnam, Ms. Hoang Thi Them can see without her eyes.

5. RELATIONSHIP BETWEEN THE FIVE AGGREGATES AND THE NERVOUS SYSTEM139

The Five Aggregates, namely Form, Feeling, Perception, Mental Formation, and Consciousness, play a significant role in the transformation of information into CS in the nervous system. This relationship can be understood based on the sequence in the process of Object.

- Form: connection /nerves/peripheral nerves: Retina for vision, Olfactory bulb for taste, Inner ear for hearing, Spinal cord for touch and Hippocampus for general CS
- Perception: Thalanus except Hippocampus for Olfaction
- Feeling: Sensory Cortex, level 1 (direct information from Thalamus) and Superior transmission pathway
- Formation: Solar Cortex and Inferior transmission pathway
- CS: Mind. Omniscience

SOUL is closely linked to sensory and motor synapses, the above event is the basis for the following phenomena:

Conversion Neurosis: a patient loses the sense of sight, hearing or touch in the skin, is deaf and blind but has normal physiology and neurology, the patient often has psychoneurological symptoms. The cause is that the Soul is no longer connected to the synapses of the sensory organs in the retina, inner ear or spinal cord

Liberation of the Soul: Buddha points out that when the Form is purified, the Form part of the Soul can be free to see and hear, smell..

When afraid, people can become immobile in the whole body or part of the legs because the motor connection does not have the Soul working at the corresponding synapses. In Akinetic Mutism (no intention or cannot move according to the will, however the soldier is not paralyzed, because the motor Soul does not work. Conversion Neurosis: the patient loses the sense of sight, hearing or touch in the skin, is deaf and blind but has normal physiology and neurology, the patient often has psychoneurological symptoms. The cause is that the Soul is no longer connected to the synapses of the sensory organs in the retina, inner ear or spinal cord

When afraid, people can become immobile in the whole body or part of the legs because the motor connection does not have the Soul working at the corresponding synapses. In Akinetic Mutism (no intention or cannot move according to the will, however the patient is not paralyzed, because the Soul does not work at the motor synapses

---

CONFUCIUS' RICE POT

Confucius and his disciples were forced to live on only vegetables and porridge, and there were many days when they had to go hungry .

On the first day of arriving in Qi, a wealthy family gave them some rice. Confucius assigned Zi Lu to lead his disciples into the forest to find vegetables. Yan Hui was in charge of cooking the rice. Confucius was at ease reading the books upstairs.
Confucius, upon witnessing Yan Hui's actions, was filled with disappointment. He realized that his most precious student was secretly eating from his teacher and his friends, a realization that weighed heavily on him.
After that, Zi Lu and his disciples picked vegetables and brought them back. Yan Hui boiled the vegetables. Confucius still lay there, suffering. A moment later, the vegetables were cooked. Yan Hui and Zi Lu brought the meal to the upper house; all the disciples clasped their hands and invited Confucius to eat.

Confucius sat up and said: "We have traveled from Lu to Qi, a long journey of thousands of miles. I am very happy that in the midst of chaos and hunger, you have remained sincere. On the first day of arriving in Qi, I was fortunate enough to have a meal. I felt sad and missed my hometown and parents. I want to scoop some rice to offer. Do you think it is appropriate?"
Except for Yan Hui who remained silent, the disciples, out of deep respect for their Master, clasped their hands and said: "Yes, Master, it is appropriate!"
Confucius said again: "But I wonder if this pot of rice is clean or not?". At that time, Yan Hui clasped his hands and said: "Yes, Master, this pot of rice is not clean."
Confucius asked: "Why?" Yan Hui replied: "Yes, when the rice was cooked, I opened the lid to see if it was cooked evenly. Unfortunately, a gust of wind blew in, and soot and dust from the house fell and soiled the rice pot. I quickly covered the lid again, but it was too late. Then I scooped out the dirty rice, intending to throw it away... but then I thought: there was little rice, and there were many brothers, if I threw away this dirty rice, I would lose a portion of food, and my brothers would have to eat less. Therefore, I took the liberty of eating the dirty rice first, and left the clean rice to offer to you and all my brothers... Master, so today I have already eaten... now, I ask permission not to eat any more rice, I will only eat the vegetables. And... Master, the rice that has been eaten first should not be offered!"

After hearing Yan Hui's words, Confucius looked up to the sky and lamented, "Alas! It turns out that in this world there are things that one sees clearly with one's own eyes but still cannot understand the truth! I, Confucius, almost became a fool!"

---

6. Conversion Disorder. The disorder is believed to be an emotional problem converted to a somatic disorder. The pathophysiology is poorly understood. Lifetime prevalence in the general population is quite variable, at about 11-22 cases per 100,000 people up to 50 per 100,000, the second most common cause of a neurological outpatient visit after headache. Rare in young children, with onset generally from late childhood to early adulthood, and seldom seen after the fourth decade of life. Study of pediatric patients: The prevalence is estimated at approximately 5%–14% of general hospitalized patients, 5–25% of psychiatric outpatients, and up to one-third of neurology outpatients. The annual incidence rate is 2.3–4.2/100,000 children, Female to a male ratio of 3:1, average age of 13.7 years (range 8–18))

The neurological deficit usually does not fit into a clinically defined neurological disease and neuroanatomy.
- Abnormal movements, such as tremors or difficulty walking.
- Loss of balance.
- Weakness or paralysis.
- Difficulty swallowing or feeling "a lump in the throat".
- Seizures or episodes of shaking and apparent loss of Consciousness (nonepileptic seizures).
- Episodes of unresponsiveness.

Signs and symptoms that affect the senses may include:
- Numbness or loss of the touch sensation.
- Speech problems, such as the inability to speak or slurred speech.
- Vision problems, such as double vision or blindness.
- Hearing problems or deafness.
- Cognitive difficulties involving Memory and concentration.

Failure to recognize the soul with attachment to the Original Mind is a problem for medical researchers.

Clinical and imaging investigations have often revealed non-specific changes that may account for disorders associated with conversion disorders..
The findings and suggestions of mechanisms such as:
overactivity of the limbic system, dysfunction of brain networks that give the movement the sense of voluntariness, and increase of activity in a cluster that included the right supplementary motor area (SMA) and another that included the right temporoparietal junction (TPJ)
CD may have a neuropathic pain component, and there may be a deficit in central nervous system inhibition

Since Buddhahood/ Original Mind is uncreated and imperishable, overwhelmingly present, the disconnection with the Original Mind is always temporary. Therefore, the Conversion disorder always resolves by itself with time but can be associated with relapses.

IX. VISUAL CS

More than 30% of the Cerebral cortex is related to Visual CS. The Retina comprises two types of cells: The parvocellular component for color/ contrast BW and the Magnocellular (Low contrast BW). The visual information is sent to the LGN, the two V1 (striate cortex),V2-V (extrastriate cortex)

X. NEUROANATOMY OF ICS (Fig 5.14,15,16)
A. Thalamus

The Thalamus is a large structure of the Diencephalon, composed of two nuclei 4x2.5x2.5 cm at both sides of the third ventricle. They are interconnected by Interthalamic Adhesion. The volume and the central location are consistent with the crucial role of the brain in communication. Thalamus does not store the MM or CS but play an essential role in CS formation

a) Thalamic Reticular Network / TRN (Fig 5.11)

The anatomic distribution of the TRN, with its wrapping around the Thalamus, and the juxtaposition of the sensory organ-specific part of TRN over the corresponding sensory organ-specific part of the Thalamus deserves special attention. All inputs to the Thalamus are moderated by the Thalamic Reticular Network, acting as a filter or noise canceling system as in a headphone case. In the attention, the TRN is useful in discarding unrelated inputs. For the four types of input (except the olfactory), TRN is divided into respective areas overlying the corresponding sensory of the Thalamus (Ahrens, S. 2015, Aizawa H2011, Akeju, O 2014.),

Crick proposed that TRN plays a role in attention due to the capability to discard unwanted information, like a filter or a noise-canceling system of a headphone. Alteration of TRN contributes to the pathogenesis of schizophrenia. The Thalamus is critical in the mechanism of wakefulness.

The Thalamus and other grey nuclei in the midbrain are the relay center for information required for making the CS. TRN plays the role of a filtering system to discard unwanted data

b). Except for the Olfactory sense, the informative input comes to the Thalamus before being sent to the cortex. The connection is bidirectional with the cortex.

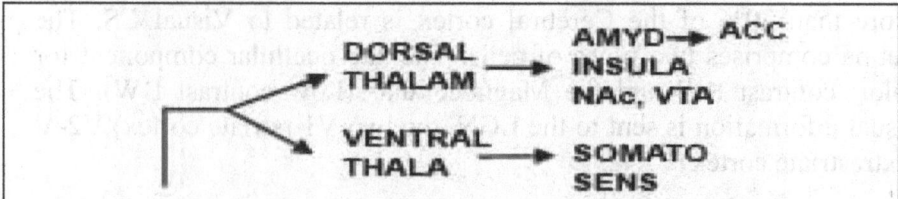

Fig 5.12 Diagram showing the pathways connected to the Thalamus composed of: - Dorsal pathway to Nucleus Accumbens/NAc, VTA (reward, joy), Insula (internal percepting) AMYGD (fear) to ACC (error detection for comparison with ICS)
- Ventral pathway to sensory cortex

c) The division of the Thalamus into various areas facing the connected regions of the remaining brain. For example: Anterior Thalamus, connected to the N, Reuniens HIPPO, PFC, Mammillary body. The DorsoMedial Thalamus (DM) to the limbic system.

i. It is of interest in a study (Scheinin 2020) on the Thalamus-Cingulate/ Limbic and Angular Gyrus systems. stable even under general anesthesia (Jin 2021 Timbie, 2015; Wolff 2015a) and essential in the emotion expression. Medial dorsal Thalamus: connected to AMGD (Fear, Anger), Olfactory cortex (Attention, behavior).

ii Anterior connected to Hypo, Mammillary body (spatial MM), and Basal ganglion (involuntary movement).

iii. Ventral receiving inputs from below.

iv. Posterior, Pulvinar and Medial Geniculate body, MG and Lateral LG connected to visual and auditory inputs.

v. Superior Colliculus/ **SC** divided into lateral for visual input posterior connected to intralaminar nuclei have an important role in the wakefulness incited by visual and auditory inputs(Wurtz)

d) Intralaminar nuclei mainly related to the wakefulness

i. Anterior Intralaminar Nuclei for Saccade movement of the eyes and wakefulness the injury of Anterior Intralaminar Nuclei will cause an unconscious state and Hemispatial neglect (Cortex - Hudetz, 2012, VanderWerf,2002),

ii. Posterior Intralaminar Nuclei role in the attention (related to sensory inputs) Minamimoto 2002, Kinomur 1996).

iii. Midline structure connected to Medial PFC, Hippo subiculum and Entorhinal Cortex, related to MM (Aggleton 2010).

e) Thalamocortical Radiation Thalamocortical and Corticothalamic, consisting of four components: anterior (PFC), posterior (parietal and occipital) superior (cortices Pre and Post central) and inferior (Insula, temporal and inferior frontal). The frontal component with GABAergic activity in Parkinson's disease, social interaction, schizophrenia Myoclonic. This component is also associated with Acetylcholine activity, resulting in pathophysiologic relationships with autism, Pick's disease, Anorexia Nervosa.

On the other hand, the Thalamus is important for the wakefulness. With the prominent thalamocortical radiation and cholinergic and adrenalinergic neurotransmitters, it is significant to distinguish CS from wakefulness. (Banerjee S2019 ,Bayley PJ 2005, Maguire EA 2005. 2018Chadwick MJ, Barry DN 2018, Andrews-Hanna 2010, Ressler KJ. Dias BG2014., Andero R 2014, Alberini CM 2011,Maroun M, Akirav I 2006,Amemiya T 2017,Ajina S2016,)

B. Prefrontal Lobe And CS (Fig 5.17,18)
The **PFC does not play an essential role** in the wakefulness
The evidence is the removal of the PFC except for the Premotor cortex: there is no alteration of wakefulness. Removal of the posterior parts of the cerebral cortex (the Parietal, Temporal, and Occipital lobes), may cause loss of wakefulness. The above changes denote the importance of the posterior half of the brain in keeping humans vigilant. However, vmPFC and cortex BA6 are critical in judgment, experiences, and attention management, influencing CS formation.. (Bricker 1952, Boly 2017 Koenigs 2007, Koch 2016b, Odegaard 2017. PFC manages social interaction and emotion but is not related to wakefulness (Dixon 2017 Berridge 2015)
Neurohormone is Norepinephrine.

Role in morality, executive decision of contentious plans, high–order thinking, with connection to the parietotemporal cortex, social interaction behavior in creation, attention, moderation with initiation versus inhibition, planning, estimating, aborting termination,
coordination, overriding, and evaluation, specifically:
- **dlPFC** is responsible for the management of most activity of the brain, influenced by the vmPFC and the Insula in decisions regarding social interaction and volition

Lesion of the dlPFC results in the inability to set the logistic sequence, initiation, and maintenance of a plan. The symptoms are apparent in the cases consisting of multiple levels of management.
The choice of the options is *automatic contention scheduling with perseverance of behavior* instead of under the control of *supervisory attentional management*.

- **mPFC**: evaluation of other people in en-motion interaction
- **Lateral OPFC**: evaluation of the feeling from peripheral inputs by connection with the Amygdala, Hypothalamus, and Periacqueductal gray.
- **Inferior OPFC**: involvement in the reward system
- **Rostral mOPFC and lateral**: BA10 containing)
- **pgPFC+ sgPFC**: feeling, emotion about fear, sadness, visceral function. sgPFC related to cardiovascular function.

dMCC: error evaluation, sharing the role with ACC. Role in anxiety in decision making, responsible for lying, inhibiting telling the truth. Stimulate the dlPFC inhibiting from dlPFC in telling the lie. VmPFC is the latest phylogenetic structure of the central nervous system that holds a vital role in the management of emotion-related sensory organs, primarily visual, auditory, and olfactory. Furthermore, it is critical in the control of the level of morality of each individual. To fulfill the above function, vmPFC is endowed with the capability of MM retrieval. Each part of PFC is involved in the respective function that is still under investigation. In general, PFC has the following functions:

- The upper part of PFC: General executive function.

- The lower part: management, planning, and thinking with an emphasis on emotion and social interaction.

- Lateral aspect; related to outer sensory organs

- Medial aspect: related to MM and inner sensory organs such as viscera

- Posterior aspect: Predictive comparative function to detect the errors in comparison with the information contained in the ICS.

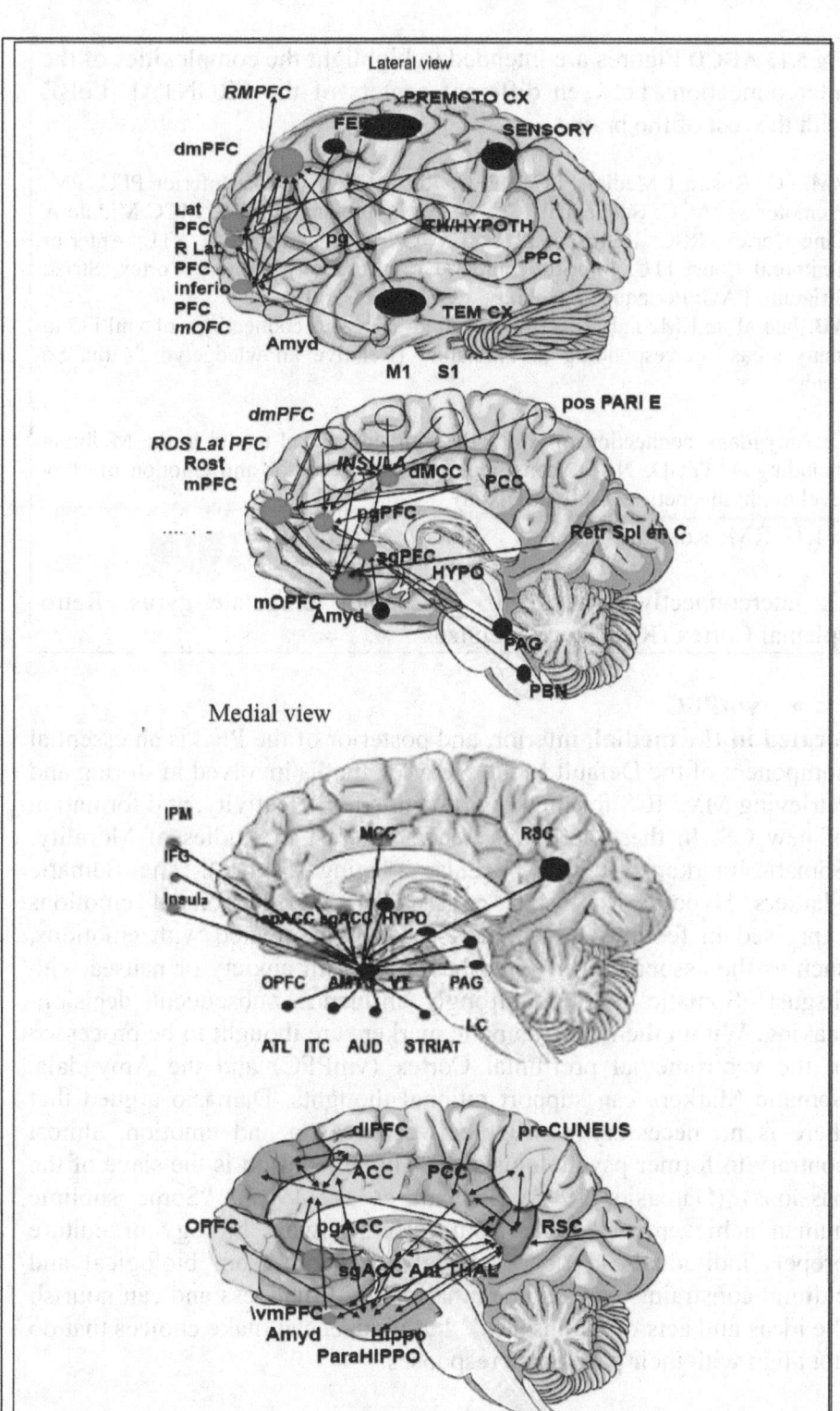

> **Fig 5.13 ABCD** Figures are intended to highlight the complexities of the interconnections between different centers of the FRONTAL LoBE with the rest of the brain
>
> RMPFC: Rostra l Medial PFC, RPFC: Rostral PFC, IPFC: Inferior PFC, PM: Premotor , sgACC: SubgenualACC, pgANN: Pregenual ACC, MCC Middle A Cing Cortex, RSC: Retrospenial Cortex, OPFC Orbitary PFC, ATL: Anterior Temporal Lobe, ITC: Inferior Temporal Lobe, AUD: Auditory Cortex, Striat: Striatum, PAG: Periaqueductal Gray, LC: Locus Ceruleus
>
> **AB**: lateral and Medial aspect of the Brain showing connections of vmPFC to many areas corresponding to Paratantra (Relative knowledge/vn: Y tha Sở Tánh)
>
> **C:** Amygdala/ connection of AMYGD with subcortical nuclei in the Midbrain including AMYGD, NAc... consistent with the behavior and emotion of low level of phylogenetic evolution (Dixon)
>
> H4.17 RM: Rostral Medial
>
> D: interconnectivity between PFC with Cingulate gyrus, Retro splenial Cortex (RSC) and Precuneus

- **vmPFC**

located in the medial, inferior, and posterior of the PFC is an essential component of the Default Mode Network that is involved in storing and retrieving MM/ ICS for thinking, imagination, creativity, and formation of new CS. In these activities, recent interest in studies of Morality, Somatic markers, and FTT (False tagging Theory). The Somatic Markers Hypothesis (SMH) consists of a collection of emotions expressed in feelings in the body that are associated with emotions, such as the association of rapid heartbeat with anxiety or nausea with disgust. Somatic markers strongly influence subsequent decision-making. Within the Brain, somatic markers are thought to be processed in the ventromedial prefrontal Cortex (vmPFC) and the Amygdala. Somatic Markers can support rational thoughts. Damasio argued that there is no necessary conflict between reason and emotion, almost contrary to former psychologists like Hume (Reason is the slave of the passions) (Damasio, 1994; Bechara et al., 1997). "Some sublime human achievements come from rejecting what biology or culture propels individuals to do" —though "freedom from biological and cultural constraints can also be a hallmark of madness and can nourish the ideas and acts of the insane." "Individuals can make choices that do not align with their passionate responses.

However, according to most Buddhist scriptures, the overall factor is not the SMH/ **Somatic Markers Hypothesis** or reasoning but the setting up of the Mind on the quest for the truth and can experience happiness in the Dharma instead of in the five worldly pleasures (arising from the five senses)." Dharma is the cosmic law of proper behavior, conduct, and thinking, such as boundless kindness composed of compassion, pity sharing, joy sharing, and detachment/abandonment

.vmPFC does not directly influence the formation of the CS. An example is the case of Phineas Gage with injury to the vmPFC. After the accident, Gage changed their social behavior, but his CS remains reasonable for professional work. dlPFC is the center for management of working MM ad for attention. The connection between vmPFC and dlPFC is necessary for the functioning of dlPFC. Therefore, vmPFC is indirectly involved in the CS. ACC, the center of error detection, is part of the cingulate Cortex. ACC is adjacent to and connected to vmPFC, Hippo Thalamus, Insula, and Amygdala. Due to the connection between ACC-vmPFC and Insula-AC-, Amygdala, ACC (rostral ACC), takes part in emotion management. In aged persons, the ACC -Hippo and Thalamus decrease is consistent with the decline of CS. However, the connection of ACC with VMPFC, Amy, and insula is increased, which is consistent with emotional instability with fear and anxiety due to lack of information (Cao 2014, tern,2010).
vmPFC is activated with joy and happiness.

* **OPFC.**
In many studies, OPFC is included in vmPFC due to its close location and role in morality and social interaction. However, vmPFC is more involved in social interaction, whereas OPFC is more concerned with righteousness.

* **Other areas of PFC.**
   - Left Ventral PFC for inhibition of overreached movement
   - Frontal eye field BA8, above dlPFC, for movement of the eyes for Creation of the 3D image in the visual Cortex.
   - BA10
   -IPS (fig 5.18)

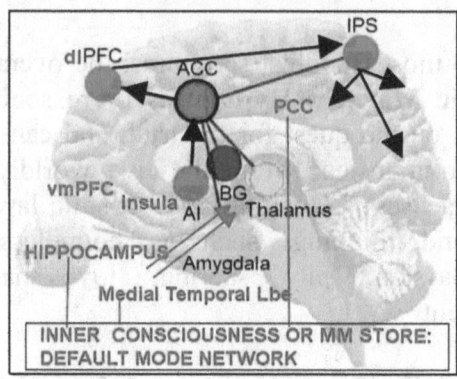

Fig 5.18

IPS and dlPFC form the executive network in the management of sensory and motor centers by the intermediate of the IPS. In addition, Via the ACC the IPS is connected with the Basal gangia, Thalamus, amygdala

XI. Emotion and CS

From antiquity, it was believed that the Brain is the site of CS, but emotion belongs to the body outside the Brain. This impression is likely based on the observation of the involvement of all body parts n expressing the emotion. Aristotle preferred to term emotions as pathos, which is considered a passive state located in the metaphysical landscape in the body and outside world. Parts of the Brain involved in emotion are located at the base of the Brain: Amygdala (fear), Ventral Striatum (Joy), Insula (Deep feeling), OFC (Empathy), ACC (Error detection, pain), mPFC (social interaction), Midbrain/Peri-Aqueductal Gray (facial /bodily expression, visceral reaction).

Along with the involvement of the anatomical structures, batteries of biochemical mediators like epinephrine (flight, fighting), DOPA (controlling movement, pain feeling), serotonin (joy), and endorphins (pleasure) are appropriately released.

In Embodied Simulation (Barsalou 2003,2008), the image of the body is represented in the Brain (in the ICS). As a result, the expression of the emotion is coming from within, in ICS (Barrett and Simmons, 2015; Chan and Barrett, 2016, Seth et al., 2012; Seth, 2013; Pezzulo et al., 2015; Seth & Friston, 2016, Barrett and Bliss-Moreau, 2009; Barrett, 2017). All living beings, including unicellular organisms, have within them an Inner Model of their body and the environment. The Inner Model serves as the rational and economic reaction mechanism for survival and adaptation. As previously discussed, this Inner Model or ICS is used as a reference model for labeling the CS and the expression of emotion through interoception and simulation. For CS, a piece of information similar to a part of ICS is recognized. For emotion, each stimulus from the environment is associated with a stereotyped emotion contained in the ICS.

Unlike the sensory cortices, the cortex of the ICS is characterized by the absence of the granular layer corresponding to layers 2 and 4, which play a role in processing and integrating sensory information. The role of ACC in predicting the information in the formation of CS and emotion is related to the capability of error detection. In this context, error detection compares the new data stored in the ICS and identifies the difference. Furthermore, the connection of ACC with vmPFC, pregenual PFC, and sensory-related cortices is attributed to the above function. According to the classic view, emotion is formulated according to an innate model according to the evolution of species and tradition engrained in family, society, and culture and hardwired in subcortical nuclei. For a long time, it was believed that the emotion is somewhat prototyped according to the genetic code, modulated by preexisting conditions, therefore representing the expression mode that is automatic and involuntary. The feeling is a secondary event of the cortices activated by the subcortical nuclei. The thoughts are the results following the cortical activities.

In the new theory of Constructed Emotions, the mechanism of the construction of Emotions is the opposite of the classical view of emotion. Emotions are not reflexes to the outside world. Like the CS conceived in this book, emotion is developed as the result of experience and the learning process by forming models for each group of inputs according to personal choices, personal social experience experiences, and genetic background. Inputs without a corresponding inner model become noise of the organism, including the Brain and body, which have to update the preexisting model. The inner model dictates the effect as pleasure, sorrow... and interoception as bodily manifestations with cardiac, and respiratory rates, hormones, facial features, and body gestures. Manifestation of emotion is finally impacted in an automatic system called body budget, which includes sleep, depression, nutrition via eating, drinking, and dressing, working, and talking habits.

Therefore, the difference between the old and new concepts resides in the ICS. (Damasio AR 1983,1999,1994., Van Hoesen GW)

XII. CS and Philosophy

A. Theory of Predictive Mind (Fig 5.18)

1. The theory is developed by Jakob Hohwy University Monash, Australia, in the early decade of 2010. The Brain is a prediction box activated by the initial sign of incoming sensory input. The Brain uses the ACC to look into its contents and then sends out to the periphery a guess of the events (TOP-DOWN). This may quickly get the information and eventually save energy. The response from the sensory organ will be sent up and then assessed by ACC to identify the similarity or to detect the differences (BOTTOM -UP). The information is then sent to the motor cortex for action to pgACC and vmPFC for correction of the errors and to update the ICS (Barrett and Simmons, 2015; Chan and Barrett, 2016, Seth et al., 2012; Seth, 2013; Pezzulo et al., 2015; Seth & Friston, 2016, Barrett and Bliss-Moreau, 2009; Barrett, 2017).

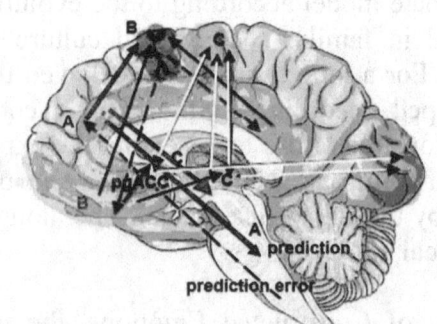

Fig 5.19 Three phases in the formation of the CS: interplay between ACC, Motor cortex (for action), and PFC (for moderation/ advice) with the participation of PCC and Sensory cortices (Vision, hearing, touch...) (for recognition of the reality)
A) Solid thick arrows: after receiving the information input, ACC sends the signal to the cortex. A, peripheral organs, and Hypothalamus for prediction and automatic emotional reaction
B) Discontinuous arrow to vmPFC, PCC for consultation, and solid thin arrow to the motor cortex for adjustment
C) pgACC: Clear arrow for the last prediction. Final information is sent to sensory cortices

The theory is based on the Bayesian theorem for the statistical analysis that permits to calculate the probability of occurrence of an event E : P(E)%,
P(E)%, that is influenced by opinion/thoughts/ concepts B.
P(E|b)%: is the result assessed by B, Knowing that B is only correct in P(B)%,
Therefore
P(E)= P(E|b)% divided by P(B)%,

Example: P(E)= 0.4%: 40% = 0.1%.
In the above example, P(E)% represents the chance of correct guessing, B represents the ICS
In summary, opinion and concept usually decrease the probability of the prediction/ survey. (Hohwy J. 2013)

2. The embodiment of the emotion as expressed by the interoceptive Awareness, originating from within/ interoception (ICS) Füstös, Soc Cogn Affect Neurosci. 2013 Dec; 8(8): 911–917.

3. Influence of emotion expands to the environment of society with the involvement of cultural and technical components..

B. Bayesian Brain and Phantom Perception (Fig 5.19).
The brain works as a box of prediction. To achieve high accuracy in the prediction, the brain has to minimize the input corresponding to the principle of Shannonian free Energy (Free Energy Processing=FEP). But limiting the peripheral input can cause the ACC Insula to work relentlessly searching the source of information in the ICS. As a result, illusion develops because the retrieved information from ICS may not be related to the expected context (BOTTOM -UP). The information is then sent to the motor cortex for action, to pgACC and vmPFC for correction of the errors and to update the ICS (Rao 1999; Friston, 2002, 2008; Lee and Mumford, 2003; Knill and Pouget, 2004; Yuille 2006;Summerfield et al., 2006; Bar, 2009; Friston 2009; Rauss 2013).
(Friston 2006, Kesner 2014, De Ridder 2014)

C. Embodied Consciousness, Simulation, Extended Consciousness, GROUNDED COGNITION) commonly referred as body language
To psychologists, the brain is composed of multiple modules specialized in different types of CS like visual, auditory... and other subtypes of general CS. The brain manages to store the input information in different modules according to the properties of the respective cortices and sites in the ICS, like storing Memories. The CS is reconstituted by interconnecting the modules based on the same connection before consolidation and storage, with connections between modules but in the opposite direction. As a result, there is an interplay between different modules in the brain, such as areas of sensory, auditive, visual, word writing, and word spoken and motor and premotor cortices. The reconstituted data represents the information that can generate an overall CS after being compared with the ICS. In this stage of processing the CS, the ICS represents the model or

background for comparison. After the comparison, the CS is labeled. This phenomenon is comparable to Grounded CS (referred to ICS). As a result, the simulation is accomplished through the grounding process because, without grounding, CS can not be generated.

The simulation process was often used in motor simulation theory (MST), which is a mental rehearsal of the body's movement or body parts without actual movement. This involves the cognitive activation of the part of ICS storing the ICS of the related action, followed by the rehearsal of the motor system offline through the simulation mechanism (O'Sheas Motor Simulation Theory Explain). In addition, deep breath may be associated with efficient retrieval of MM:

In the theory of Embodied Consciousness, in addition to the CS grounded in the ICS in the brain, evidence also suggests that there is an embodiment of CS in the body outside the brain (see below). (Beauregard M 2001 Apps MA.2013Anderson MC 2022,20011).

1. Respiration and the Brain
   a) Respiration modulates the coordination arm, fingers, and eye movement. Respiration provides oxygen in exchange for CO_2. The lung, chest wall muscles, diaphragm, and other accessory muscles of the abdomen, larynx, mouth, and nose are involved. In addition, CS is related to the brain at the areas of the brain storing MM, mainly in the sensory cortices and the DMN (Beate Rasslerr). In addition to the brain, the manifestation of CS can be seen in the respiration and movement of extremities. Holst, in 1939 demonstrated that arm and leg movement could coordinate with temporal changes in breathing. The entrainment of cyclic breathing can be synchronized with the rhythmic movement of cycling, walking, and running (Rabler B 2000). Forced expiration, especially against a closed glottis (closed upper part of the respiratory tract: Vasalva maneuver) is commonly associated with increased flexor muscle strength in the fingers and hands (Li &Laskin 2006). Movement, coordination, and precision of the index finger are found to be dependent on breathing. Therefore, the movement of the fingers and the respiration are intercorrelated, especially in meticulous tasks requiring skill and accuracy of movement, such as threading a needle or artistic task like drawing a picture (Rabler B 1996 Rassler, B. 2000). (Of interest, there is no such evidence of a correlation between pianist's fingers and respiration). The correlation may be attributed to the neuronal interaction and modulation of the central nervous system with the participation of visual input as rowing stroke and respiratory cycle at the integer ratio rate of 1:1.

b) Respiration influences brain activity and CS. During respiration, cortical activity is detected and forms oscillations of different frequencies of gamma type that can modulate neuronal responses to sensory perception (Arieli et al., 1996; Poulet and Petersen, 2008; He, 2013). The rhythmic respiratory activity likely sent the interoceptive and proprioceptive input to the cortex. The evidences are:

- In mice, the smell activates the olfactory bulb and the olfactory cortex. During the quiet wakefulness state, the brain's cortical activity is independent of respiration. Ito has demonstrated that there are links between the lung and cortex. Respiration causes modulation in the somatosensory cortex with respiration-locked delta oscillations and gamma waves (30-100Hz) (Ito et al., 2014). In addition, these electrical changes are considered to be linked to the CS and other high cortical functions like emotion, pain perception, motor control and attention (Detlef H. Heck12017). According to Porges' Polyvagal Theory, Jerath suggested that respiration in Meditation also activates the Periaqueductal Gray that is responsible for widespread muscular, vascular, and visceral activities. In mice, sharp wave ripples from Hippo are linked to MM sending vmPFC to Memory for consolidation and are associated with respiration. Whisker barrel cortex of awake mice producing Delta and Gamma oscillations are synchronized with respiration. Removal of the olfactory bulb abolishes this synchronization (Ito 2014 Nat Commun).

In summary, the function of respiration is not limited to O2 and CO2 exchange. Still, it is also associated with the modulation of high-function cortical activities such as CS, emotion, pain perception, and movement of limbs. The airflow through the nose with activation of the olfactory nerve is suggested as a mechanism of this cortical modulation(Detlef H. Heck12017).

2. Role of the peripheral nervous system
Like the olfactory nerve, other peripheral nerves may influence the CS. Example: rowing eyes is also accompanied by increased gamma waves in the occipital cortex. Presenters, actors, or singers usually exhibit changes in faces, tone of voice, and gestures with arms and hands to supplement the expression of some content of CS. Therefore, all body parts can express CS and emotional pain perception in a manner comparable to the central nervous system: The phenomenon is considered an embodiment of CS. CS is considered part of the Mind or the Soul, that is the metaphysical part, as opposed to the body and the brain which represent the physical part. The CS, as discussed above, is

linked to the DMN and sensory cortices of the brain. Medical practitioners or psychologists still need to address the metaphysical part of the body (head, trunk, four limbs). However, there are some states of "metaphysical form of energy, so far poorly understood and poorly defined like Qi in Chinese medicine, Chi Gong in Chinese martial art and possibly "bioenergetic flow in Kunladini Meditation technique" though different points of Chakra on the body. The difference between Qi / Chi Gong and Chakra is that Qi and Chi Gong are more materialistic, while Chakra energy is more spiritual. Qi, Chi Gong, and Kulanini likely represent an aspect of the Soul linked to the bodied CS. An embodiment of the Soul will be discussed later..

3. Extended Cognition.

Cognition based on a physical structure outside the brain and the body is considered extended CS. Examples: CS obtained in magazines, books, computers... As a result, the entire world containing information known to each individual is the extended CS.

XI Consciousness in Christianity, Medial Doctrine of the Mahayana Buddhism (Basic Doctrines of the Sole Consciousness or Idealism) (skt:Vijnanamatra).

A) Discriminative Mind.

The case of Adam and Eva in the Eden Garden of the Old Testament exemplifies a discriminative Mind. Adam, the original man, was commanded by God that he could earn every fruit in the Garden except the fruit of the tree in the middle of the Garden since eating that fruit would bind you to know good and evil. Adam lives with his wife. Both are naked, but they are not ashamed. One day Eva listens to a serpent saying they can eat fruits from every tree. She convinced Adam and, after eating the fruit from the tree in the middle of the Garden, their eyes opened, and they realized they were naked. That was the original sin; because of eating the forbidden fruits, they were punished for acquiring a discriminative Mind and had to leave the Garden and cultivate for their food. Since then, the world gradually developed and expanded, carrying the Original Sin with them. The discriminative Mind represents the original Consciousness Vijñāna.

B) BUDDHIST EPISTEMOLOGY

Unlike Christianism, in Surangana Sutra, Buddha revealed to the meeting of 1250 Bodhisatsavas, Arahats and disciples that CS is generated from the Original Omniscience, not the thinking mechanism and the five sensory organs. As described in chapter one in the paragraph on the veil of ignorance, objects are not recognized by living sentients as such. Still, they are modified by the environment, sensorial organs, and the mechanism of data integration in the brain (6x3=18 fields or filtering systems) (known as inference by some authors). The most significant filters are within the brain. The data is compared with data in the inner consciousness. The comparison process is particularly highlighted in the Lankavatara sutra, by Nagarjuna in his middle-way discourse Asanga and Vasubandhu. In Tibet, a similar epistemology was developed by Dharrmakirti. The notion of Original Omniscience (Pragma) is almost ignored by Theravada scholars as well as most Westerners. Of interest, the role of attention in the formation of the CS was only mentioned in some Buddhist sutras, likely representing the insufficient recording of this step.

There can be no doubt that all our knowledge begins with experience, For how should the faculty of knowledge be called into activity, if not by objects which affect our senses and which, on the one hand, produce representations by themselves or, on the other, rouse the activity of our understanding to compare, connect, or separate them and thus to convert the raw material of our sensible impressions into the knowledge of objects, which we call experience.

But though all our knowledge begins with experience, it does not follow that it all arises from experience, for even our empirical knowledge may be a compound of that which we perceive through impressions and of that which our faculty of knowledge (incited by sense impressions) supplies from itself, a supplement which we do not distinguish from that raw material until long practice and rendered us capable of separating one from the other.

According to E. Kant, therefore, there is a question that deserves closer investigation and cannot be disposed of at first sight: Is there any knowledge independent of all experience and even of all impressions of the senses? Such knowledge is called 'a priori' and is distinguished from empirical knowledge, which has its source 'a posteriori', in experience..." Such questions remain unanswered if the components of SubCS, UnCS and store karma are called into play. In other words, inner CS (that includes SubCS and UnCS) and karma of previous lives likely play a role

C) Buddha nature, Original Mind, Ultimate Omniscience/ vn:Bản Giác, Wonderful/ Miraculous and Ultimate Awareness/ vn: Tánh Giác Diệu Minh, Emptiness/ vn: Chân Không.

These nouns are almost synonyms to describe the ultimate and marvelous/ miraculous state of equality, non-discrimination, and Emptiness. It has a neutral nature that is neither good nor bad, misleading/ wrongful/ false, or correct/ rightful. From this state of Ultimate omniscience/ Miraculous Emptiness and without causes/ dependent origination (Hetu-pratyaya) or conditions arises a false and transient thought/ skt-p :Vitathavitakka/ vn:Vọng niệm. Since the transient false thoughts are a mixture of good and bad states, resulting in continuous conflicts that represent discrimination with loss of equality defined as the state of ignorance. This state of ignorance obliterates the Original Mind or Buddha nature, therefore, is called the veil of ignorance,

In Lankavatara Sutra/ vn: Kinh Lăng Già composed of the discourse of Buddha explaining to Mahāmati Bodhisattva/ vn: Bồ-tát Đại Huệ that the root of all phenomena is Mind.

Cittamatra (skt)—Mind-only or
Mind itself—Nothing exists apart from Mind.
vn: Tâm ngoại vô sở kiến
Hết thảy các pháp đều do Tâm/ Nhất thiết duy Tâm tạo, không có pháp nào ngoài Tâm.
Buddha added: "Those attached to the notion of duality, object, and subject fail to understand that there is what is seen of the Mind."/ vn: ai còn vướng mắc vào *nhị nguyên, chủ thể và vật thể thì* không hiểu được Tâm.
It will be easy to understand that this Universe's Creation resulted from incidental thoughts from the Original Mind. The incidental thought first creates the original CS or Ultimate Awareness, as outlined above

D) The Relationship of the CS and the Five Aggregates: FORMATION OF VEIL OF INORANCE
/ vn: ngũ Ấm/ Uẩn/ skt: Panca, khandha)
The five aggregates make up a living being and consist of
- Form/vn: Sắc, the physical component as perceived by the five senses.
- Metaphysical component that is not perceived by the five senses but manifests with effects through the physical Form, i.e. the Mind or Soul, which consists of:
    - Perception/ vn: thọ.
    - Feeling/ vn: tưởng.
    - Formation of the CS/ hành.
    - CS/ vn: thức.

Perception, feeling, and formation constitute the three steps before the Creation of CS from the environment.

The Original CS, at the beginning, splits into the duality of physical components or Form and Metaphysical components. The Form is created because the Miraculous Emptiness is also the origin of the current Form. Therefore, according to Madhyamika Saska (or Medial Doctrine of the Mahayana Buddhism/ vn: Trung quán Luận) the five aggregates are created from the Original Mind. Thus, they are associated with the law administered by the circle of birth and death or coming into Existence and ceasing to exist. Furthermore, because of the above reason, the five aggregates are egoless. But because of **the veil of ignorance**, living sentiments have a false impression that they own their body and the Mind/ Soul. The result of this ignorance is the suffering. The metaphysical component is represented by the Soul.

In Buddhism, ignorance is the fundamental and the most serious sin. This coincides with Christianity in which the discriminative mind (development of CS, also called veil of ignorance) represents the most serious sin, leading to the eviction of Adam and Eva from Eden garden, for harsh labor work penalty and loss of eternal longevity (death penalty)

XII. THREE DHARMA SEALS
p: Tilakkhana.
Emptiness is associated with Ultimate Omniscience. Due to the incidental Though /False Thought, thr UO obliterated by **the veil of Ignorance**. The veil of ignorance is Dependenr cause of the
Three Dharma Seals/ vn: Ba PhápÂn—Three marks of Existence, or three characteristicsof all phenomenal Existence:
- Impermanance/ p: Anicca.
- Non-Ego/ p: Anatta —No-self—Egoless— Impersonality.
- Suffering versus Nirvana

The three Dharma Seals are the most essential doctrines in Buddhism that govern the practice of the spiritual path. The concept is simple if:

- As discussed above, the Universe is impermanent and perishable because a false thought from the Original Mind creates it. The latter is also the Ultimate Emptiness that exists by itself because Nothingness or Emptiness can be considered as the Creator of itself. As pointed out by Parmenides, "What exists is uncreated and imperishable, for it is whole and unchanging and complete. It was not or nor shall be different since it is now, all at once, one and continuous."

- Ones can consider themselves children of the Creator, that is, in this case, the Original Mind, the Ultimate Emptiness. Since they are created, they are perishable. Furthermore, as children of the Creator, they are not entitled to own anything in the world, including their body and Mind. This ownership is not unconditional but limited to the necessity for their existence and their work on behalf of the Creator. They have to follow the commandment and do not have the liberty to do their own will except those conforming to it. This means that they are egoless.

-What are the commandments? The commandment consists of four wonderful Minds/ Tứ Diệu Tâm and Egolesness. Practically, that means the five precepts in Buddhism or the ten commandments in Christianity.

-However, humans usually forget or are not guided to realize that they are children of God. They misunderstand that they own their body, Mind, and the propriety they acquire during their lives. In fact, as children, dependent on their parents, they are given some privilege to use many means enough for their living, development, and work for the amelioration of nature. According to their fruitful work, they are commensurably rewarded. The remaining benefits from their work belong to the Creator. The non-proportional benefits belong to God and are considered extra giving to be used to benefit nature, communities, and other beings.

The extra giving can not be used for personal inappropriate entertainment. Abuse of the extra giving sows the seeds for bad Karma. Failure to recognize the impermanence, the egolessness, and the limit of privilege is the manifestation of ignorance. This failure is similar to that committed by Adam and Eva and is considered a sin that will be followed by punishment and suffering.

The Creator or the authentical owner of the Universe is God according to Christianity, Alla in Muslimism, and The Original Mind/ Wonderful and Miraculous Emptiness in Buddhism. There is no question that the above are uncreated, unchanged, and imperishable.

To understand the mechanism causing the ignorance, impermanence, and egolessness resulting in sufferance, humans from antiquity relentlessly search for the ultimate cause of the Genesis of the Universe.

This concept of three Seals of Dharma is not always acceptable in the common opinion, even in some religions. In Hinduism, with the concept of homology between Atman and Brahman, with Brahma being a perfect god, humans are perfect. Sufferencae is the problem of Humans lacking recognition of Brhama. The deviation causes bad karma, leading to suffering with a circle of samsara birth and death

XIII. NON-EGO, EGOLESS/ Emptiness of a Self/ Non-Personality/ skt: Anatma).

Since God/ Original Mind/ Emptiness is the Creator, all sentient beings are children of God. As in a family, children are dependent of their parents, and they have the privilege of having benefits commensurable to their needs to follow the guidance of their parents but have no free will. As a result, all humans are egoless and have no free will or ownership. They are not independent, self-existing, or sovereign. The proprieties and belongings that exceed their needs should be used for the benefit of others, not for unjustifiable, accessible, and unrestrained expenditures or uses.

A. The Self and Illusional Egoism or Social Egoism

As discussed above, only God/ Original Mind is the absolute Creator and the Owner, and has the Ego and Self. The illusional ego consists of Memory stored in the inner Consciousness and emotions that have the neural correlate as the DMN (Inner CS), namely PCC RSC, MTL, vmPFC, Precuneus (different types of MM), and related cortices like HIPPO, DMPFC, DLPFC, and ACC (indispensable for MM formation). This social egoism has the instinct to maximize one's own welfare in society and at the expense of others. The illusional ego is often intentionally or unintentionally covered up for individuals with an illusional ego and developed Consciousness. Many times, this covering-up is the attempt to lower the self or practice the Four Immeasurable Minds/ sk:t Catvariapramanani/ vn: Tứ Vô Lượng Tâm for its consolidation. Therefore practicing egolessness is a difficult task. Understanding the mechanisms of building up the illusional ego plays a central role.

Social egoism is often proportional to Consciousness, education, and high social class and inversely proportional to Awareness. The veil of ignorance limits Awareness. The Ultimate or Original Awareness is only present when the veil of ignorance is cleared from the obstruction of the perception of the sensory organs. Knowledge is considered as Understanding/ Consciousness + Awareness.

By adopting Einstein's formula in Ego=1/Knowledge and from the above discussion, an ego varies from 0-1:

$$EGO\ (,1) = \frac{CS}{KNOWLEGE} = \frac{CS}{CS + AWAREN} = \frac{1}{1 + \frac{AWARENESS}{CS}}$$

The false ego is commonly associated with preconception (vn: chấp trước).

Scalabrino distinguishes different disorders of the False Ego:
   - Self-constitution, the related disorder is Psychosis
   - Self-manifestation, the related disorder is Borderline Personality disorder
   - Self-expansion, the related disorder is neurosis

These three layers of Self is based on the true Self or Original Mind, Emptiness/ Buddhahood. A disorder related to the relation of the DMN with the Original Mind is Cotard syndrome.

Krishnamurti said: *The Self can never be anonymous; it may put on a new robe, and assume a different name, but identity is its very substance. This identifying process prevents the Awareness of its own nature.*
And
Anonymity is humility; it does not lie in the change of name, cloth or the identification with that which may be anonymous, an ideal, a heroic act, a country, and so on. Anonymity is an act of the brain, the conscious anonymity; there's anonymity that comes with the Awareness of the complete. The complete is never within the field of the brain or idea.
and
Nobody can teach you about yourself except yourself, so you must be the guru and the disciple and learn from yourself. What you know from another is not valid.
And
Meditation is not a practice; it is not the cultivation of habit; Meditation is heightened Awareness.

Therefore, to cultivate egolessness without spiritual practice is very difficult to achieve the goal logically.

Egoism has to be differentiated from Personality. Personality is the expression of the Ego (self-manifestation and self- expansion) that can be categorized into:
   - Neuroticism - Emotional stability.
   - Extraversion.
   - Openness to experience .
   - Agreeableness and,
   - Conscientiousness.

The personality is influenced by the heredity (Tellegen 1997, Plomin 1997, Kendler 1993 p341) and social interaction, education.

In post-traumatic stress disorder/ PTSD, the social Ego is altered. The traumatized Ego responds by the development of a mechanism of survival, manifested by two modes of Personality that can be instantly switched from the other:

· Emotional Part of Personality/ EP with Positive Dissociative symptoms. Therefore, the patient presents many symptoms, such as anxiety and pain. As a result, EP is associated with positive dissociative symptoms.

· Normal Part/ NP: Negative Dissociative symptoms According to Freud, the EP is repressed in the Consciousness. The patients appear normal. They appear normal however may present with visceral symptoms by the intermediate of the central nervous system with nuclei such as the Amygdala, ACC, HIPPO Insula mPFC, and finally, activation of the sympathetic and parasympathetic systems.

Egolessness can be categorized in:
  - Incomplete egolessness, such as in the case of Arahat,
  - Partial Egolessness with the decrease of three poisons: Greed/ skt: Raga, Anger/ Aversion/ Hate/ skt Pratigha/ Dosa and Ignorance, stupidity/ Delusion/ skt: Mudhaya/ p: Moha and development of four innumerable Minds.
  -Masked Ego usually occurs in an educated person with some lack of spiritual practice.

Buddhist Dharma distinguishes two types of Egolessness: Being type (or five Maras composed of Form and No-Form) and Dharma type (Awareness of the creation of Maras).

B. ALTERATION OF THE SELF: Depersonalization, Derealization Disorders/ DDD.

The neural correlate of the Self or the inner CS is the DMN. DMN can not work without connecting with the Orginal Mind or Buddhahood. Alteration of the inner CS differs from cleaning/ clearing the veil of ignorance. Contrarily, alteration of the inner CS may increase the obliteration of the Original Mind.

For example, an educated patient suffers cranial injury with subdural hematoma over the inferior Right Parietal area. After the removal of the hematoma, the patient develops Asomatognosia (loss of Awareness/ feeling with seeing the outside world) consistent with the derealization/ DR.

Another patient has the Right Subdural Haematoma in the Parietal Lobe due to a cerebrovascular accident; after the recovery from Left hemiplegia, the patient complaints agnosia involving the Right arm and leg with a feeling of loss of the Right leg arm and the body. There is no sensory loss of these parts of the body. The changes are consistent with Depersonalization/ DP.

Derealization and depersonalization usually occur when the lesion involves the right Parietal lobe. The Right hemisphere plays a role in artistic activity, comprehension, and emotional feeling.

The DR and DP do not represent the state of enlightenment because the loss of the respective connection with the Original Mind also accompanies the loss of the Self or part of the Self. The perception of the body part or external world is not lost, but the body parts and external world appear strange or unreal. The brain, in this case, is similar to a camera registering the images, but the photographer does not examine the images. This type of disorder has recently been recognized more often than before because medical practitioners increasingly recognize the entity.

C. DISSOCIATIVE IDENTITY DISORDER/ DID.
This disorder is also more and more recognized than in the century before. In DSM V (2013) the disorders are defined as followings:
    i. Two or Multiple Personalities composed of the Core (Principal) Personality and Alter Personality that replaces the Core Personality that is traumatized
    ii. Gap of Memory
    iii. Disturbance in the emotion and in social interaction
    iv. Disturbance in religious belief
    v. not associated with drugs or medication

The Alters personality can be difficult to recognize. Therefore in DSM V, the criteria of a positive diagnosis of DID can include the switching between Core and Alters personalities, such as a change in the manner of dressing, Personality, thinking, and emotional state. The Alters personality is responsible for temporary mental behaviors and can bypass the principal Personality and control an individual in routine daily activities for a short time. When the Alters personality controls the inner Consciousness (the Self), symptoms of DR or DP can develop.

DID is often associated with and masked by anxiety and depression neurosis. Of interest, the patient usually displays symptoms of numbing, withdrawal/ absorption, *Agarophobia, Amnesia)*

1. Dissociative identity Disorder (DID), Multiple Personality (MPD) Disorders of Extreme Stress not otherwise specified (DESNOS)

The DID is composed of at least two identities, the core Personality and the Alters Personality.

The two types are separated by Memory gap amnesia. The Amnesia gap is significant for the diagnosis but often difficult to identify. Switching between the two identities is frequent and can occur without obvious reason. The gap in Consciousness accompanies the Memory gap.

The incomplete Memory gap or Consciousness gap and inclusion of features of DR or DP characterize atypical DID. DIDNOS is another variant such as Fugues and DID with inapparent alert personality. Post Traumatic Stress Disorder PTSD may be associated with DID.

2. Borderline Personality may be associated with DID, DR, or DP.

3. Cotard's syndrome (Walking corpse syndrome) (Bermúdez 2021)

Cotard's syndrome is a rare disorder classified as a neuropsychiatric condition in which the patient denies the existence of parts, including internal organs or the whole of their body, and believes that he or she is dead. Hence, the patient may present with delusion, depression, and feelings of unnecessary eating leading to starvation.

4. Physiopathogenesis of DR, DP and DID.
  a) DP/ Depersonalization, DP/ Derealization.
   i. Theory of the Body image.

This theory is based on the existence of the inner Consciousness or the Mind that contains the body image. This image also stores the model of emotion that is continuously updated in life. According to the level of alteration of the body image, DR, DP or Cotard syndrome develops. Experimental injection of Stophanyl into the Thalamus can induce DP. **Hoff 1931: Zeitschrift fur die gesamte Neurologie und Psychiatrie 137,722-734).** The Thalamus is also part of the neural correlate for the Self and participates in the DMN.

The current concept tends to include DIP in the spectrum of DR and DID; the continuum of symptoms of DR, DID, nevertheless, DR and

DID are distinct from DID due to the presence of anxiety/.depression component.

ii. Other theories in a small number of cases
- Psychodynamic views.

DR and DP are manifestations of Libido investment in the subconsciousness level reactional to the Ego.

As a result, the disorder originates from exaggerated Narcissistic Gratification. The formation of the Self in infancy is the root of the problem.

- .Alteration of sensory perception.
- Alteration of the Memory with the impression of Déjà Vue, False Memory of double personality/ Double Consciousness.
- Alteration of Affect, such as in Melancholia Anesthesia, there is a lack of emotion in integration, the feeling of being imperfect and underappreciated, as in cases of obsessive-compulsive disorder.
- Phobic Anxiety.

b) DID

It is currently believed that DID is not causally due to anatomical lesions but due to stress.

- According to Pierre Janet:

Each individual represents a whole entity and may use the entire or part of the Consciousness with respective Memory. One part of the CS represents the core personality. The other part represents the Alters personality.

i. First law: each part of CS is associated with the specific Self. Therefore there is no amnesia but the switching from one to another personality.

ii. Second Law: The repression of others creates each type of personality. The objection to the mechanism of repression proposed by Freud is a surprise to his contemporary researchers.

iii. The two personalities can coexist at the same time, accounting for some automatism.

iv. Alters personality can completely replace the core Personality.

v. The person with DID can not control the switching of the Personality but can know that there are two or multiple identities. As a result, the subject "we" is used to design the status of multiple personalities. The manifestation is different voices, writings, and unrelated answers to the questions (Ganser's syndrome).

vi. The loss of Memory may last for various lengths of time. The loss only involves the explicit and implicit Memory (short and long term) but does not involve the procedural Memory is stored in the basal ganglia and cerebellum *(van der Hart 2001)*.

vii. DID often occurs in persons with High **susceptibility,** the personality of absorption, absent-mindedness.

c) Unstable Personality, Alternation of Mind.

- Pierre Janet did not indicate the mechanism of the existence of multiple personalities. Probably he is a clinician and experimenter but not a theoretician on the matter of metaphysic.
- The disorder can be successfully treated with hypnotherapy.
- PFC (vmPFC) may play a role in physiopathogenesis (Spiegel 2013, van de Hart 2001, Dell 2005)
  - Freud with suppression of part of the Memory: suppression of the uncomfortable Memories is the autonomic mechanism of survival, a mechanism of defense.
  - Current concept.

PTSD originates from a history of childhood abuse.
The physiopathogenesis is entirely unknown to medical professionals. Therefore, there is no effective therapy. Clinical investigation and imaging techniques do not help unlock the mystery of the disorder.. The above fact suggests that the mechanism needs to be more concerning in the context of the present medicine. On the other hand, one can conceive that some sentient beings may harbor more than one Soul/ Consciousness. In this world of duality, at the time of the ensoulment of the embryo, more than one Soul can enter into the embryo which is easily understandable. This phenomenon is similar to one ova fertilized by two spermatozoa in the case of molar pregnancy or partial mole. As a result, in individuals with two or more than two Souls, one Soul takes control of the physical body, others are repressed.

As in the section on Memory, the retrieval of explicit episodic and semantic Memories requires the specific Soul corresponding to the particular Memory.
For implicit Memories, the retrieval does not require a specific Soul. Therefore, the absence of a specific Soul is the cause of the amnesia, whereas the non-requirement of a specific Soul for implicit Memories facilitates the individual to meet the basic needs for life

5. Neuro Anatomical Biochemical Changes and Therapy of DID/ DR /DP

DP often occurs in patients with temporal lobe epilepsy, tumors in the inferior parietal gyrus, and Angular gyrus area related to out-of-body experience. The frontal lobe is otherwise inhibited in DP.
The above changes suggest the parietotemporal junction as a zone important in the integration of the images and emotion formation.

In Cotard syndrome, in addition to the TPJ as anatomical lesions, the fusiform gyrus of the Temporal bole (Face areas) has been suggested

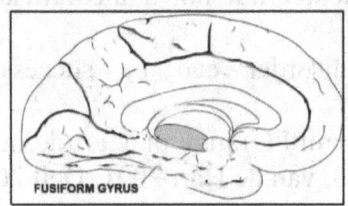

Antagonists of NMDAR like Ketamine ("Special K") Marijuana and Hallucinogens (LSD, DMT, Psylocibin, Opioids) are known for their dissociative effects. Lamotrigine reverses this action may cause psychotic side effects and inhibits NMDAR.

The sympathetic system with epinephrine is inhibited in DP, as occasionally seen in sexual abuse cases.

HPA/ HIPPO-Ptuitary-Adrenal Axe decreases the sensitivity in DP, as evidenced by the benefit of Benzodiazepine. But SSRI treatment is not effective in the control of the disorder.

XIV. THE CLEARING/ CLEANSING/ REMOVAL OF KARMA, TRANSFORMATION OF THE DESTINY

In the removal, cleansing, or clearing of Karma, the concerned Karma is completely removed, partially removed from the path of spirituality.

Karma is all the deeds each individual created in the past.. These deeds are registered in Consciousness in the form of Memory. The Memory is represented by a set of neuronal synapses in different parts of the ICS / DMN. Karma in the present life is mainly stored in the PCC, Karma in the previous lives is stored in store Consciousness, likey in the RSC. In addition, deeds in the form of information that does not receive enough attention are stored in the subconsciousness level in the present life or in the previous lives, Buddha said in Nikaya sutra that one could not hide Karma even when he is in the upper level above the earth, in the mountain or in the sea. Even the intentional repression of the Memories by focusing in other matters will not remove the Karma but only bury the Karma deep in the Consciousness or the subconsciousness and make its retrieval more difficult. (https://www.psypost.org/2021/01/neuroimaging-study-sheds-light-on-how-to-clear-thoughts-from-your-brains-working-Memory-system-59077, Wang 2019, Kim 2020).

It is noteworthy that the concept of Store CS is distinct from the concept of repository CS of the externalists (Laṅkāvatāra Sūtra)

> *"Thus, Mahamati, if the intrinsic of our Store Consciousness and the unfolding aspect of Consciousness were separate, the repository Consciousness could not be its cause. But if they were not separate, the cessation of the unfolding aspect of Consciousness would also mean the cessation of repository Consciousness. And yet, its intrinsic aspect does not cease. Thus, Mahamati, what ceases is not the intrinsic aspect of the Store Consciousness, only its karmic aspect. For if the intrinsic aspect of Consciousness ceased, repository Consciousness would cease. And if repository Consciousness ceased, Mahamati, that would be no different from the nihilistic views proposed by followers of other paths (externalists).*

(meaning: Store CS= past Karma+ Buddhahood, Repository CS of externalist = only past Karma)

The removal of Karma can occur in Meditation. In Meditation, one can look into the ICS by retrieving the Memory. After the retrieval, the Memory, , i.e. neuronal synapses, become more labile. If not reconsolidated, the synapses will be prone to destruction by the mechanism of neuroplasticity (destruction and repair). Neuroplasticity (destruction and repair) is the phenomenon of the removal of synapses by the phagocytic system.
accompanied by the regeneration of new synapses. In the cleansing and removal of the Karma in Meditation, purification is activated as one can progress from:
- Dhyana stage 1, (Srotapanna/ vn:Tu Đà Hoàn, the person will never again regress in his cultivated path and is guaranteed to reach Arhathood after, at most, seven more times of rebirths among Heaven and Humans).
- Dhyana stage 2: Sakrdagamin/ vn: Tu Đà Hàm a state of coming back once more.
- Dhyana stage 3 Anagamin/ vn: A Na Hàm—One who does not return to the earthly world—One exempt from transmigration—Who attains the third stage of Sainthood.
- Dhyana stage 4; attaining the level of Arhat.

Of course, this cleansing is also initiated with a voluntary commitment to not creating new Karma.

In Surangama, Buddha said to Ananda in tome 9 &10:
Ananda, in the cultivation of samàdhi, when the fourth aggregate of discrimination (saüskàra) comes to an end, the subtle disturbance in the state of clearness (that is, the functioning of samsaric Mind), which is the mechanism of birth and death, suddenly explodes and exposes an outlook completely different from that of the profound Karma of pudgala (, i.e. all beings subject to transmigration). This is when Nirvàda is about to dawn, like the cock-crow that heralds the first light of the day in the east, when the six senses are void and still and no more wander outside. Within and without, there is only a profound brightness reaching the root of the life of all beings of the twelve forms of birth in the ten directions of space wherein nothing can be further penetrated. This contemplation of the essence of basic clinging (, i.e. the fifth aggregate of Consciousness) releases the practiser from all attraction (by old habits and new Karma). It immunizes him from further transmigration in saüsàra for he has realized the identity of Mind with its self-created externals everywhere. As the nature of Consciousness now manifests clearly, he will discover its hidden depth. This is the fifth aggregate of Consciousness which conditions the practice's Meditation.

CHAPTER VI: THE SOUL.

I. THE SOUL.
A. Review of Concept of Soul in Literature.

From antiquity, most researchers have concurred that Memories and CS are seated in the brain. The brain manages specific types of MM and CS in specific areas, such as the Nucleus Accumbens with joy/reward, vmPFC, and OPFC with social interaction, and morality. Penfield described a case while performing a surgical procedure on the brain, stimulation of certain regions incited the MM in the past. In near-death experiences /NDE and in out-of-body experiences/ OBE, the CS can be present out of the brain, suggesting the Soul exists outside the body.

The Soul, known as the Soul or psyche by the Greeks, Amina by the Romans, and Atma by the Hindu, carries significant historical and cultural significance. These different names reflect the diverse perspectives and interpretations of the concept of the soul across different civilizations.

The Soul is often associated with religion, myth, and divine authority. Philosophers and scientists have been actively involved in the debates surrounding the soul, offering diverse perspectives and interpretations that have shaped our understanding of this complex concept.

It was commonly believed that after the death of the living beings with a physical body, the Soul leaves the body. This belief suggests that, from that moment, the Soul no longer belongs to the brain.

Moreover, the Soul is also considered an active force of life in animals, humans, and unicellular organisms (Pereira 2015).

In early Western civilization, the heart is believed to be the seat of the Soul, as today one may say, *"Words from the bottom of the heart,* or *"If I create from the heart, nearly everything works; if from the head, almost nothing." (Marc Chagall)* (this means that the head creates the Mind/ Consciousness, which is usually wrong. The heart means the true self/ Awareness/ close to the original Mind, which is always right) Since Hippocrates (5th century) and after that, the brain is commonly believed to be the seat of the Soul. According to Pythagore, intelligence, reasoning, and passion belong to the heart and the rest of the Soul to the brain; therefore, the Soul spreads from the heart to the brain. According to Aristotle (384 -322 BC) in the book entitled De Amina (on the Soul), a human is born with an empty Soul (Tabula rasa), which will be added up later by experiences and concepts. This concept is equivalent to that of Mencius, a great Chinese philosopher who also said *'Man is born virtuous /with right morality in original nature"/* vn: "Nhân chi sơ, tính bổn thiện".

Fernel (1496–1558) proposed in his book Physiology (Tart 1998) that the brain acts as a filter, cleansing the Soul from the physical aspects of the body, a concept that stimulates intellectual curiosity.

For Lansi (1654–1720), the Soul is located in the corpus callosum propagating along the nerve of Lanson, the dorsal aspect of the corpus callosum to the entire body (Di Ieva 2007).

Knowing that the pineal gland develops at the embryonal age of 3rd week is noteworthy. The gland contains a small number of nerve fibers and MNDAR. Importantly, this is the endocrine gland making Melatonin, which induces sleepiness/drowsiness. It also produces N-dimethyltryptamine with pharmacodynamic action as a hallucinogen related to the state similar to that in the Meditation (also called spirit molecule) David E Nichols (Chair in Pharmacology at Purdue University) called this reference illogical. In some animals, the pineal gland is related to the geomagnetic force that guides the direction of flying birds. The gland usually degenerates in old age and becomes calcified. The scientific relationship with Consciousness is minimal. Similarly, the reference to the pineal as the third/ wisdom eye is also baseless. Diseases of the pineal gland are not associated with psychological or psychiatric disorders. (Mittal 2010, Carson 1997, Mordecai 2000).

According to Sherrington (1857–1952), the CS is not only based on the cortex but also in the deeper part of the brain, separated from the sensory and motor cortices. For Pinker, CS is localized in the neurons and disappears after death (Pinker 2003).

More recent researchers agree that there are no reliable answers regarding the Soul's location, but the Soul is loosely related to the brain. In the case of craniopagus parasiticus, the human newborn with two heads and two brains (in Bengal in 1873) displays two different facial expressions and emotions (Allen 1989). Gazzaninga, in studies of patients with callosotomy (longitudinal splitting of the Corpus Callosum for the treatment of intractable epilepsy, reported a few cases displaying different emotional states when the corresponding hemisphere was exposed to different situations. In callosotomy, there is no intercommunication of separate information (Gazzaniga 2016). However, in a majority of cases studied by

Sperry (Nobel prize in 1990), patients with callosotomy only display one emotional state and do not feel that they have two separate minds. Sperry cannot logically explain the physioneuropathogenesis of this phenomenon.

Electromagnetic waves from the cellular metabolism may behave like the Soul and vice versa. This may create the axis Cell-Soul pathway(Arnold 2016). The objection to this theory is that the Soul is not associated with significant electromagnetic activity

B. **Features of the Soul.**
1. Soul is an entity- specific for each sentient being.
2. Pervasive effects on the baryonic and probably on the nonbaryonic matter or the Emptiness.
3. Absent or minimal effects on electromagnetic energy. No interaction with living beings if the nervous system is not involved.
4. Likely having the gravitational and attractive force. In the study of Dr Duncan McDougall in 1901 the human Soul may weigh 21.3g. However, the study is considered non-scientific according to scientists' preconceptions about the existence of the Soul.
5. Adherence propriety to different tissues of living organisms.
6. Able to remain still or move at a breakneck speed,
  receive information on sensory organs, create Consciousness with emotional sensitivity, think, store and retrieve information in the form of Memory outside the nervous system.
7. Based on the Buddhahood to create the CS. Without the Buddhahood, the formation of the CS is not possible.
8. Obliged to be associated with a physical structure for motor function like movement and motility for interaction with other sentient beings.
9. Able to ensoul in the embryo to complete the formation of a living being. The embryo constitutes the form, one component of the five Maras/ Aggregate of Grasping.

The Soul is responsible for the formation of the metaphysical part of the living being, Four of Five Maras/ Aggregate of Grasping, namely: The Form, Perception, Feeling, Mental formation, and Consciousness

By analogy to the Chinese Ultimate Pole, the body/ form constitutes the Form/ Positive/ Yang part. The Souls with four remaining Maras constitute the Yin/ Negative part. Such division is artificial, aiding examination and study because this division is never complete. In the body / Yang part, there is a minor Yin that symbolically represents the site of attachment of the Soul to the body. In the living human, the Minor Yin represents the inner Consciousness. The Yin or Major Yin represents the Soul,

After death, when the Soul leaves the body, leaves behind it, the remainder of this minor Yin represents the memoir of the person. In addition, the Soul carries with it the minor Yang, the essential component of the Memory, likely the imprint of synapses activated by the semantic Memory.

C. Relationship between the Soul and the Brain.
In the embryo and after the ensoulment, the Soul controls the development of the central nervous system. After forming both hemispheres, the Soul is likely and mainly seated in the DMN, where the inner Consciousness is stored, while the store Consciousness is likely stored in the retrosplenial cortex..

In the proposed structure of the Soul (please see page 280, 281), the Soul is not electronically charged, but after the ensoulment, neutrinos in the Soul are bound to electrons (-). The electronically charged Soul is attached to the electronically charged (+) synaptic membranes. When the synapses actively code the Memory, the synaptic membranes become depolarized and electronically (-). The Soul is, therefore, temporarily detached from the synapses. This change may represent the imprinting of the MM onto the Soul as a second copy of MM.

As a result of the above-proposed phenomenon, and in view of the pervasiveness property, the Soul likely attaches to all parts of the brain but mainly in the DMN where the critical component of the MM, the semantic MM, is stored....

During life, the explicit Memories (episodic MM and semantic MM are stored in the DMN, particularly the PCC, where the stored MM is called autobiographic MM.

The short-term MM is transitorily stored in the HIPPO before being transferred to vmPFC for consolidation. The implicit MM is stored in the basal Ganglia and cerebellum for procedural MM. Other MM for space and location are variously stored in the HIPPO and Precuneus ACC.... At the same time, with encoding (for storing) of MM., the autobiographic MM is imprinted onto the Soul. The mechanism of imprinting is unknown and has not been investigated but is proposed as above.

In addition to the storing, the Soul is also necessary for retrieval of the MM, as evidenced by the amnesia in DID because the repressed principal Soul is not available in this disorder. The mechanism of the soul as guiding the neural transmission in encoding and in retrieving MM is discussed in chapter I, paragraph **PARADIGM OF SOUL AS A QUANTUM COMPUTER (Figure 1.16), page 126.**

The storage of the MM in the Soul may account to some extent for the relatively intact MM in some children with severe hydrocephalus.

II. SOUL OF THE BODY OUTSIDE THE BRAIN.

As previously suggested, the Soul is a part of the wholeness of a structure generally conceived to be composed of two parts: one part, the form, is accessible to the five sensory organs, and the other part, the formless, is not accessible to the five sensory organs. The brain's Soul is responsible for Consciousness enabling living beings to think. By analogy, the Soul of the non-neuronal part of the body can develop the strength of the tendinomuscular system and allows the skin and viscera to develop sensitivity to stimuli and motility .... These capabilities are well known by the Far Easterners as Chi or Qi in acupuncture or martial art.

In Samahdi with near complete extinction of sensation and thought/ Kenosis/ vn:Diệt Tận Định, the Soul may play a role in maintaining the integrity of the body.

For example, the martial arts practitioner can use the cubital side (side of the fifth finger) of the hand to break one or several pieces of brick without suffering damage to the skin or bone of the hand. If F^0 is the force applied to the object and is the force of reaction, Fbreaking, the equation will be: $F^0 = F^R + F^{breaking}$
Commonly $F^R > 0$

The significance is that between the hand and the brick, there must be a medium sustaining the F^R. This medium is the Qi or the Soul of the non–neuronal part.

The numbness distribution is non-neuronal, therefore suggestive of the alteration (stagnation) of the Qi/ Soul.

The Soul of the non-neuronal body accounts for the Qi/ Chi in acupuncture, which controls the pain of different body parts. The site of needling in acupuncture and areas of controlled pain do not correlate with the nervous system distribution.

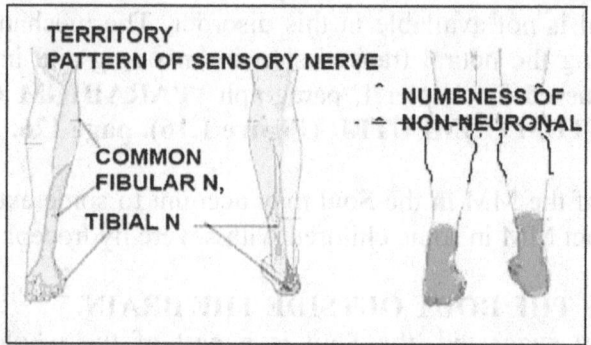

Fig 6.1 In Meditation, sitting with crossed legs usually causes numbness in the lower leg. (Fig 6.1).

1. The connection of the Soul to the non-neuronal body and System of Acupuncture Channels and Points.

The Soul of the non-neuronal part is probably attached to the correlating part. The acupuncture channels under the skin's surface could represent an important attachment part. The acupuncture channels are usually 1cm in depth under the skin and follow the path of the vascular and nervous systems. The points of acupuncture are usually located near joints and tendinomuscular junctions.

Phenomenon of Fractal in Acupuncture and Homeopathy (Fig 6.2).

Fig. 6.2, A Fractal is a rough or fragmented geometric shape that can be split into parts, each of which is (at least approximately) a reduced-size copy of the whole. Simply fractal is self-similarity at a larger or smaller scale.

In acupuncture, the procedure can be performed at restricted areas of the body to achieve a similar result when performed on the acupuncture type of the whole body. For example, ear, face, palm, and plantar acupuncture. This phenomenon again supports the concept that acupuncture operates not based on the nervous, vascular or endocrine system.

Homeopathy therapy is another example of the Fractal phenomenon: in this type of therapy, one uses an infinitesimal quantity of medicine. Smaller amounts can have more potent effects on the illness. Probably the mechanism is likely a metaphysical aspect of the medicament.

As previously discussed, the Fractal phenomenon seen in Acupuncture and Homeopathy is consistent with the feature of homogeneity of the Emptiness.

2. Soul of Aminal of low phylogenetic level.
Based on the concept of the Soul as an unidentifiable part of living and non-living beings, there is gradual progress in the complexity of Souls of the minerals or organic compounds. The complexity will be followed by the ability of the Soul to form Consciousness for perception and storage and help to communicate with others.

3. Soul of the mountain, river landscape, country.
Is the summation of the components like soil rocks, plants, rivers, and mountains....
The city has a face; the country has a Soul. - Jacques de Lacretelle.
(Likely the inhabitants of the town often change, but the landscape rarely changes, therefore the Soul of the town is stable).

4. Soul as an entity in the Universe and Proposed Soul Architecture

Table 1: Forces and Matters in Universe

Six organs of perception	Force/Form	Particle/ matter	Connecting force	Range	Force
Perceptible, non-emptiness or physical /Form	Weak force,	Quarks, leptons	W,z (decay, radioactive)	Very short	Weak
	Strong force (quarks forming Protons and Neutrons)	Quarks, Gluons binding Quarks	Gluon/Boson (nuclear binding)	Short	Very strong
	Electromagnetic incl Gamma Ry	molec binding)	Photon /Boson(	Infinite	Strong
Non perceptible like Emptiness/ Formless	Gravity (mass binding)	mass	Graviton not yet determined	Infinite	Very weak
	Dark force of 5th force	Universe expansion	? Dark photons	Infinite	Strong

The Universe is likely composed of Matter:
- Baryonic Matter, accessible to vision, touching, hearing, smelling, tasting
- Non-baryonic Matter or dark matter not accessible to the five sensory organs but has caused mutual attraction between all things

HYPOTHESIS OF STRUCTURE OF THE SOUL.

Dark Force, Dark Matter, and Neutrino, enigmatic entities in the Creation, bear intriguing features of the Soul. The Dark Matter, a potential cornerstone for the architectural arrangement of Neutrinos, adds a layer of mystery to this cosmic relationship. After ensoulment, electrons are attached to

Neutrinos; as a result, the Soul, a fundamental entity, becomes electrically charged. This revelation of the Soul's electrical nature carries profound implications. Outside the body, data can cause an architectural change in the arrangement of Neutrinos.

In the brain, the presence of electrons in the Soul renders the Soul the capacity to attach to Synaptic membranes, which are positively charged. In

During memory encoding, the depolarisation of the synaptic membrane triggers the detachment of the Soul from cell membranes. This process of attachment and detachment could be a mechanism overseen by the Ultimate Omniscience within the Soul.

After the death of the brain or in the Meditation with out-of-body experience, synapses lose the electrical potential, .allow the Soul detached from the brain.components of the constituent of the Universe, given the fact that the Soul does not have electromagnetic energy but is capable of moving and gravity. Furthermore, in the brain, the Soul may have the additional capability to attach to neuronal synapses, which are polarized when the synapses are activated to create Memories. Therefore, the Soul could be composed of dark matter and neutrino. Neutrino is chargeless but can be attached to an electron with a negative charge. As a result, the Soul can be attached to the synapse with a positive charge when the synapsed is activated.

HYPOTHESIS OF STRUCTURE OF THE SOUL.

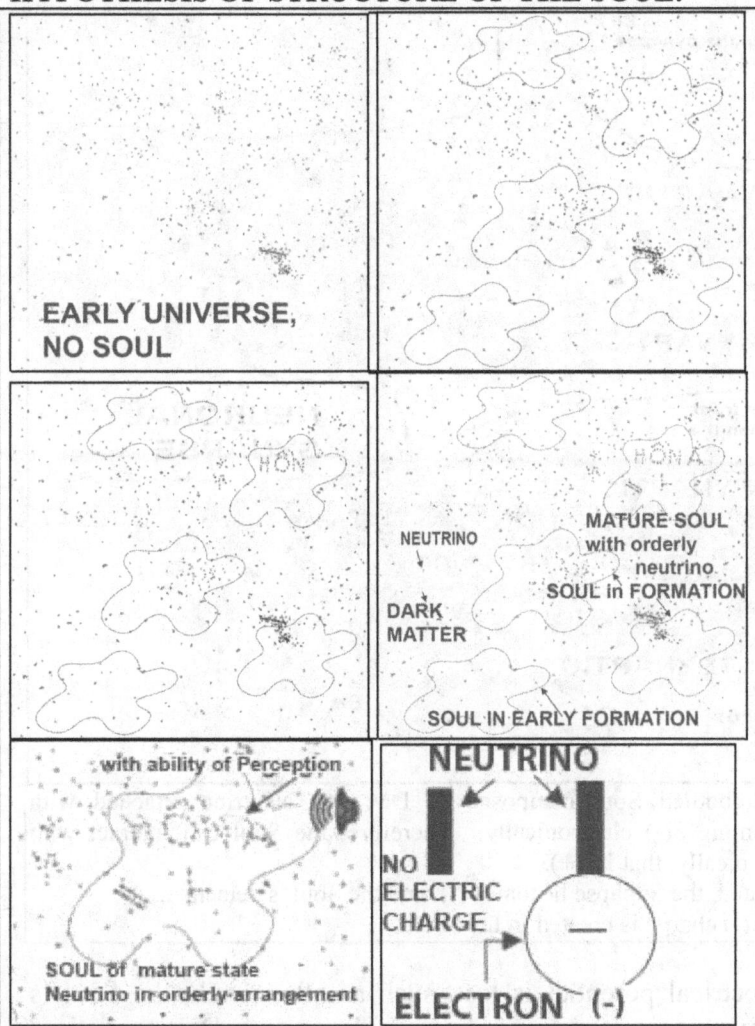

Fig 6.3 A,B,C,D, E,F
By analogy with Memory Computer Disc composed of billions of transistors (comparable to ON/OFF switches) arranged in many layers to register the electromagnetic changes:
Each particle of Neutrino, with or without attachment with an electron of a negative charge, renders the Soul to have an electric potential and electromagnetic field. For the information input, light, and sound...) exert momentum on the neutrino, therefore creating a change in the structure of the Soul. The architecture of the Soul changes with each sentient being.
A: Early Universe with Neutrino, Dark Matter and Dark Force.
B, C: in simple, low-level sentient.
D: higher level sentient.
E: Soul with Neutrino affected by the sound (speaker), without interference of the Brain.

F: Neutrino attached by an Electron becoming charged with (-) potential.

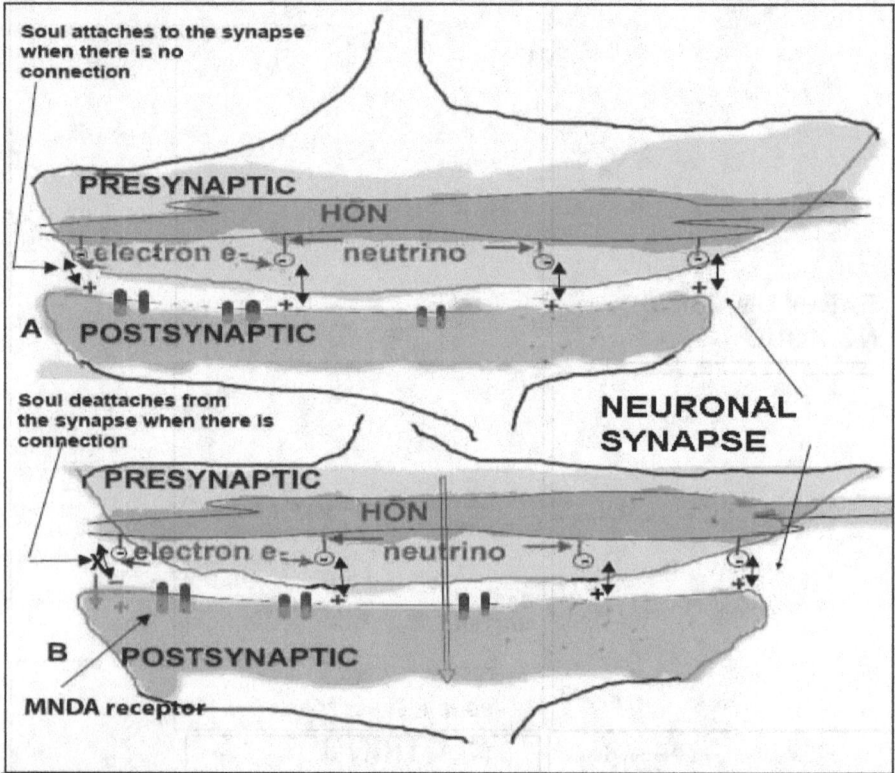

Fig 6.4 A: Embodied Soul composed of DM and Neutrino attached with Electrons becoming (-) electronically. Therefore, the Soul can interact with Synapse electronically, that is, (+).
B: when activated, the Synapse becomes (-), and the Soul is detached. As a result, the (-) charge is created in the Soul.

When the electrical potential is lost, such as after death, the Soul is detached from synapses and loses electrons. Long-term Depotentiation/ LTD occurs when the synapses are activated and represent the same phenomenon with the detachment of the Soul (likely associated with a change in the architecture of neutrinos in the Soul). In Meditation with OBE, the synapses become refractory stimulation with membrane potential >-55mV).

NEUTRINO AS POSSIBLE PARTICLE OF THE CONSCIOUSNESS

In view of the role of the Neutrino in the Soul, which plays a role in registering, storing, and retrieving memories (see page 122), Neutrino is likely the elementary particle of Consciousness.

III. ENSOULMENT

The complexities in structures of organs with high capabilities of absorption of nutrients characterize animals. Moreover, the essential differences are the nervous system enabling animals to react to the environment and communicate with other animals through acquiring information and developing the Consciousness stored in the central nervous system and in the Soul.

The Soul and the Body are separate parts of each sentient being, conforming with the concept of duality in the process of ensoulment. The embryo with stage 1-4 cells is undifferentiated and is not dissimilar to vegetal cells because of the absence of nerve cells. The Mind is the most essential and sacred part of each sentient. Phylogenetically the development of the nervous system is paralleled with the progression of the Soul and Mind. In the duality of the Soul and body, the Soul's attachment to the body can be conceived as the attachment of the negative and the positive electrical charge to each other. As previously outlined, the separation of an entity as the wholeness into two separate parts can never be complete, and vice versa; the combination of the Soul and the body is ingeniously intimate.

The transition of the embryo from belonging to the mother to becoming a separate individual is a significant stage in the development of the Soul. The non-fertilized ovum shares the Soul with the mother's non-neuron Soul. After fertilization, as the embryo progresses to the stage of Morula, it begins to form its own identity. The implantation into the gestational endometrium marks the point at which the embryo becomes a distinct individual, separate from the mother.

1. Christian concept of ensoulment.

The ensoulment occurs at the same time as fertilization. However, in cases of non-identical twin or quadruple or higher order multiple pregnancies, each embryo developing separately from each blastular cell, it is obvious that the fertilization of each subsequent blastular cell occurs later in the embryonal development.

Aquinas, a priest, philosopher, and theologian (1225 – 7 March 1274) agreed with Aristotle that sperm power generates the human Soul. This Soul only enters the embryo in the later stage of development (40 days after the conception of male conception and 90 days in the female embryo. Aristotle claimed that God made intellectual Souls at the beginning of Genesis. The Soul undergoes three periods of transition: vegetative (plant Soul), Sensitive Soul (animal), and Rational Soul, which is distinct due to its intellectual character. This Intellectual Soul is only present in humans many weeks after conception. (Aquinas)

Isaiah 44:24 says, *"This is what the Lord says- your Redeemer, who formed you in the womb: 'I am the Lord, the Maker of all things, who stretches out the heavens, who spread out the earth by myself.'"*
Or
Psalm 139: 13-1 says, *"For you created my inmost being; you knit me together in my mother's womb. I praise you because I am fearfully and wonderfully made; your works are wonderful; I know that full well."*

Again, this is consistent with God's spiritual power, which is sophisticated, ingenuous, and elegant.

The non-fertilized ovum belongs to the mother and shares the Soul with the mother's non-neuron Soul. After fertilization until the stage of Morula (stage of the embryo having more than eight cells) with the formation of the cavity and embryonal disc, the embryo starts the implantation into the gestational endometrium. After the implantation, the embryo is a separate individual and does not belong to the mother

. 2. Ensoulment in Artificial Fertilization of high-level multiple pregnancies.

In artificial fertilization and identical twin pregnancy, Rose Koch-Hershenov has demonstrated that, until day 4 of Morula (up to 8 cells before embryonal disc formation), each cell is an individual stem cell, having the capability of developing into a separate embryo (Smith 2018, Koch-Hershenov 2006). Therefore if the ensoulment occurs at fertilization, each cell of Morula has an individual Soul. The corollary is that in artificial fertilization or in natural pregnancy, each living being developed at the expense of the other Soul, which has no chance to develop into a sentient. This appears to be illogical as God designs such a paradigm.

3. Buddhism and Ensoulment

Buddha pointed out to Ananda in Surangama, chapter 10,
"Your body owes its existence first to your parent's thought of giving birth, but had you not thought (of being born), there would have been no chance for your incarnation in their thought."

4. Proposed Ensoulment.

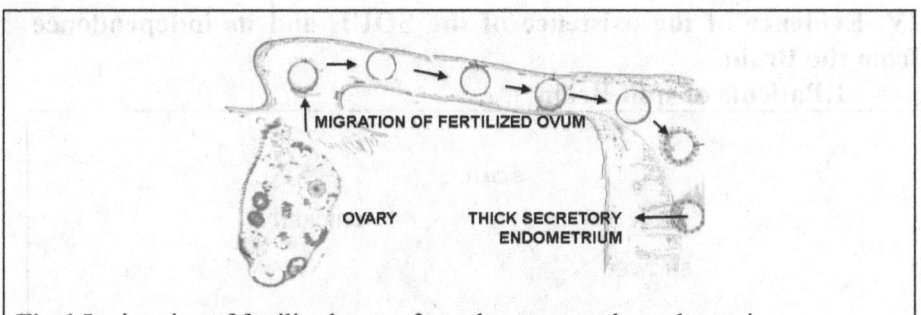
Fig 6.5 migration of fertilized ovum from the ovary to the endometrium

Building on the above findings, this book proposes a significant timeline for ensoulment. It suggests that ensoulment is likely to occur after the implantation of the fertilized ovum into the endometrium, specifically between day 4-6 of the embryo (stage of blastula). This timeline is crucial as it marks the transition of the embryo into a living sentient being. From the 6th to the 30th day of the embryo, ensoulment will be associated with the nervous system's development, marked by the successive formation of the spinal cord, brain stem, and cerebral hemispheres. Before ensoulment, the fertilized ovum or embryo shares the mother's Soul.

Another intriguing question is whether the Soul can guide the development of the central nervous system. The Soul demonstrates a progressive and continuous transition from a vegetative to a sensitive and rational Soul. This transition suggests that the human (rational/ intellectual) Soul may appear at the beginning of the ensoulment. The manifestation of the intellectual Soul only occurs when the cerebral hemisphere develops, shedding light on the intricate role of the Soul in the development of the nervous system.

According to philosophers Donceel and Pasnau, the Soul only appears when the brain develops with signals in EEG. It is also reasonable to believe that the Soul is only present when the human form is apparent after the gestational age of the 20th week. This reasonableness provides a strong foundation for the proposed timeline for ensoulment.

The Soul can only manifest when a neural tube with neurons are formed. Therefore, according to scientists, the ensoulment happens after 3-5 weeks.
This suggests that abortion can be allowed until seven weeks of gestation. Since 30 days of pregnancy, the Thalamus is present but cerebral hemispheres have not been developed.

IV. Evidence of the existence of the SOUL and its independence from the Brain.
1. Patients of split Brain

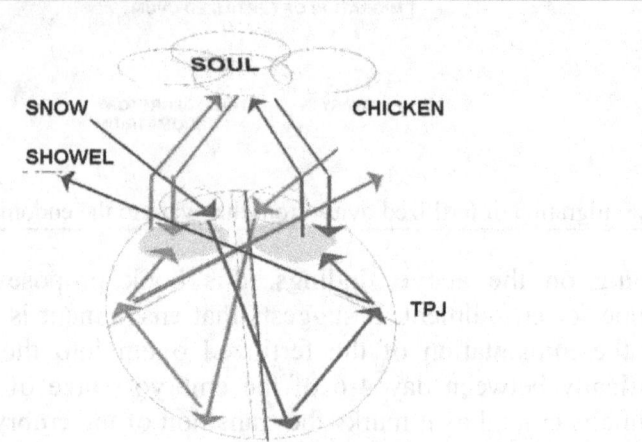

Fig 6.6: a) Path of conduction of the data in the split brain. When a chicken claw is shown in the right visual field (corresponding to the Left hemisphere, the patient reasonably acts by choosing the chicken and understands the action logic because the Left hemisphere is the predominant side with complete CS function.

When a shower is shown in the Left visual field corresponding to the Right hemisphere, the patient appropriately chooses the snow but does not understand the action logic. The Left hemisphere has limited CS, mainly for art and emotion. In the split brain, the information received by the Right hemisphere is not shared with the left side due to the severed connection between the two hemispheres. As a result, the CS still needs to be completed.. Due to the soul not being divided in the split brain, the patient still has one soul.

    b) In an experiment using hands for a specific maneuver in case of a person with split brain and if the action is first performed with the right hand, then completed with the Left hand without difficulty:

    c) Repeat the experiment b) With blindfolding of both eyes, the patient cannot act with the Left hand.

This is because the Left brain learns to perform action after seeing. This is called self-cueing. The Right TPJ plays the role of imagery after seeing the left hand performing the movements. According to Voltz (2017) and Gazzaniga (2013), learning is related to vision, the CS, and subconsciousness. However, for other researchers (Pinto 2017, Lamme 2003), more than cross-cueing is needed in the explanation of all phenomena in the split brain. The perception is likely split, but the CS can not be split.

In treating intractable epilepsies a few decades ago, callosotomy by splitting the brain along the corpus callosum appeared to control many cases. The left hemisphere plays the sensorimotor, reasoning, and Consciousness role. The right hemisphere plays a role in conception, art, and abstraction,

In the experiments of a patient with callosotomy, the patient only expresses one Consciousness despite each hemisphere receiving separate information. The physiopathogenesis has debated the controversies regarding different mechanisms as follows:

- Most researchers agree that there is no evidence of interconnection of hemispheres through the subcortical connectivity (through diencephalon and mesencephalon)
- Cross cueing between hemispheres without the need for a single integrated conscious control (Volz and Gazzaniga, 2017a,b,c): "pictures and objects in the left visual field cannot be named through overt speech, they can be matched with written or spoken words, matched to conceptually related items, stored in short term Memory for matching with subsequent probes: (Gazzaniga, 1995
) Moreover, high-level cognitive judgments of the right hemisphere can initiate appropriate and accurate motor responses without the knowledge of the speaking hemisphere (Gazzaniga, 2000

). This is difficult for other researchers to be able to happen.
- Therefore, it is likely that the Consciousness of both hemispheres exchange the information through the no physical mode, namely the Soul.

Pinto and Lamme proposed that Consciousness plays a role in keeping the patient with two separate hemispheres displaying one CS still using the subcortical nuclei. However, the HYPO and tegmentum do not play a significant role in CS. As neuroscientists only recognize CS is associated with the brain, since they do not accept that CS can exist without the cortex, the problem of unlocking the mystery of splitting into two hemispheres with one CS has remained unsolved.

The above experiment demonstrates that a patient's Consciousness with two separate hemispheres is still connected even though the two hemispheres are separated. This phenomenon is similar to EPR Paradox/ Einstein-Podolsky Rosen Paradox, in this phenomenon, two electrons or photons are separated but still communicate with each other due to quantum entanglement or non-locality/ interconnectedness. This is also called Local realism. The phenomenon will be discussed in Chapter XI, paragraph VII, Free Will.

God doesn't play dice. In atomic-scale physics, the physicists "come to praise Bohr and decry Einstein," but they need the world of Newton and Einstein on a large scale. In fact, as Heisenberg echoed, the elementary particle is not fundamental; they form the world of potentialities and possibilities rather than one thing of fact. So, as Buddhists believe, there is no reality at the quantum level. The wave function of light and sound represents the wave of probability, not the oscillation of matter. The wave function is simply the mathematically calculated probability that a particle will be discovered in the condition set by the experiment. The above said one can feel that our macroscopic world is, in fact, illusional. Furthermore, some scientists believe that Consciousness does play a role in quantum mechanics and accounts for the fuzziness in subatomic particles. Even if the machine registers the results of the experiments, the interpretation always relies on Consciousness, on the human cortices.

Furthermore, the realities are multifaceted, while Consciousness, if limited in potential, is narrowed down to focal areas. Therefore, Awareness, the Original Mind, must be used without any technology to discover the realities.

In Lotus sutra, the omnipresence of Buddha or Bodhisattsavas is characterized by the presence of multiple responding bodies/ skt: Sambhogakaya/ vn: Báo thân, at the same time at different places, planes, worlds, performing acts and speaking or teaching the same things.

> The Buddha's body fills the cosmos, Appearing before all beings everywhere-
> In all conditions, wherever sensed, reaching everywhere,
> Yet always on this seat of enlightenment. In each of the Buddha's pores
> Sit Buddhas as many as atoms in all the lands,
> Surrounded by masses of enlightening beings
> Expounding the supreme practice of the universally good.
>
> Buddha, sitting at rest on the enlightenment seat,
> Displays in one hair oceans of fields; The same is true of every single hair,
> Thus pervading the cosmos.
>
> He sits in each and every land, Pervading the lands one and all;
> Enlightening beings from everywhere gather,
> All coming to the enlightening scene.
> (......)

In addition. knowledge of the Buddha is overwhelming everywhere, at any time, all inconceivable. One Buddhra produces infinite Buddhas, thunderous sound pervading all lands In each hair pore are infinite light sounds pervading all quarters

According to the principle of indeterminism from Copenhague's Quantum Mechanics Interpretation, the nature (Path and Momentum)

of a particle is undetermined until it is observed. Therefore in Quantum Mechanics, the particle is "not known as it is but as observer is".

However EPR paradox, which is later demonstrated in "reality", particles display features and behave in a fashion similar to Emptiness with Unreal Locality (**Not Locally Real**) and Interconnectedness. Furthermore, particles, due to the infinitesimal size close to zero are consistent with the concept of Emptiness with homogeneity, and Non-discrimination.

2. Attachment of the Soul to the Brain.

The body and the Soul are commonly considered as two separate entities. For the inherent Soul, the ensoulment in the intrauterine embryo has been described before. The attachment is likely by the intermediate of the gravitational attraction and the electromagnetic force developed at the synapses. Along with the attachment, electromagnetic changes at the synapses may be accompanied by electronic changes in the Soul. The changes will follow this in the subatomic particles in the Soul. The changes are specific for each electrical potential and are comparable to the encoding of the Memories.

In addition, in some cases extrinsic Soul can be attached and adherent to synaptic neuronal junctions in a mechanism similar to the inherent Soul.

The attachment/ adherence of the Soul to synaptic junctions plays a role not only in Memory recording to the Soul for storage but also in retrieving the Memories from the Memory-storing cortices of the DMN and the sensory cortices.. The evidence is from the DID. Patients of this disorder have at least two Souls. When the principal Soul is suppressed, the explicit and implicit Memories related to the principal identity remain intact but can not be retrieved. As a result, subsequent amnesia develops. However, procedural Memories are not affected by the disorder. Therefore, retrieval of explicit and implicit Memories, not procedural Memories, requires specific Soul.

3. Phenomena of out-of-body experiences and near-death experiences and other supernatural events.
a) Near Death Experience.

i. Near-death experiences have been recognized from antiquity, recorded by Plato, Egyptians, Tibetans, north-American Indians…Since the development of cardiopulmonary resuscitation, there has been an increase in cases of near-death, and the NDEs are more frequently seen than before. The book Life After Life in 1975 by Raymond Moody, then a medical student, marked the beginning of the proliferation of publications on the topic of NDE .

The incidence of NDE is challenging to estimate, ranging from 6-23%, because of the facility to report the experience due to religion, culture, personality, and psychological problems. NDE is more prevalent in suicide 15-47%, accidents 23%, chronic illness 60%, and delivery 7%. The high prevalence increases with the frequency of dreams in sleep, fantasy, a Memory of previous life, and emotional instability.

ii. Characteristics of NDE.

- Out of Body Experiences.

The Soul is elevated to the top of the head. The leaving out of the body is commonly associated with varying phenomena ranging from

- The welcoming radiant light with a feeling of liberation and happiness, strange musical sounds.
- Distasteful, unpleasant, dreadful, like literal hell.

- OBE can be accompanied by looking back at the body (autoscopy).
- Flight with high speed in the tunnel of light.
- Encounter of angels or religious leaders.
- Vision clearer than before.
- Review of the significant and vivid events in life in the past, more in Asians.
- Encounter of deceased relatives.
- Encounter people unknown before for sending a message to their relative when returning to life.
- Returning to the body, forcibly, voluntarily, or by unknown cause .

b) Autoscopy, Heautoscopy, Out Of Body Experiences, Supernatural Phenomenon Or Mental Disorders.

i. OBE can be accompanied by autoscopy. The person is conscious that the Soul leaves the body with:
- Floating under the ceiling and looking back at the body.
- Flying in the open sky, looking down at the landscape below with a vivid vision.

ii. Autoscopy: a vision of the face and upper part of the body. There is no feeling of the Soul leaving the body, floating or flying.

iii. Heautoscopy: intermediate between 1) and 2). Seeing only the upper part of the body.

All the above cases can be accompanied by the audition of sounds and voices.

iv. Meditation with visual experiences and OBE.
- Out of Body Experiences.

Similar experience in NDE.
Also reported in epilepsies, stimulation of the temporal lobe with a needle or by neoplasms, mindfulness Meditation, and in rare cases without an obvious cause in normal persons. OBE can be accompanied by autoscopy. The person is conscious that the Soul leaves the body with
- Autoscopy, Heautoscopy.

In the Surangama, chapter 9-10, Buddha pointed out that
In the meditation Mind becomes pure (and cleansed from Karmas), one can see the external world, temples , buildings, people, demons multiple Buddha (as numerous as snds of Ganges river)

Buddha warned to the Assembly that these vision result from the intermingling of the one of the five aggregates (in this case :form/skt: Rupra/vn:Sắc) with the Mind (Buddha means the Inner Consciousness retrieved by the Mind represented by the Awareness/vn: Trí Tuệ /Janan and Prajna, not Trí Huệ /transcendental Awarenss/ Prajna-paramita .skt)

In Surangama Sutra, Chapter 4, Buddha said
"Ananda, do not you see in this assembly Aniruddha who is blind but sees, Upananda who is deaf but hears, the Goddess of the Ganges who is noseless but smells, Gavampati who does not taste with his tongue and the God of Sunyata who has no body but feels touch.

...... And Mahakasyapa, who is here, succeeded long ago in rooting out the organ of intellect thereby realizing perfect knowledge which does not derive from the thinking process".

Meaning: the CS in one of five layers of the veil of ignorance. Meditation clears the karma of the CS and makes it detached fron the brain (becoming transparent), so the Original Mind can receive the data as it is.

c) Mechanism of NDE: NDE is not a Supernatural Phenomenon
Phenomenon (Mobbs, D. and Watt, C. 2011 Holden JM2009, Tanne JH 2007: Tucker,JB 2021).

i. Psychological reaction to the fear of death. The incidence of Tunnel of light in NDE is variable, according to the culture. However, this incidence remained the same after the NDE became more recognized to the public after Raymond Moody's report.

ii. A manifestation of depersonalization. However, in DP, the feeling is frequently unpleasant as compared to the pleasant feeling in OBE, in NDE.

iii. Dreams, fantasies, and illusions are different from NDE by the incoherent, illogical stories in dreams, lack of clarity in fantasies and illusions.

iv. Bood biochemical changes before death:

- Change of blood O_2/CO_2 before death, causing the narrowing of the visual field with an illusion of a light tunnel.

Pilots of fighter jets commonly experience a momentary loss of Consciousness followed by a pleasant impression of a tunnel of light and even an impression of OBE. However, loss of CS, orientation, and transitory paralysis are usually not experienced by the person with OBE. In addition, when meeting relatives or acquaintances, angels are not reported in hypoxia.

- Increased blood level of endorphin, NMDAR can cause a pleasant feeling but not a long-lasting feeling in NDE.
- Ketamine in anesthesia can induce illusional or nervous feelings due to the mechanism of imagery nearly identical to feeling in NDE.
- Cerebral cortex TPJ is an area that can cause OBE feeling with needle stimulation or by neoplasm. These cortical areas are adjacent to the auditory and visual cortices and receive vestibular input from the cerebellum. In exercises like jumping, drawing pictures composing music, or practicing a speech, this TPJ uses long and short-term Memory to create different scenes for the practitioner to review.

In the state of near death, there are changes with the increase of monoamines (like DOPA and norepinephrine), serotonin (Morse 1986) can mimic NDE.

fMRI of a person with voluntary OBE demonstrates activation of the supplementary motor cortex (area for planning movements, TPJ, inferior temporal gyrus, cerebellum, middle and superior OPFC (related to the ICS).

For Greyson, the above reports of cerebral and humoral changes with mental changes are not similar to the cognitive changes in NDE, which are more orderly, logical, specific, and characteristic in NDE. The Soul

(Penfield 1958) (Fig 6.7).

> Further reading
> Lateral aspect of the Temporal lobe.
> -26-year-old female, Temporal Left: at various points, the patient reported seeing herself giving birth to a baby many years ago.
> - 14-year-old female with epilepsy, having an illusion in her childhood, walking on a beach, being followed by a man saying that there was a snake in the bag, she ran away. Thereafter, the story was repeated in many dreams.
> Penfield was able to cause the illusion of seeing relatives and fears in a patient by the stimulation of brain areas with a needle.
> When stimulated, different locations marked by a number are associated with specific stories.

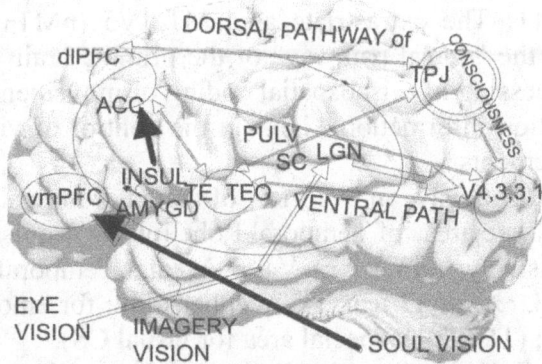

> IFG: Inferior Frontal Gyrus, FG: Fusiform Gyrus, IPS:IntraParietal Sulcus
>
> **Fig 6.7** Three mechanisms of vision using the same pathway but with different directions.
> -**Eye vision**: → Lateral Geniculate body (LGN) →Occipital Visual cortex --: dorsal stream→PFC, and → Ventral stream → Temporal Lobe .
> - **Soul**: or vision without eyes →PFC → Occipital Visual cortex.
> - Imagery Salient network (Dorsal ACC –Insula) stimulating PFC /dlPFC then using Soul pathway.

has the impression of liberation with happiness, traveling to distant, scenic areas, and encountering people sending messages to living persons. The later cases, are difficult to unlock the mystery using the scientific mechanism (Smith 2014, Carmona-Bayonas 2016, Bos 2016 Greyson 2012, 2014, van Lommel2014, 2002, Blanke 2005).

Penfield reported that the stimulation of the middle and inferior temporal lobes could trigger images and sound more precisely than in a dream, suggesting that each area of the temporal lobe corresponds to individual videos of life in the past.

The distinction between the different mechanisms of imagery in the brain

One can imagine and visualize images in the brain by using experiences of exposures of images seen in the past. Retrieval of these images in the ICS has been studied (Dijktra 2017, Mechelli 2004). The retrieval pathway is similar to storing these images when seeing with normal vision or without using the eyes. These pathways involve the PFC, TPJ, inferior Temporal cortex, and the occipital cortex

- ***Auditory Pathway, additional.***
  d) Visual pathway (Fig 6.8).
- Dorsal stream- *CS/Mindfulness) V1>hMT>PPC>motor cortex>Frontal Eye Field:* for eye movement (HOW?) for defense reflex. hMT: The extrastriate area MT [V5 (hMT+) in humans] located in the "dorsal pathway" of the primate brain is specialized in the processing of visuospatial coding of movement direction for visual motion information , i.e., in the control of visually guided hand movements.
- Ventral pathway: an overview of the environment *(Awareness)*, involving large areas of Temporal Lobe for the acquisition of CS:
  - Successively from V1 → V4 → Medial Temporal Lobe (MTL) →TEO,TE, →Amyg/ fear, N. Accubens/joy: for emotion in vision (WHAT?); (TE, TEO: special area for visual CS).
  - → Pulvinar /Superior Colliculus for location (WHERE).
  → >> Finally the information transferred to Inner CS (Carruthers 2005; Rosenthal 2006; Lau and Rosenthal 2011).

IFG: Inferior Frontal Gyrus, FG: Fusiform Gyrus, IPS:IntraParietal Sulcus, TEO, TE: Particular cortices of Temporal Lobe specialized in shape.

Visual information in the brain consists of:
- Eye perception: eyes→ LGN → Visual cortex → inferior Fusiform Cortex of Temporal Lobe IFC→ IPS/IntraParietal Sulcus → → Dorsal and Ventral pathways.
- Imagery Dorsal/TopDown): in the absence of eye vision, dlPFC activates IFG/PFC →V1,+MTL/Inferior Fusiform Gyrus +TPJ/Temporo Parietal Junction (storing the images of implicit and explicit MM) →IPS/IntraParietal Sulcus then TopDown.

The clarity of the images is due to *IntraParietal sulcus/ TPJ and MTL/ Inferior Fusiform Gyrus,* → *PFC/PreFrontal Cortex* : mechanism: image from the imagery compared with images in the ICS → IPS (TPJ) --)Fusiform Gyrus (temporal Lobe).

In addition AMYGD, Pulvinar also contributes to the process of imagery, accounting for the emotional and configurational components of the sensory input (Pessoa 2010, Walla,2013).

e) Hallucinogens and NDE.

Some chemicals or drugs are strongly associated with the phenomenon of OBE and NDE. They are Ketamine, Psilocybin, DMT, MDA, and LSD (injectable or oral) and share a common mechanism of action by binding to neurotransmitter receptors at the synapses (Ketamine: NMDA for Glutamate-, LSD, DMT: receptor for Serotonin 5HT, MDA: receptor for Serotonin 5HT, norepinephrine, dopamine) involving in Memory formation, pleasure, attention.. The experiences induced by these drugs are comparable to the natural NDE, which includes encountering external entities. Attaching these chemicals to the receptors may render the Soul detached from the neuronal synapses and free. As a result, the Soul becomes non-local, as in the case of Emptiness, and can be freely connected with other external entities. This phenomenon is equivalent to free flight, seeing and hearing without being obstructed by the brain. The underlying mechanism is that the Soul in the brain is coupled with electrons (-) that help the Soul attach to the synaptic membranes electronically (+). In the presence of Glutamate (for Memory), serotonin (pleasure), or hallucinogens, the membrane is depolarized (C^{++} ions enter into the cytoplasm) and is detached from the Soul. This detachment is overwhelming in the case of voluntary consumption of hallucinogens. In Meditation with OBE, or higher levels of experience, the clearing of Karma destroys synapses through the mechanism of neuroplasticity (MM in the inner CS is retrieved but is rejected by the mechanism of attention operated by the IPS, therefore retrieved MM and related synapses become unstable, are not reconsolidated) The experiences induced by drugs are temporary. In contrast, those in Meditation are permanent or long-lasting. Therefore, using medicines or chemicals is forbidden in most major religions.

f) Neuroscience of the OBE

Since OBE can be associated with needle stimulation, neoplasm, and epilepsies involving the sensory cortices. The TPJ is especially well known for areas related to OBE and is consistent with its function associated with integrating audiovisual and vestibular inputs. Alterations of this latter area of cortex with vestibular disturbances may be accounted for the images and the motility (Ridder 2007, Greyson 2008, Blanke 2005,2004,2002, Tong 2003, Sang 2006 Mobbs, D. and Watt, C. 2011)

. Alterations of PFC, Temporal Lobe, cerebellum. Occipital is also revealed in OBE (Easton 2008, Braithwaite 2010, Terhune 2009). Since dreams associated with REM intrusion can be seen in the state of wakefulness before Near-death states, the relationship between OBE and illusions occurring in dreams associated with REM intrusion could be accounted for NDE (Nelson 2006Mobbs 2011, Greyson 2012.)

g) ROLE OF THE SOUL293

Under normal circumstances, the Soul is tightly attached to the Synapses connections, preventing it from leaving the Brain. This is due to the embodied Soul is associated with a Negative (-) charge where electrons stick to the Neutrino of the Soul, combining with the Positive (+) charge of the Synapses. However, during altered states such as when using psychedelics, electric stimulation, tumors, Meditation, near death, or when the body dies, the (+) area of the synapses decreases or diminishes, and the synapses connection decreases when the Karma is reduced. As the (+) charge decreases, the Soul is no longer attached to the Synapses, granting it the freedom to have Non-Locality, which is synonymous with astral projection. The Soul leaving the body carries with it all the Karma of the human life and many lives. Moreover, the Soul retains its ability to hear, see, think... as it did when it was still attached to the Brain. This freedom from the pentagon and the Brain transmission and control system allows for a more visible flow of information.

Buddha emphasized the transformative power of meditation. He suggested that the Form part of the Soul corresponding to Five senses and body is the first part of the Soul to become free and become pervasive, allowing the Soul to penetrate every part of the body or the external scene. This profound insight is beautifully captured in the following passage from the Lankavatara Sutra at the end of this section:

Therefore, when near death, the Soul can begin to separate from the brain. This is due to its all-pervasive nature, allowing for self-seeing or the seeing in the out of Body Experience /OBE to become obvious.

The Sound or Light in Near Death is akin to that experienced in Meditation, originating from information from the Inner Consciousness copied to the Soul. Without attention, the retrieved Inner CS data result in a pure sound with little or no detail. The Inner Consciousness from the Right hemisphere can produce mystical or magical realms (please see section.

h) Mechanism of Hearing Music (Fig 6.9).

Sound, particularly music is an essential attribute of the Original Mind/ Buddhahood. Therefore some kind of music may carry with it transcendental pleasure. Auditory cortex has a connection with reward system composing of VTA and Ventral Pallidum/Nac (DOPAergic system).

i) Discussions.

The role of the TPJ in some clinical cases of OBE is essential and can not be deniable. In addition, the TPJ cortex is close to PCC and RSC, which are cortices storing immediate and remote autobiographic Memories. Injection of hallucinogen chemicals into PCC can also induce OBE with floating or flying experiences. Similarly stimulation of different areas of the fusiform cortex of the temporal lobe also induces the OBE. Knowing that the Occipital visual cortex only stores visual MM for a short time.

However, the neuroscience changes can not be accounted for many OBE and NDEs, such as:

· AS and OBE associated with lesions or stimulation of the TPJ appear to be distinct from the OBE or AS in NDE or mediation. The events appear orderly, logical, and harmonious in the latter circumstances. The mental status is usually happiness, pleasantry, and peaceful Mind. Images in AS and OBE in pathological and provoked conditions appear to be more similar to those in dreams, with illogical connections between different steps.

· No vestibular disturbances in the case of pathological and provoked OBE.

· Some cases of provoked OBE are associated with images and scenes of worldly lives as opposed to the supernatural appearance of OBE in Meditation or NDE.

· Flying high and fast.

· Encountering dead relatives or acquaintances who would like to send messages to living persons.

· Lesion of PFC may be associated with religious phenomenons.

Neuroanatomical changes can account for some cases of AS and simple cases of OBE but can not account for all features of most cases in Meditation or NDE. As a result, AS and OBE in pathological or provoked conditions and AS, OBE in NDE, and Meditation likely use the same neuroanatomical mechanisms of clinical cases of NDE and OBE. In Meditation and NDE, OBE, and AS may represent extracerebral extensions in which the Soul plays an essential role with its base in the brain.

In living sentients, the soul is attached to neuronal synapses to register the electrical changes when the data enters the brain. In Meditation or NDE, the Soul is detached from the neuronal synapses and re-acquires its original free state (without attachment of electrons). Due to its pervasiveness and non-locality, the soul can get in touch with any parts of the body (i.e. face and head) or external landscape. As a result, it can see what is called OBE or NDE. In addition, in NDE, the retrieval of internal data (memories) in the Inner Consciousness, without attention, generates light and sound, as is often reported in OBE or NDE. The retrieval of the Right Inner Consciousness can account for some bizarre visual contents (please see paragraph of Content of Dreams). This is supported by the revelation of Buddha when discussing Experiences in Meditation in the Surangama sutra.:

> *Ananda, when you sit in Meditation, **if your thoughts are wiped out, the state (of your Mind), now free from them**, will be clear, and will not be changed by either stillness or disturbance. In this state, **both remembrance and forgetfulness are one undivided whole**. While in it and before realizing samā-*
> *dhi, you are like a man whose eyes are clear but who is still in the dark, for though your Mind is clear, it does not yet shine. This is the aggregate of form that conditions your Meditation. If your Mind radiates, you will clearly perceive all the ten directions of space...... .As your **Mind becomes pure and clean, its uttermost purification causes you to see suddenly the great earth, mountains and rivers in the ten directions change into the Buddha.s (pure) land adorned with all sorts of precious gems whose radiance is all-pervading**. You will again see clearly Buddhas as countless as the Ganges. sands with beautiful temple buildings filling the whole space, with the hells underneath*
> *and deva palaces.*

From Buddha's saying, one can understand that in NDE or OBE, the Soul/ Mind acquires vision and hearing by:

- The intrinsic capability or perception of information

- Clearing of the Form (the physical body which obliterates the Soul since the ensoulment, Therefore, the Soul can hear or see as such the body (the skull) is transparent.
- Detachment of the Soul from its attachment to the brain (Default Mode Network of the brain storing the CS) when there is extensive depolarisation of DMN.

According to Buddha, Meditation helps to cleanse the Karma. Progressively, the five layers of Maras are cleansed. In the first stage, the Form (in this case, the brain, skull, and scalp) is clear to the Soul because the Soul can penetrate baryonic material (all physical forms). Scientifically, the body is "opaque" to the Soul because the Soul is obliterated by electrons that tie the Soul to the synaptic membranes. Therefore, clearing the Form in Buddha's sermon means that, in normal conditions, the Soul is connected to the brain. *Death or Meditation depolarizes the cell membrane of synapses enough for the Soul to traverse or leave the brain*.

j) Soul and DID.

As previously outlined, the physiopathologensis of DID remains controversial unless the role of existence of two or multiple Souls in same individual is taken into account. Since the Soul not only plays the role of storing but also retrieving Memories. All the souls share the same implicit MM bur individually specific for explicit MM. Faces, shapes, colors, and motion are stored in the temporal fusiform cortex. (Dell P 2009.) Under stress the principal soul is suppressed, one of the other soul may take the control of the body

V. Is Buddhism Pantheic Or Atheic.

a) Concept of God worship or Theism/ Pantheism.

In this concept, the superpower/ spiritual power comes from God/ Deignity, who creates the Universe. In The Old Testament, 2. .... God's active force was moving to and fro over the surface of water...God proceeded to say; *Let light come to be* ..... However, nowadays, God in The Bible is represented by personalized power. God is represented by The Father, The Son, and The Holy Spirit. This is the spiritual power which has all the creativity, miracles, and ingeniosity.

b) Concept of Atheism and Non-Theism.

Non-Theism is the lack of belief in God/ Gods. One does not believe in the divine Creation of the Universe since God or Deity has not been physically demonstrated.

The Creation of the Universe is an incidental or accidental event of the Big Bang, which is followed by natural selection according to Darwinism. Furthermore, many scientists believe that the Universe is created by the eventual development or incidental by chance since the Big Bang, despite the fact it has been known, doubtful, or not known that the Universe can not be developed by chance or by eventual evolution. Atheists may believe in spirituality, the existence of a metaphysical realm, and soul or spirit.

As previously discussed, it is very difficult to expect that beneficial accidents can occur multiple times in the Creation. It is often believed that the optimal location of the Earth in the galaxy and the Solar system is for the existence of life on the Earth. Should the location of the Earth be misplaced, the change in the temperature would be unsuitable for life.
Moreover, the perfect nature of the Earth and the knowledge of human beings are exceptional without the intervention of spiritual power.

In non-Theism, there is no belief in spirituality, in the existence of the soul or the metaphysical realm. One believes in the sole existence of the actual life and world. They would like to only accept the physical and objective realm with the doctrine of existentialism and humanism

c) There is no personalized God in Buddhism. Therefore, most Westerners, including Western philosophers, consider Buddhism as Atheism. This is mainly due to the need for more understanding of Buddhist scriptures, particularly the Mahayana scriptures, which are considered to be composed of Buddha's proclamation of dharmas in major meetings. Nevertheless, Suangama, Diamond, Ornament Flower. Avatamsaka is a scripture that reveals the root of life and Creation. According to Buddha's teaching, The Emptiness/ Original Mind represents the Original Power that is manifested in three forms called Trikāya/ vn: Tam Thân Phật, similarly to Trinity/ vn (Thân:kāya) : Ba Ngôi in Christianity

Dharmakaya (skt)/ vn: Pháp *Thân*, Nhu Lai thân (*Tathāgatakāya*) and Nhu Lai (*Tathāgata*): appearing in a way beyond Consciousness (acintya/ vn: bất khả tư nghì), in billions of billions of forms with power of perception like the mirror or pure, still ocean that is vast without shore but without identifiable energy or actions. It is uncreated and imperishable.

Sambhogakaya (skt)/ vn: Báo *Thân* and
Nirmāṇakāya / vn: Hóa thân.
In comparison to Christianity, there is no difference in the concept of Creator/ God in Buddhism.

Buddhas in "ten directions" is the manifestation of Buddhahood represented by the Emptiness that is miraculously existence. Perception of dharma, 'Subhuti, as the Tathagata has taught no-perception. Therefore it is called 'perception of dharma'. Or simply **Even dharmas should be re-relinquished, how much the more so no-dharmas** (vn: Chánh Pháp cũng cần buông bỏ huống là Phi Pháp)

d) The difference between Buddhism, Christianity, Hinduism and Eastern philosophy (I Ching)

resides in the standpoint of view, resulting in the belief in the existence of God and the Emptiness/ Buddhahood in the Creation of the Universe with all living beings sharing the same Origin. In Genesis, God manifested as the Holy Spirit or God's Active Force. The Holy Spirit is often personalized as in the Michaelangelo (an Italian sculptor, painter, architect, and poet of the High Renaissance) art painting:

The Earth was without form and void, and darkness was upon the face of the deep. And the Spirit of God moved upon the face of the waters

There are analogies between:
- The Earth without the form or Void and darkness, in Christianity
- The Void, formlessness and fuzziness in Lao Tsu's and Chuang Tsu's Taoism and
- The Emptiness, non-discrimination, and stillness in Buddhism

In any circumstances, personalization of the creating source was not necessary. In Buddhism, the Ultimate Creation Source is represented by Buddha appearing on Earth or in other dimensions in the Universe. These dimensions of the formless realms are represented by different worlds of "ten dimensions" of the Universes's spaces occupied by the Dark Force and the Dark Matter. In both religions, the creator power is manifested in three forms called.
- The Trinity/ vn: Ba Ngôi) in Christianity: God the Father, God the Son/ skt: *nirmāṇakāya*/ vn: Chúa Giêsu and God the Holy Spirit,
- The **Trikāya**/ "three bodies"/ vn: tam thân in Buddhism and the Trinity in Christianity. (Fig 6.11)

The significant difference is God in Christianity that is represented by the only Son (skt: *nirmāṇakāya* Ứng Thân/ Hóa Thân, by numerous Buddha such as, *and many past Buddhas such as i. The Father is equivalent to the Dharma Buddha/ skt:dharmakāya/ vn: Pháp Thân; in Buddhism is represented by Emptiness. According to* Nagarjuna, *the 14th Patriarch in Indian Buddhism, this Emptiness is characterized by* three (original) own natures/ self-nature/ Prakriti or Svabhava *or its noumenon of*

· *Being non-created (no birth no death),* Self-existence, ownership, free will.
· Permanence.
· Oneness.

ii. Buddha, (The Son/ skt: nirmāṇakāya/ vn:Ứng Thân/ Hóa Thân. Jesus Christ), the current Buddha/ *Shakyamuni Buddha, future Buddha Maitraya.*
The Buddhahood represents the Holy Spirit/ skt: saṃbhogakāya/ vn: Báo Thân, Thụ dụng Thân endowed with thirty-two excellent physical marks

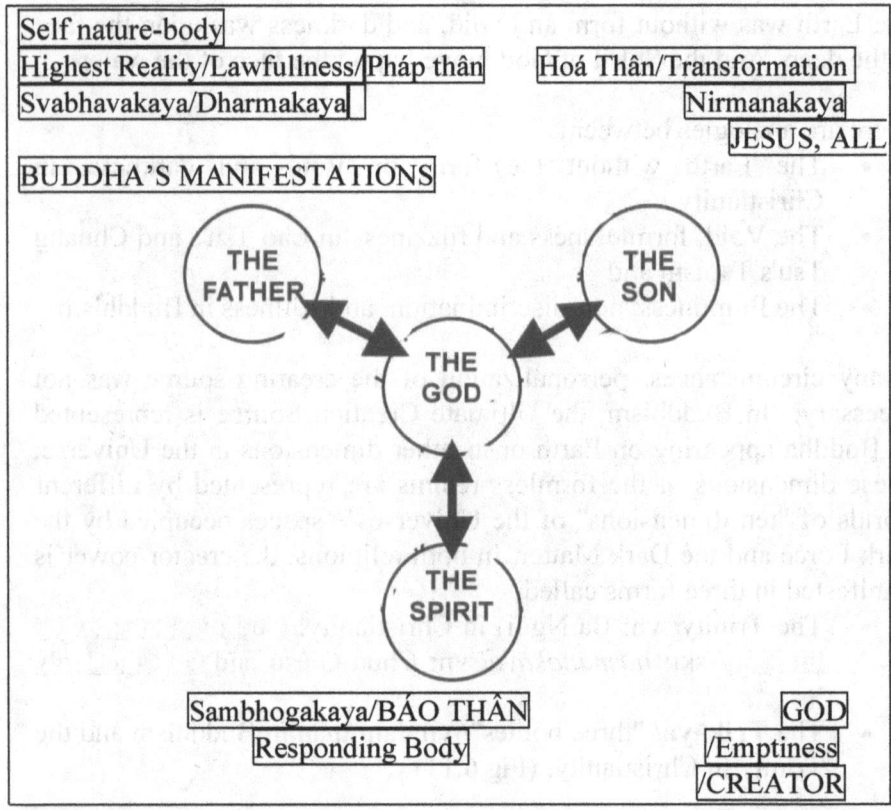

Fig 6.10

iii. The reward body of bliss or enjoyment or Celestial body or bliss-body of the Buddha, personification of eternal perfection in its ultimate sense.

302

The Holy Spirit only exists in the metaphysical realm. *For example, Amita Buddha is in the Pure Land.* Dipankara Buddha/ vn: Phật Nhiên Đăng who predicted/ vn: thọ ký *Shakyamuni Buddha. In addition, when Shakyamuni Buddha proclaims the Buddha truth/ dharma/ vn: hoằng pháp, multiple Buddhas, presiding in remote different worlds, often attend the Shakyamuni Buddha's grand meetings.* Despite different bodies,

All Buddhas are spiritually identical but manifested in different Transformation bodies having different characters to accommodate living/ human sentients for education, guidance, and saving. As a result, the number of Buddhas is as numerous as the number of sand grains of the Ganges river. Likely the difference is in the method of proclaiming the truth by Jesus and Buddha.
Jesus being The only Son, is exactly similar to Buddha, representing the Emptiniss that is also unique.

Therefore, the difference between Theism and Atheism / Polytheism lies in the way God is expressed, not in the concept or belief. God is omnipresent and can, therefore, appear in different forms according to different circumstances and functions. In Hinduism, Brhama, Visna, and Shiva represent the three forms of God for Enlightenment/Creation, Maintenance, and Destruction, respectively.

In addition, as Buddha pointed out, the identical Brhaman and Atman reflect Brhama as God in the regulation of the samara circle. The common question often posed by eternalists (Hinduism) to Buddha or even by Buddhists in some Sects
·*Uninerse is permanent or impermanent*
· *Universe is limited or unlimited*
·*Buddhaia eternal or not*
·*Soul and body are identical or not*

In Hinduism, the concept of Creation is viewed from a late stage perspective where Emptiness is not fully recognized or understood. This perspective leads to the belief that life is happy if one behaves correctly in the light of Brhama. The reasoning behind this is straightforward: if Brhama is perfect, his Creation must be without mistakes. Buddhism, on the other hand, does not accept the notion of a perfect Brhama and places him lower in the Creation hierarchy, not as the Creator. *The sufferences of sentient beings are due to the veil of Ignorance following the False Thought that subsequently leads to the creation of the Uninverse/Multivers. The False Thought is an incidental event of Emptiness is not an intention of the Creator (see Figure 1.4A).*

In Far Eastern I Ching, The Ultimate Pole is equivalent to Emptiness. However, the Creation is represented by 64 symbols (hexagrams) representing various aspects of the Universe of both physical and metaphysical realms. The First two hexagrams, ☰ and K'un (natural response) ☷ (similar to Father and Mother respectively or Adam and Eva)) forming initial steps of the Creation. After the creation of Ch'en and K'un the successive steps are Chun (difficult beginning), Meng (Inexperience)....Chi Chi (After the End) ䷾ and Wei Chi (Before the End) ䷿. The sixty trigrams are theoretically not arranged in a linear, but in the circular diagrams; there is no end or beginning in I Ching. Hexagram Chi Chi, the 63rd of the 64 hexagrams, symbolizes the after End that hexagram 64th, Wei Chi represents the beginning. This reflects the work in this Universe: The End is only illusional in this world as long as one can eliminate the illusional self to merge with Emptiness/Nirvana. Therefore, the meditation appears to be an endless process, until the ego is completely eliminated. The circular diagram of 64 hexagrams is comparable to the Causal Dependent 12 links in Buddhism

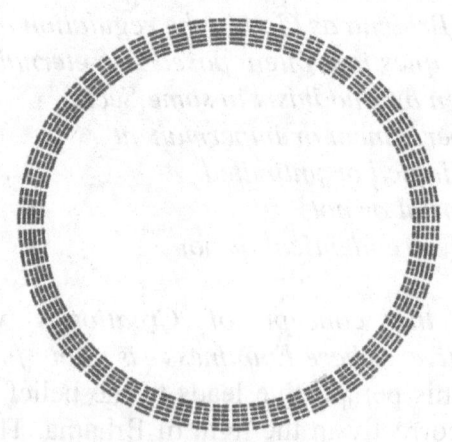

Circular arrangement of 64 hexagrams

Chapter VII: ATTENTION and MINDFULNESS

Without mindfulness:
There is no seeing when looking
There is no understanding when hearing
There is no tasting when eating
...Within the Tao, knowledge would be miraculous
Tâm bất tại yên:
Thị nhi bất kiến,
Thính nhi bất văn,
Thực nhi bất tri kỳ vị.
...Thường cư đạo trung, Huyền diệu khả kiến
(Đại Học, Tu Thân)

The Ancient's saying above reflects the importance of mindfulness in daily activities. The essential step in the integration of incoming information is attention. Mindfulness is the method to calm the Mind, which sentient beings use to investigate the environment and the body as long as the sentients live and are awake. Therefore, the reason is always active: *"The Mind is like the monkey, and thinking is like the horse"*. (Wolfe 2019, Gobulic 2019, Treisman 1992. Therefore, understanding the mechanism of attention is critical in mindfulness Meditation.

In life, attention is not just a prerequisite for Consciousness/CS but a key to understanding. Information without enough attention is only recorded in the unconscious and cannot give rise to understanding. This is shown in the saying:
Peace of mind is the gateway to understanding. If the mind is not at peace, it cannot see, hear, or understand and cannot taste.
(in Vietnamese :Đại Học, Từ Thân,)
In Meditation, Attention is not just a tool, but a catalyst for transformation. It is the beginning of limiting CS to a phenomenon or a simple point of observation. As will be presented later, CS is the veil of Ignorance that limits Awareness, in the process to develop Awareness or knowing. Awareness is the goal that Meditation wants to achieve, not just observation/attention.

Theory of Searchlight
The background plays an important role in the search, which is easier in a simple than in a complex background. In photography, the bokeh effect blurs the background to enhance the subject's portrait. The examples in portrait photography indicate the presence of two systems

in the vision: the search system for the background and the attention focusing on the target.

More significantly, the examples highlight the essential role of reducing the noisy environment that interferes with mindfulness.

II. Systems of Attention

A. attention is the observation based on the perception by the five senses and the Consciousness

Attention is paid to the target or to the background to identify the target. When used for the background, a mechanism exists to eliminate/ inhibit the noise or unwanted/ irrelevant information. The search process of quick scanning of the entire system usually precedes the inhibition. F Crick discovers the neuroanatomy of the scanning process to be the Thalamic Reticular Network. The characteristic anatomic distribution specific for each type of sensory input is very suggestive of the inhibition of unwanted inputs. Logan 1992, Robertson 2003, Li 2017, Fritz 2007 Healey et al., 1996).

B. Role of the Preattention that precedes the attention

The preattention is subconscious, because the preattention is not intentionally used in the process of attention.

Information in the subconscious usually helps the search by increasing the search speed and efficiency. The information is stored in the subconsciousness. For example:

- The preattention favors the choice in the previous training (ex: green color), the neuroanatomy of the preattention, in this case, is the dorsal pathway (Folk 2006, Tollner 2010, Schnitt 2018 Van der Heijden 1996).
- Or, it is easier to find a relative in the crowd if the person is known before or wears a particular or previously known type of cloth.
- The green ball in the background of the grey ball is easier to find than in the background of the red ball because of the dominant red color. The preattention, in this case, is to favor the red color. The neuroanatomy is the ventral pathway.

C) The mechanisms of attention.

There are many types of attention.

i. A. Unconscious Attention.

In an experiment, the subject is shown scary faces or unthreatening faces for 250ms, below the required period for Consciousness generation. However, fMRI demonstrates the changes when the menacing faces are shown. (Vuilleumier 2001, Tamietto 2102). The findings reveal the subconscious formation.

    ii. **Involuntary attention** is associated with quick initiation, lasting only 3-5 secs. For example, stimulation by light or sound. The Ascending activating the Reticular network plays the primary neuroanatomical role.

    Attention is secondary to **alertness/ arousal** due to changes in the environment. Since the person is *unprepare*d for the event, the attention is idecreased by mental and physical status like sleepiness, tiredness, and mental depression. The neurotransmitters are different: Alertness is related to norepinephrine (Locus ceruleus) and serotonin, and arousal is related to acetylcholine (thalamic intralaminar nuclei).

    iii. **Voluntary attention** slow in initiation, following emotion, Working Memory, and experience. This management occurs even in the absence of the incoming information because the management is under the control of the Salient network of Insula-ACC, which ultimately initiates all brain activities. Subsequently dlPFC TPJ, and Angular Gyrus play the executive role in connection with different areas of the cortex. For example, in an attention using vision, the dorsal stream V1→motor cortex →FEF (Frontal Eye Field) : for movement of the face and eyes. In addition, working Memory is crucial in connecting the FEF with the lower pathways of attention:

The short Memory involved in the work involves reward, fear, and other types of emotion. (NAc, VTA, Amydala) (Matsumoto 2022, Lindsay 2020, Rinne 2017, LaCroix 2020, Kaya 2016 , Maunsell 2015, Pozuelos 2015, Zhao 2017, Strait 2011, Vossel1 2013). Working Memory is usually limited to up to four items.

    . Selective attention is related to voluntary attention in the presence of multiple incoming data.

    . Attention with vigilance. The volition often controls the vigilance to maintain the attention since the short attention span usually limits it. The neurotransmitters acetylcholine (Basal Forebrain, brain stem: LDT, PTT), DOPA and thalamic intralaminar nuclei.. Of interest, the thalamic intralaminar nuclei play the role of boosting the arousal state to alertness. Alertness is mainly attributed to the Right hemisphere.

    . Furthermore, attention enhances the perception, as evidenced by the fact that the same sound is heard more clearly when the vigilance of attention is exerted.

    iv. **Divided attention**: Attention is split into two targets. It is easier and more natural if the targets are of different types of sensory input like auditory and visual, but difficult if they are the same type..

1. Neurochemical mechanism

Locus Ceruleus/ LC/ Blue spot is a small nucleus in the midbrain producing nor-adrenaline to initiate and maintain wakefulness. The wakefulness is characterized by the alpha wave of 8-10 Hz/sec (EEG) in the occipital cortex (or in REM as seen in the PFC). The alpha waves disappear when LC is stimulated (in Laboratory animals) and correspond to the state of attention with dilated pupils (Dahl 2022). The underlying mechanism is the Thalamus, which is wrapped around by the thalamic reticular network (TRN), playing the role of inhibiting the noisy incoming information

Other neurohormones pathway are acetylcholine system (BasalForbrain, PPT/.Pontopedunculat Tegmental nucleus and LDT/.Lateral Dorsal Tegmental) and Serotonin/ Dorsal Raphe.

2. Neuroanatomical Mechanisms
Role of the Thalamus
a) TRN (Fig 7.1)

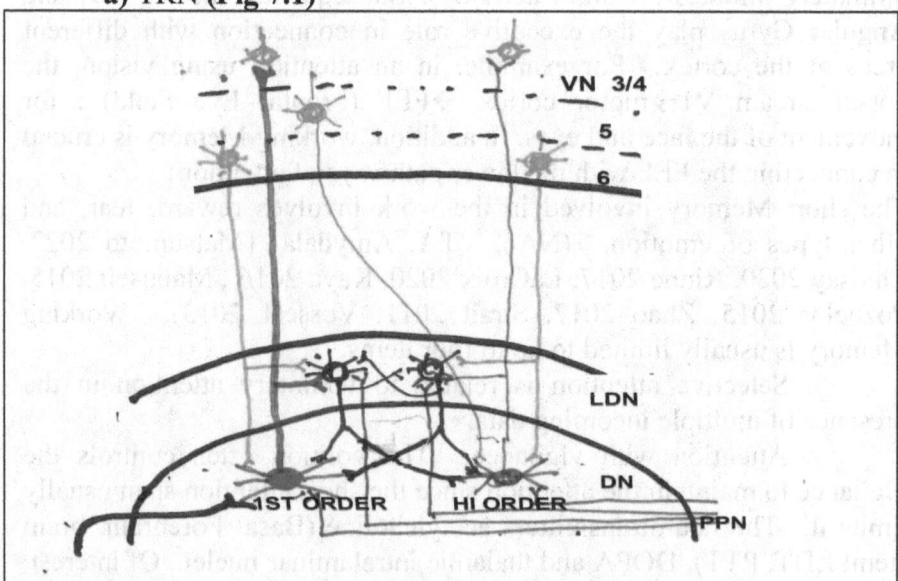

Fig 7.1:
Thalamic Reticular Network plays the filtering role in inhibiting the noise of the information. (Sherman 2016). Pedunculopontine nucleus (*PPN*)
Incoming information to the Thalamus, First Order, and High Order send signals to level 3 và 4 of the cortex.
The cortical neurons of levels 5,6 send back signals to the thalamus associated with PPN neurons for filtering purposes.

Further reading

TRN/ Thalamic Reticular Network is indispensable in filtering the incoming information to the Thalamus by discarding unwanted information (noise canceling) to maximize attention. The location of the network to respective sensory nuclei is suggestive of the specificity of the filtering process. Parallel to this inhibitory process, the PNN/ Pedunculo- Pontine Nucleus plays a role in wakefulness and motility.
· Two mechanisms of filtering information.
- Chemical mechanisms with GABA inhibit the pathway (Thalamocortical): This occurs when the TRN receives the input from the cortex.
- Synaptic mechanisms: Tonic and Bursting in sleep, Spindle (7-14HZ) manifest the bursting mechanism from TRN necessary for the MM. (Halassa 2016).
· Role of PFC in the function of TRN: TC (Thalamocortical) is coordinated with PFC(Pre Frontal Cortex} by the intermediate of TRN (mediodorsal) (Zikopoulos 2007).
· Similarly, the connection between TRN and BG in regulating the movement (Ahrens et al., 2015)

The attention is initiated when being cued. This mechanism is critical when the milieu is affected by unwanted noises (high noise). Therefore, attention affects the quality of the CS. CS is modified and may become distorted by voluntary input from the cortex. Furthermore, to increase the CS, the sensory cortex areas are restricted (Receptive field) in V4, Middle Temporal Lobe, and lateral intraparietal Lobe.

In the attention, repeated scanning of the object, the interaction between nuclei of TRN and those of the Thalamus to coordinate the filtering information.

In the Thalamus, adjacent synapses form groups of synapses to compare the information of the object in search and those in the ICS.

Groups of synapses not consistent with those in ICS are
i. discarded and destroyed by the neuroplasty mechanism (Crick 1984);
ii. stored in the SubCS, with feasibility for retrieval, or
iii. integrated and stored in ICS.

With the involvement of the cortex, particularly the PFC, the filtration of the information can be controlled by the CS.
For example, one can voluntarily discard unwanted information in the attention. Therefore, with experience, the meditator can voluntarily focus only on the objective of the attention.
- **Zona incerta and anterior pretectal nuclei also play a role in TRN in attention through the cortex to enhance the TRN in the filtering mechanism to increase** GABA-B. GABA-B from TRN only inhibits the Dorsal of the Thalamus) and, subsequently, the sensory cortex and emotion (Austin 2011)
- **GABA inhibits** the Occipital and Parietal cortices to limit the incoming sound and light, increase attention, and control emotion through the connections PFC-Amygdala-Cingulate Cortex.
- Since CS is only formed when attention is involved. Improve attention will eventually improve the CS. In Meditation, alpha waves decrease all over the cortex. Beta waves decrease in the RT frontal cortex, anterior part of the middle cortex, and posterior cortex.
In **Meditation, changes are seen** more in the RT cortex than in the LT. Parasympathetic (ventromedial HypoThalamus) is activated in Meditation.

> - Meditation and the Reticular Activating System. TRN, ARAS. (Saggar 2015, Austin 2013, Guglietti 2013 Kovalzon 2016)
> Meditation enhances the activity of the PFC, TRN to increase GABA (with inhibition of the sensory cortices and the emotion)

b) **Role of the Superior and Inferior Colliculus.**

The SCC plays the role of maintaining eye contact when the head or the object moves away from the initial position. This means that the Superior (for vision) and the Inferior Colliculus play a role in directing attention. In the patient with Supranuclear palsy, it appears that the patient is blind. Of course, this is different from the hemineglect syndrome.

c) **Role of the Pulvinar.**
As a reference center between different frames of vision that eventually help an individual to decide the target of attention

d) **Cortices in the Attention .(Fig 7.2,3,4,5) .**
Involving IPS and FEF. Upon receiving input from
- The visual/ other sensory cortex, or from
- Volition under the control of the Salient network.
dlPFC-IPS network initiates IPS to connect to:
    i. **Top-Down Pathways.**
    ii. **The Bottom-up pathway.**

3. **Sensory system used in the attention.**
4. **Default Mode Network/ DMN** > DMN is the site MM and CS storing; in attention, the retrieval of MM is minimal. Therefore, DMN is inactivated while other regions, especially the executive network dlPFC-IPS are activated.

III. SPECIAL CASES OF ATTENTION.
A. **Unconscious Attention.**
See above

B. **Hemispatial neglect.**

The RT Hemisphere represents RT and LT visual fields, but the LT only denotes the RT visual field. As a result, a specific lesion of the Temporal Lobe of the Rt hemisphere causes the blindness of the Left space but does not abolish the reflex reactional to the object in the LT visual field. The visual cortex is normal. The ventral visual stream is affected by this syndrome. Often, the emotion component of this ventral remains intact, as evidenced by the expression of the emotion despite the blindness of the hemispace.

C. Attention Deficit Hyper Activity Disorder/ ADHD (Fig 7.2)

This is seen in children with decreased social interaction, eye contact, and restricted or repetitive behaviors or interests, skills, and speech. ASD/ Autism Spectrum Disorder has many etiologies with alteration of multiple genes. Developmental **restriction** of **visual** attention may play a critical role in the early years of development.

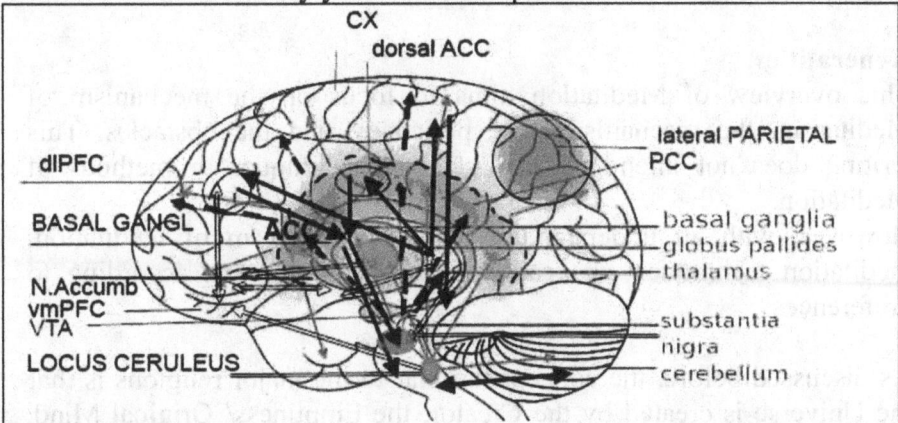

Fig 7.2 Hyperactivity.
Alteration of Norepinephrine/ NE from LC/ Locus ceruleus and other nuclei NE with widespread connection to the Cerebral cortex, ICS, (Cingulate gyrus, Basal Ganglion Thalamus, Cerebellum, DOPA system from SN/ Substantia Nigra.
The reward system Nucleus Accumbens/ NAcc is connected to vmPFC-dlPFC and Thalamus, LC, SN.
Motor system: BG connected to dlPFC, SN, Thalamus, Cerebral Peduncle to PFC. In AHAD, the involved cortices and nuclei are PFC, vmPFC, Parietal Thalamus, dlFPC, IPS, ACC (emotion), BG, Nucleus Accumbens/ NAc, Caudate nucleus, Putamen: decreased activity. (Faraone 2018, Douma 2019)

Table summarizing ASD

	Genet	Neurobiol changes	Symptoms
Matern /Infect /Tox	IntraUterine Post Natal Maturation	Cortex/Synapses/Modules Networks of the Brain Brain Growth Autisn Spectrum	Socio cognition Language/Motor

Chapter VIII: MINDFULNESS MEDITATION.

Hear, O Children of immortal bliss
You are born to be united with the Lord
Follow the path of the illuminated ones
And be united with the Lord of Life
Swami Vivekananda

Generalities.
This overview of Meditation aims to focus on the mechanism of Meditation, the mechanism of experiences, and the obstacles. This writing does not intend to introduce the techniques or methods of Meditation.
However, with an understanding of the mechanism of meditation, meditation techniques are readily planned and chosen according to preferences.

As discussed before, the concept familiar to the major religions is that the Universe is created by the Creator, the Emptiness/ Original Mind, according to Buddhism. The Thought in Buddhism or the Will in Christianism represents the beginning of the Creation. Subsequently, upside-down thinking/ phenomena or discriminative minds generate the different aspects of the Mind of today's World. The original process was written in the Heart Sutra/ The Prajnaparamita Sutra with "The Emptiness miraculously creates The Form, and The Form creates The Emptiness". Along with false thought, which is the origin of the upside–down, inversion, or discriminative mind, the development of science and technology has tremendously improved the quality of the physical world environment and the body, all representing the form of the Five Maras. This was developed at the expense of the Spirituality. Therefore, the present worldly life is imbalanced. Supplementary attention to the spiritual aspect is critical. Meditation or other alternative methods guide sentient beings to return to their homeland. The Emptiness, Nirvana, is the goal. The practice of the Noble Eightfold Path/ vn: Bát Chánh Đạo/ Astangika-marga (skt) for improvement of personal sufferance, mental conditions, and social upside down. The stillness of Mind, harmony of society, and peace of the Universe are original states; therefore, returning to the homeland should be a simple, easy, and natural pathway. Meditation is a healthy, happy, stress-free living method for most Meditation beginners. Meditation has no goal. It is a natural and decent course of living, just like a worker returning home after a long day of labor. However, due to the veil of ignorance, **incomplete and deviated CS, according to**

Gödel's incompleteness theorems) lasting for billions of years, many humans forget this natural pathway.

Mindfulness Meditation uses attention to focus on certain areas of the body or sensory information that will ultimately lead to inner Consciousness. The inner CS represents the ego, the actual veil of ignorance (incomplete and deviated CS) of each individual.
Meditation can cleanse this veil of (ignorance (incomplete and deviated CS), which is nothing but the Memory of the present and multiple past lives. Memory is represented by the synaptic connections, which can be subjective to the neuroplasticity (destruction and repair), a process of degeneration and regeneration turnover (elimination of old Memories and regeneration of new Memories).

Another potential of mindfulness Meditation is to put aside and exclude the veil of ignorance. This can be done by narrowing down the Consciousness to expose the Awareness.
According to Suzuki:

> *"We say concentration, but to concentrate your Mind on something is not the true purpose of Zen. The true purpose is to see things as they are, to observe things as they are, and to let everything go as it goes. This is to put everything under control in its widest sense." Zen szellem, a kezdő szellem.*
>
> *"When we first hear that everything is a tentative existence, most of us are disappointed; but this disappointment comes from a wrong view of man and nature. It is because our way of observing things is deeply rooted in our self-centered ideas that we are disappointed when we find everything has only a tentative existence. But when we actually realize this truth, we will have no suffering." Zen Mind, Beginner's Mind.*
>
> *"For Zen students, the most important thing is not to be dualistic. Our "original Mind" includes everything within itself. It is always rich and sufficient within itself. You should not lose your self-sufficient state of Mind. This does not mean a closed Mind but an empty and ready Mind. If your Mind is empty, it is always ready for anything; it is open to everything. In the beginner's Mind, there are many possibilities; in the expert's Mind, there are few.", Zen Mind, Beginner's Mind.*

After the False thought, human beings get lost in the ocean of ignorance. The decision to return to Buddha nature is not conditional, not secondary to a cause, dependent origination Hetu-pratyaya (skt)/ vn: nhân duyên but Mental initiation or initiative—to make up one's Mind.
Buddha said to Ananda (in Surangama, tome IV):

> *Even after the ideas of both Creation and destruction have been considered as natural, this idea of being natural is also of dharma of Creation and destruction, not the Bodhi Dharma (not the reality of being uncreated and imperishable). Even after the ideas of both Creation ……… of being uncreated and imperishable is considered as spontaneous. This is like the mixture and fusion of various worldly materials into a composite compound which implies its opposite, the uncompounded. With no more thought of practice and realization. This kind of reasoning is illogical.*
>
> *The Absolute which is neither original nor unoriginal, neither mixed and united nor not mixed and not united, and neither apart nor not apart from union and separation, is above and beyond all sophistry.* **Bodhi and Nirvana are still far away and cannot be attained without aeons of practice and experience.** *Even if you (succeed in) memorizing the twelve divisions of the Mahayana canon taught by all the Buddhas and the profound and perfect doctrines countless as the Ganges, and sands, this will only increase sophistry.*

The reality/Emptiness of being uncreated and imperishable should not be considered as spontaneous. Because the Absolute is neither original nor unoriginal, neither mixed and united (Please see the characteristics of Emptiness) Buddha said in Surangama sutra: with this concept of spontaneous effort or diligence in spiritual practice, **Bodhi and Nirvana are still far away and cannot be attained without aeons of practice and experience.**

The meaning is that **"the being natural "** is not the reality of being uncreated and imperishable. As a result, spiritual practice is not natural.

Going home, the Original Mind or Emptiness/ Nirvana where the turbulence or disorder of the Emptiness is nil/ zero. Turbulence or disorder is analogous to the Entropy of a physical system. According to the second Thermodynamic law, Entropy (degree of chaos) can only increase but cannot decrease. Therefore, turbulence/ disorder tends to increase naturally with time. To go back home to Nirvana is very challenging unless voluntary effort is exercised. This difficulty represents the basic difference between spirituality and worldly life since in spirituality, going back to Nirvana is the natural path..

I. Time and Space in Buddhism.
Time is an important component of life because it can define the past, present, and future.
Time combined with space and Consciousness are essential in the Creation.

In 1905 and 1915 Einstein developed the theory of Relativity and General Relativity, respectively. The equation E=mc2 (E: Energy, c speed of light 300,000m/sec and m= mass) represents the simplest and most famous formula. The equation ds2=dx2+dy2+dz2-c2t2 (xyz) measures the three dimensions of the space, d the change, and space that can be
represented as Consciousness, t time. Space-time represents Consciousness (Space and Time áe recognized by the CS; Consciousness/ CS is specific for each individual.

In the ample space of the Universe, the presence of the material generates the curvature of the space.

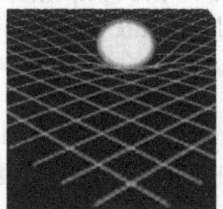

In **The Avatamsaka Sutra,** 5, chapter 31 Eternity 325
the time changes depending on the location in the revelation of the observer (Consciousness) as followed:
- A day of Brahma equal to 4.320.000.000 years in the World of endurance).
- One day in Ultimate Bliss World/ vn: Thế giới Cực Lạc is equal to one Kalpa/ Aeon/ kiếp in the World of endurance/ vn: Thế giới Ta Bà, (refers to our world, filled with sufferings and affections yet gladly enjoyed and endured by its inhabitants).

In the case of Meditation, the time length may vary depending on the mental status of the mindfulness. When time is the focus of attention, the time shown up in Consciousness is longer than when the time shows up in Awareness.

Ksana (skt)/ vn: sát na is the shortest of time, a moment or an instant. Ninety ksana is sufficient for one thought to occur, equivalent to 13 - 16msec. In a video production,. the minimum interval between two consecutive frames is 1/60 sec= 16 msec

Time and Space in Meditation.
For the Emptiness before the False Thought, or Universe before the Big Bang, there is no time space and Consciousness. "This is here now. There is no thought and no self". This is comparable to the concept od

Singularity as the beginning of Big Bang or Singuar point in Black Hole. **Ksana of Thoughtlessness** is a very short instant in earthly life when one can feel oneself without involvement in any mental or physical activity from inside or outside; as such, one can feel like one is in Nirvana. However, the instant is too short to be significant..

II. Volition in Meditation.
Volition is the most essential factor in Meditation.
The Creation of the Universe is not spontaneous. Still, it is inherent to the development of the Right and False thought with subsequent generations of ignorance of all events and the inappropriate interrelationship between parents, children, and ancestors. (note: Thought is an inherent nature of the Emptiness. Thought could be Right or False. The Right Thought creates Emptiness; only False Thought creates the Universe.)

The initiation of the volition must be followed by:
    Continuous, unremitting, and increasing effort.

  — *Jiddu Krishnamurti said about* <u>*The Real Crisis*</u>
The crisis is there. The crisis is not in the world, it is not the nuclear war, it is not the terrible divisions and the brutality that is going on. The crisis is in our Consciousness, the crisis is what we are, what we have become."

Spiritual practice is like going in the opposite direction of the stream of life.
In Surangama Sutra, volition is important, as seen below:
The Buddha took pity on the sravakas/disciple and pratyekabuddhas/self-enlightened disciples in the assembly whose minds set on enlightenment were still not at ease and (also) on future living beings in the Dharma-ending age who will want to develop their Bodhi minds and to tread the Path of the Supreme Vehicle.

Buddha points out two decisive factors in the development of your Mind in spiritual practice by the realization of
  i. the Mind and the World as illusional
  ii. the volitional making up Bodhi Mind

Ananda, as you decide to give up the state of a sravaka to practice with the Bodhisattva Vehicle to possess the Buddha's All- Wisdom, you should see clearly if the cause ground (used as) a point of departure and its fruit-ground (, i.e. realization) are compatible or not. Ananda, if you use your worldly Mind as a causal point of departure, you will fail in your search for the Buddha Vehicle beyond birth and death. Therefore, you should inquire into all

> the creations (of the Mind) that are subject to change and destruction in this material world. Ananda, which one of them does not decay? Yet you have never heard that space can perish.
> Why? Because it is not a created thing.

An example is when the Celestrial King asked Boddhisatsava how much the merit is when a Boddhisatsava earn when initiating the Supreme Bodhi Mind for the enlightenment of the Buddha
The Bodhisatsavan Dharma-Awareness/vn: Pháp Huệ said:

> The Buddha took pity on the sravakas and pratyekabuddhas in the assembly whose minds set on enlightenment were still not at ease and (also) on future living beings in the Dharma-ending age who will want to develop their Bodhi minds and to tread the Path of the Supreme Vehicle. He said to Ananda and the assembly: As you are determined to develop the Bodhi Mind and practice the Tathagata's Samadhi tirelessly, you should first ascertain the two decisive factors in the development of your Mind. What are they?
> Ananda, as you decide to give up the state of a sravaka to practice with the Bodhisattva Vehicle to possess the Buddha's All- Wisdom, you should see clearly if the cause ground (used as) a point of departure and its fruit-ground (, i.e. realization) are compatible or not. Ananda, if you use your worldly Mind as a causal point of departure, you will fail in your search for the Buddha Vehicle beyond birth and death. Therefore, you should inquire into all the creations (of the Mind) that are subject to
> change and destruction in this material world. Ananda, which one of them does not decay? Yet you have never heard that space can perish. Why? Because it is not a created thing.
> An example is when the Cetrial King asked Boddhisatsava Dharma-Awareness/vn: Pháp Huệ ho much the merit is when a Boddhisatsava earn when initiating the Supreme Bodhi Mind for the enlightenment of the Buddha
> The Bodhisatsavan replied: The meanings of this making-up Mind are extremely profound, unspeakable, indescribable, unprovable, unbelievable, unaccomplishable, unthinkable, and inconceivable. However, thanks to the miraculous power of Buddha, I will clarify it for you. Buddhist!, if one takes all his beloved treasures to offer to sentient beings in innumerable worlds in ten directions, the train to keep the five precepts diligently. How much do you think about the virtue and merit of this person,. The Celestial King said: these virtues and merits are uncomparable to anything except those of Buddha
> The Boddhisatsava Dharma-Awareness said, Buddhist!, in comparing this virtue and merit to those of the Boddhisatsava when initiating the Supreme Bodhi Mind for the enlightenment of the Buddha equivalent to one per millions fold

The meanings of this making-up Mind are extremely unspeakable, indescribable, unprovable and inconceivable.

Huike cuts off his arm.
In one legend, Bodhidharma refused to resume teaching until his would-be student, Dazu Huike, who had kept vigil for weeks in the deep snow outside of the monastery, cut off his left arm to demonstrate sincerity.

Position and Posture in Meditation and the Volition.

Performing egolessness is always recommended in spiritual practice. This duty has to be emphasized in Meditation. Classically the sitting position with crossed legs is recommended. Lotus or semi-lotus are optimal. These positions are neither physiologic nor natural, requiring a lot of effort and volition translated in the position and posture: staying still, having a straight spine of neck and back, face upward or forward, hands holding each other, touching the palate with the tip of the tongue clenching teeth and with a determined Mind. Sitting on a chair, walking, or lying will dramatically decrease the efficiency of the Meditation because these positions reflect low volition.

Egolessness in Meditation.
Performing egolessness is always recommended in spiritual practice. This duty has to be emphasized in Meditation. Egolessness is automatically associated with mindfulness of no space and time.

This includes all moral obligation like Four immeasurable Mind of Virtue, The Noble Eightfold Path, Renuciation to Five Attachment (tree poisons/Greed-Aversion- Ignorance, Doubtfulness/Arrogance, Erring Minds, being relaxed and Joyful, separated from four False illusions (Affliction,Ignorance, Death, Heavely Illusion), ana all 62 Perverted views (impermance versus permanence…..)

III. THE ATTENTION
Generalities
This is the first step of Meditation and spiritual practice. In all Meditation, including Asana Yoga, and other activities in life at the workplace and school, attention is necessary to focus the Mind on fixed physical or mental items.
The characteristics of attention in Meditation are to cut off or limit the information input from the outside world (five senses and narrow the field of Consciousness) to the maximum. Since attention enhances the CS, representing the veil of ignorance and stupidity, Meditation aims to reduce the CS, enhancing Awareness and eventually Omniscience. Knowing this principle, it will be evident that the attention method (as a result, Meditation) is less important than the volition. Attention is only a means to reduce CS and to enlarge Awareness.
In the Surangama, tomes 1 and 2, Buddha pointed out to Ananda the Mind of the six senses (five sensory organs and the CS). The Mind is not located in front or behind the sensory organs or within the body but only attached to the Original Mind/ Buddhahood. The Emptiness is uncreated and imperishable. In Tome 3, Buddha said that the external

information enters into the five sensory organs and is thought to be the receptive organ (that is, the brain). These are:
- **Six entrances**/ vn: **Lục nhập**/ skt: Sadayatana. Buddha said: "Ananda! Why do I say the six entrances have their origin in the wonderful nature of true suchness, the tathagata".
- **The twelve areas vn: Thập Nhị Nhập or xứ/ skt: Dvadasayatanani**. 6 entrances + 6 organs, since organs can modify the information
- **The eighteen entrances/ fields/ vn: Thập bát giới/ skt: Astadasa dhatavah**: 6 entrances+ 6 organs+6 Consciousness.

These 18 entrances constitute barriers that modify and distort external information. This distortion is commonly unknown to human beings. They are misled by their physical form and their soul for innumerable lives. Buddha pointed out that the resulting CS is the enemy, the veil of ignorance.

As discussed before, attention is essential for the generation of CS. In Meditation, attention differs from non-Meditation, like reading or working.
In Meditation, the difference is the target that will not add any or only minimal information to the Consciousness.
For example, in Koan Meditation, one of Koan is: the significance of Bodhidharma traveling in the Far East, or in Samantha Meditation, the focus is on the rhythm of the respiration and movement of the abdominal walls. There will be minimal addition to the Consciousness. The primary role is to make the CS preoccupied with the simplest thing, which is boring to try with intention and effort to tie up the running thinking to a fixed target. The purpose is not to explore the target or to learn from it

Example: Story of calming the Mind of Budhidarma.
In a Zen practice, the popular story is that Hui Ko, the second Patriarch of the Zen Sect asked his Master, Bodhidarma to calm his Mind. The Master said: *"show me your Mind"*. Hui Ko couldn't find the Mind. The Master then told him that he had calmed Hui's Mind.
The key in this story is that the Mind is the Consciousness. When Hui Ko uses the top-down pathway, he can't find the Mind, (the Manas); he gets lost in his CS, the Ego, the veil of ignorance. Then, the Ventral pathway takes over with the Awareness to realize the Original Mind.

This method of calming the Mind is similar to that Buddha teaches his monks, according to Vitakkasanthana Sutta:
The Relaxation of Thoughts:

> *Just as a skilled carpenter or his apprentice would use a small peg to knock out, drive out, and pull out a large one; in the same way, if evil, unskillful thoughts — imbued with desire, aversion, or delusion — arise in a monk while he is referring to and attending to a particular theme, he should attend to another theme, apart from that one, connected with what is skillful.*

(meaning: In the problem of unblocking an obstacle, instead of directly focusing on the obstacle, the problem can be solved by working on its counterpart (in the principle of duality).

IV. VOLITION, WAKEFULNESS and ATTENTION.

In everyday daily life, the relationship between wakefulness, attention, and CS is causal and necessary for most activities. In addution volition is also critical in the completionof a task. However, focusing attention on a fixed area in Meditation renders a marked decrease in Awareness. For the formation of CS and Awareness, attention is a critical requirement. The following paragraph in Nikaya Sutra (Long Discourse, **Mahāparinibbāna Sutta: The Great Passing,** The Buddha's Last Days) illustrated two examples of the absence of attention associated with the lack of CS and Awareness.

> 4.27. 'Once, Lord, **Ālāra Kālāma** was going along the main road and, turning aside, he went and sat down under a nearby tree to take his siesta. And five hundred carts went rumbling by very close to him. A man who was walking along behind them came to Ālāra Kālāma and said: "Lord, did you not see five hundred carts go by?" "No, friend, I did not." "But didn't you hear them, Lord?" "No, friend, I did not."
> .....
> 4.30. 'Once, Pukkusa, when I was staying at Atuma, at the threshing-floor, the rain-god streamed and splashed, lightning flashed and thunder crashed, and two farmers, brothers, and four oxen were killed. And a lot of people went out of Atuma to where the two brothers and the four oxen were killed.

The problem here is the absence of attention to the environment

Meaning: in Meditation, Alara Kalama did not see or hear even when hundreds of carts passed; it rained, and thunder killed cows nearby.

V. MEDITATION (Fig 8.1)
A. Methods
The most popular of Meditation is:

- Pure Land Sect Doctrine: Mindfulness recitation of Amitabha Buddha to gradually one can see one's own Buddha nature or to achieve the ultimate goal of Buddhahood
- Vipassana Mditation

B. Six input pathways (five sensory organs + Consciousness): Disappearance of Both "One and Six".

To explain the relationship between the Mind and the six sense organs, Buddha uses a piece of cloth (representing the Mind/inner Consciousness), and tied six consecutive knots (representing the five sense organs and general CS) (Due to the knots, the piece of the cloth become less useful.. To return to its previous state the knots should be untied).

Buddha starts undoing from the center of the knot individually and successively. After untying, the piece of fabric is restored to its original state without a knot, equivalent to Emptiness, the original state of the Mind.

It is noteworthy to mention that the Consciousness specific to each organ is connected to the others (see Figure 5.3, pge 214).

Therefore, only hearing can give some relevant information to the other five remaining organs (similarly, when one knot is untied, the other knots will be untied in the same manner). Another example of a congenitally blind person. In the fusiform gyrus of the temporal lobe, the Visual Word Form Area (VWFA; Striem-Amit 2012, Cohen2000; Dehaene and Cohen, 2011; Schlaggar and McCandliss, 2007) also called Ventral occipito-temporal cortex , Left (vOT; Price, 2012;

Price and Devlin, 2011; Wandell, 2011). *This visual cortex is activated when reading a book. The exact mechanism is applied to playing music. The musician has the imagery of the note in the song. Inversely, when seeing the notes, the musician can hear the sound of the notes. In the blind person, the visual information is enhanced due to the inputs from other sensory organs.*

As seen above, focusing only on one sense most appropriate to each individual is necessary to attain enlightenment.

The key is to attain the Origin, the Emptiness. Loss of Origin is the root of all deviation, as exemplified in the following sermon:

In Flower Adornment Sutra/ vn: kinh Hoa Nghiêm) , Book 10, An Enlightening Being Asks.
For clarification, Bodhisatsava Manjushri asked Bodhisatsava Chief of the Awakened:
Since the nature of Mind is one, what is the reason for seeing the existence of various difference, such as going to good or bad tendencies , having complete or imperfect faculties...

Bodhisatsava Chief of the Awakened replied:
....Phenomena have no function (because phenomena have no original nature).....
and have no self nature; so each phenomena do not communicate each other
and do not know one another..
Eye, ear, nose, tongue, body, are phenomenontal, do not know each other

> *Since the nature of Mind is one, what is the reason for seeing the existence of various difference, such as going to good or bad tendencies , having complete or imperfect faculties...*
> Bodhisatsava Chief of the Awakened replied:
> *....Phenomena have no function (because phenomena have no original nature)......And have no individual nature;*
> *Therefore all of them Do not know one another. Like the waters in a river, Their rushing flow races past, Each unaware of the others :So it is with all things. It's also like a mass of fire,Blazing flames shoot up at once, An Enlightening? Being Asks for Clarification Each not knowing the others : Phenomena are also thus.*
> *Also like a continuous wind fanning and drumming whatever it hits, Each unaware of the other: So also are all things . Also like the various levels of earth, Each based on another, unaware of the others: Thus are all phenomena.*
> *Eye, ear, nose, tongue, body, Mind, intellect, the faculties of sense: By these one always revolves, there is no one, nothing that revolves . The nature of things is fundamentally birthless, they appear to have birth; Herein there is no revealer ,and nothing that's revealed . Eye, ear, nose, tongue, body, Mind, intellect, the faculties of sense: All are void and essenceless; The deluded Mind conceives them to exist. Seen as they truly are, All arc without inherent nature . The eye of reality is not conceptual : This seeing is not false. Real or unreal, False or not false, Mundane or transmundane: There's nothing but descriptions.*

Meaning: **there is no difference in benefit of using different methods of Meditation because all methods, although different, originate fronthe same source, the Original Mind/Emptiness.** Data consists of many different types representing the phenomena of sensory organs, knowing that phenomenon has no noumenon (not the Original Mind). But all phenomena ultimately originate from the Original Mind/ Emptiness that is unique and non-discriminative. Therefore, bringing them back then to their origin (by spiritual practice with the realization of Emptiness) will make them know each other; this is the state of Omniscience

C. In Meditation
In the Meditation, an individual uses attention and detaches from reflexive, irrelevant thinking", to attain an awareness as large and as detailed as possible of nature/Universe. The secondary effects are the achievement of a mentally clear and, emotionally calm and stable state of mind

Generally, the acquisition of data from the environment is obtained through the five sensory organs,

In Buddhism, Buddha showed that Data enters the brain through five steps:
- Form (Data of the mental or physical object must undergo some process to reveal the data)
- Perception: reception of data by the sensory organs
- Feeling: unconditionally reactional reflex to data as a stimulus, corresponding to the Dorsal Pathway (sensory cortexàMotor cortex
- Formation: process of integration of data in the central nervous system to create data and store these data in the DMN recognizable by the Original Mind (Sensory Corte to Temporal Lbe at different areas: Ventral Pathway)
- Recognition of Data as CS

In Neuroscience, the process of the data in the brain follows two Pathways, Dorsal, and Ventral as mentioned above
\

D. Why is Attention necessary in Meditation?320
Cultivation in meditation aims to eliminate the Six Entrances (sensory organs), Twelve Realms (sensory experiences), and Eighteen Elements (mental factors), which are considered the original causes of the Ignorance veil, a metaphor for the state of unawareness or lack of complete understanding.

The Brain is equipped with a system to search for information in the Inner Consciousness to compare with new information received from the periphery through the thalamus. This system, known as the ACC, is always active except when sleeping. It plays a crucial role in managing the information from the Inner Consciousness and eliminating peripheral information by focusing on a simple goal. This process of elimination is facilitated by the dlPFC-IPS management network, which ensures that the IPS focuses on the predetermined goal and discards other information.

The Inner Consciousness is based on the Default Network/DMN. In thinking, DMN is active in recalling past experiences (MM). In Meditation, MM is discarded so DMN rests.

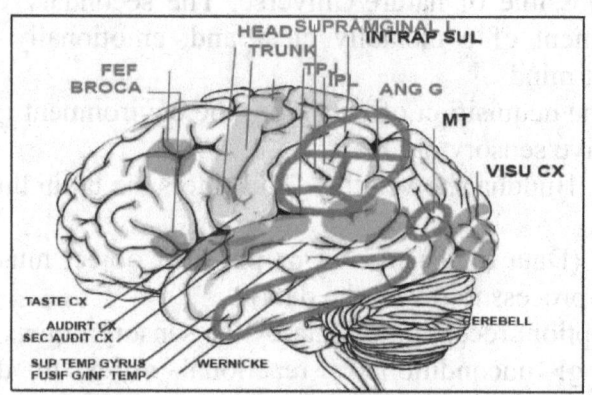

(H5.1) In Meditation: Upper Pathway - Mindfulness/Mindfullness V1> SPL (Superior Parietal Lobe) → hMT→>PPC→>Cortex Movement→>Frontal Eye Field: specializes in movement/action to master the method - there is no Knowledge without the lower pathway.

When attention in the meditation is directed to a limited target, the ACC is activated to look into the Inner consciousness. It retrieve the inner consciousness information, which is known as Karma. This information, being different from the predetermined target of attention, is not noticed, due to the lack of attention, and therefore, it is destroyed.

-(The supramarginal gyrus, plays a role in emotions and speech).
-(The angular gyrus plays a role in numbers, language, space, attention and recall of MM, guessing the minds of others. In addition, it also has the task of estimating the depth of space to create a 3-dimensional shape).

-IPS is the executive center of the dlPFC-IPS management network (for visual attention). Therefore, IPS is the center for selecting the target of attention and, more broadly, the center that makes the mind run erratic.

In Meditation: Upper Pathway - Samatha V1→ SPL (Superior Parietal Lobe) → hMT→PPC→Cortex Movement→Frontal Eye Field: specializes in movement/action to master the method - there is no Knowledge without the lower pathway

E. Dorsal Pathway In Meditation and Ventral Pathway
Dorsal Pathway Up-Down

- As pointed out above, in Samatha: V1→ SPL (Superior Parietal Lobe) (particularly IPS→hMT)→PPC→Cortex Movement→Frontal Eye Field (specialized in movement/anon conditional reflex- there is no CS because of the absence of involvement of the lower pathway.

Focusing on a limited target (to abolish or decrease other input of data) will activate the ACC to look into the inner consciousness to retrieve the inner consciousness information, which is also called the Karmic CS. As mentioned above, this information differs from the predetermined attention target. As a result, it is not noticed by the mind and, will not be processed for conscious formation and will be destroyed by the process of neurogenesis. The supramarginal gyrus is involved in emotion and speech. The angular gyrus is involved in numbers, language, space, attention, memory, and guessing other people's minds. It is also responsible for estimating spatial depth to form 3D shapes. The IPS is the action center of the dlPFC-IPS network for sensory attention. Thus, the IPS is the center for selecting the target of attention and, more broadly, the center for making the mind erring.

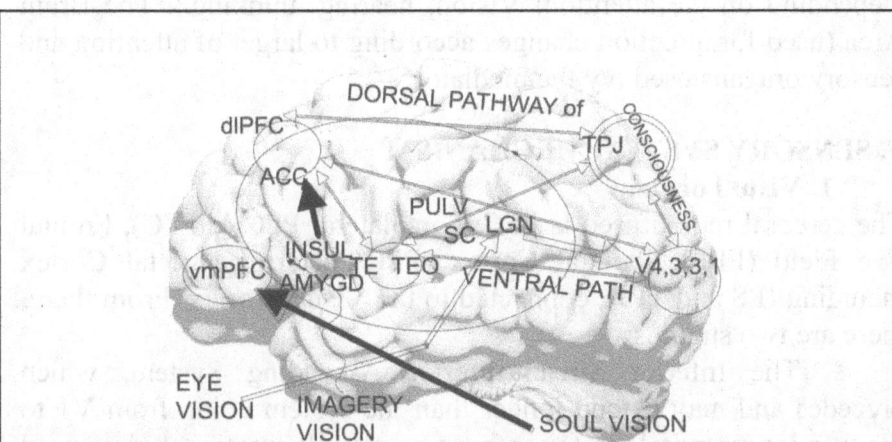

Fig 8.3: Fig 17,1 Three mechanisms of vision using the same pathway but with different directions.
 -**Eye vision**: → Lateral Geniculate body (LGN) →Occipital Visual cortex --: dorsal stream→PFC, and → Ventral stream → Temporal Lobe .
 - **Soul**: or vision without eyes →PFC → Occipital Visual cortex.
 -Imagery Salient network(Dorsal ACC –Insula) stimulating PFC /dlPFC then using Soul pathway. --> IPS/ TPJ --> Chấm Dưới lên : IFG/ PFC-->FG /

> MTL---> Chăm IFG: Inferior Frontal Gyrus, FG: Fusiform Gyrus, IPS:IntraParietal Sulcus, TEO TE: areas of Temp Lobe for form

The bottom-up system is necessary for all living beings to survive because it gives general understanding/awareness in an objective way. This understanding system needs to use Buddhahood comprehensively.

Therefore, when paying too much attention to the goal, one can reduce the Buddha Nature, but when looking at the problem as a whole, Buddha Nature is always involved. Attention can, therefore, lead to errors in CS due to the lack of objective factors and Buddhahood. Therefore, sometimes the observer needs to stop paying attention to have a comprehensive view. The above phenomenon may be the cause when the Physicists try to determine the Particle and Wave phenomenon in light. When paying too much attention to Particles, one only sees light in the form of Waves and vice versa. Again, CS and Awareness all need Buddhahood to participate in the process. Without Buddhahood as in Cotard's syndrome, there is no Knowledge and Understanding.

In Meditation, the two equivalent Dorsal and Ventral systems as above are also used depending on the method of attention.

Depending on the attention: Vision, hearing, thinking... The Brain Area (used for attention changes according to target of attention and sensory oragans used ivy the mediator.

F, SENSORY SYSTEM MECHANISM
1. Visual organ

The cerebral region used is PFC: dorsolateral PFC (dlPFC), Frontal Eye Field (FEF), Parietal Cortex (PPC/Posterior Parietal Cortex including IPS and TPJ), connected to the Visual Cortex. From there, there are two streams:

-The Inferior Stream/Inferior Ascending System, which precedes and may extend longer than the system going from V1 to the inferior temporal Cortex to have an overall emotional view used to collect information in an overview before aiming at the target.

Information from the outside world (Tran) enters the corresponding Root called Nhap (Perception as in "Luc Nhap"). Therefore, the greater the intensity of the outside information, the greater the perception, and as a result, attention is drawn to it. Therefore, in attention, unnecessary stimuli need to be blurred by the Thalamus Network surrounding the Thalamus.

- The Upper Stream goes to Cortex Premotor and to the Frontal Eye Field to pay attention to the target: starts after 120 msec and can last 300 :

2. **Tactile/Proprioceptive System similar to above: Breathing,** Vipassana Meditation upper system is the Mind Root, and lower about Self-feeling of Breath (posterior medial parietal region (Delhaye 2018))

-The Dorsal pathway connects to the Premotor/Motor. cortex

-The Ventral pathway connects to the Temporal cortex for Emotion and Knowledge (Zanescó 2013, Raffone 2010, Tăng 2015, Bernajee 2019

For paying attention to the superior Colliculus, Thalamus, Basal Ganglia, and Mesolimbic (VTA, Subthalam us, Rauss 2013, Long 2018).

When paying attention, the brain uses two attention systems: Up-Down/Dorsal and Bottom-Up/Ventral. In Vipassana Mindfulness, the Dorsal pathway is used to pay attention to the in-and-out breathing, the Ventral pathway is used to feel the abdomen expanding or collapsing or other sensations of the body. Attention is, therefore, tied up to the target. Because the Dorsal and Ventral pathways do not bring in new information, the Inner Consciousness is self-stimulated (under the direction of the ACC-IPS). The characteristic of the ACC, in this case, is "Predicting what is going to happen." Without new information, theACC-IPS looks for existing but deeper information in the Inner CS to work with.

Zen method instructs the meditator to use both the Top-Down and Bottom-Up systems at the same time. Each time one pays attention, one uses one of the three common Senses: the Sight, Hearing, and Breathing (Tactile/Proprioceptive) systems. Each system has an Upper and Lower transmission stream. These two transmission streams work in parallel and inhibit each other. When the Bottom-Up system is used, the Top-Down system will reduced or stopped working. The upper and lower streams can belong to two different systems: the upper visual stream and the lower auditory stream. The phenomenon is used very effectively with street pickpockets: one thief distracts the pickpocketed person with different tricks so that the second thief can pick the pocket.

Other Related cortices in the Meditation:
FEF, and Superior Colliculus control the eye head for attention
The Thalamus filters information to avoid interference
vmPFC, PCC HIPPO Memory is used to select targets.

G. VARIOUS METHODS TO ATTAIN NLIGHTENMENT.325

Buddha invited 25 Arahat and Bodhisattva to present their own methods of spiritual practice.

1. Kaudinya/ vn: Kiều Trần Như, first of five bhikshus, hearing Buddha's teaching on the Noble Four Truths: Contemplation on **Buddha's teaching on Dharma to reach the Original Mind/Emptiness.**
2. Upanisad/ vn: Ưu Ba Ni Sa Đà: Contemplation **on Body and Form to reach Emptiness**.
3. Bodhisattva named Fragrance-adorned/ vn: Hương Nghiêm Đồng Tử: Contemplation on **Smell to reach Emptiness** .
4. Two Bodhisattvas called Bhaiùajya-ràja and Bhaiùajya-samudgata/ vn: Pháp Vương Tử Dược Vương và Dược Thượng who were present with five hundred Brahmadevas: Comtemplation on **Taste to reach Emptiness**.
5. Bhadrapàla/ vn: Bạt Đà Bà La who was with sixteen companions: Contemplation on **Touch to reach Emptiness.**
6. Mahàkà÷yapa/ vn: , Ma Ha Ca Diếp, Golden Light/ vn: Tử Kim Quang and others (of his group: Contemplation on **Dharma of Emptiness to reach Emptiness**
7. Aniruddh/vn: A Na Luật Đà, blind because of crying and insomnia: **Contemplation on Original Mind)**
8. Kùudrapanthaka/ vn: Châu Lợi Bàn Đặc Ca: Comtemplation on **Breath to reach Emptiness**.
9. Gavaüpati/ vn: Kiều Phạm Bạt Đề: Contemplation on **taste**.
10. Pilindavatsa (Tất Lăng Già Bà Ta, Comtemplation on **Touch to reach Emptiness to reach Emptiness.**
11. Subhảñi/ vn: Tu Bồ Đề: Comtemplation on **Emptiness to reach Emptiness.**
12. Sàriputra: vn: Xá Lợi Phất: Contemplation on Original Mind of the **Vision) to reach Emptiness**
13. Samantabhadra Bodhisattva/ vn: Phổ Hiền Bồ Tát: Contemplation on the Meaning of Emptiness **to reach Emptiness**.
14. Sundarananda: vn: Tôn Đà La Nan Đà: Contemplation on the **Breath to reach Emptiness.**

15. Puraamaitràyaõãputra: vn: Phú Lâu Na Di Đa La Ni Tử: Contemplation on **Dharma) to reach Emptiness**

16. Upàli/ vn: Ưu Bà Ly: disciplining the body so that it can free itself from all restraints and then disciplining the Mind: Contemplation on **Discipline and Commandments to reach Emptiness.**

17. Mahà-Maudgalyàyana/ vn: Đại Mục Kiền Liên, returning to stillness to allow the light of the Mind to appear just as muddy water by settling becomes pure and clean as crystal: Contemplation on **Original Mind to reach Emptiness.**

18. Usschuùma/ vn: Ô Xô Sắt Ma looking into the non-existent heat in my body and Mind in order to remove all hindrances thereto and to put an end to the stream of transmigration so that the great Precious fight can appear and lead to the realization of Supreme Bodhi: Contemplation on the **Warm Inner Energy to reach Emptiness.**

19. Dharaõiüdhara Bodhisattva/ vn: Trì Địa Bồ Tát sameness of body and Universe which are created by infection from falsehood arising from the Tathàgata store, until this defilement vanishes and is replaced by perfect wisdom: Contemplation on
Non-discriminative Mind) to reach Emptiness.

20. Candraprabha Bodhisattva/ vn: Nguyệt Quang Đồng Tử contemplates the element of water in order to enter into the state of Samàdhi: Contemplation on **Water nature to reach Emptiness.**

21. Bodhisattva of Crystal Light/ vn: Lưu Ly Quang Pháp Vương Tử: Contemplation on **Wind nature, time and space to reach Emptiness.**

22. Âkà÷agarbha Bodhisattva/ vn: Hư Không Tạng Bồ Tát: **Comtemplation on Emptiness) to reach Emptiness**

23. Maitreya Bodhisattva/ vn: Di Lặc Bồ Tát: Comtemplation on **Tathàgatas to reach Emptiness.**

24. Mahàsthàma/ vn: Đại Thế Chí Pháp Vương Tử, a son of the Dharma king, six senses with continuous pure thoughts, recitation in order to realize Samàdhi: Contemplation on **Consciousness to reach Emptiness.**

25. Avalokite÷vara Bodhisattva/ vn: Quán Thế Âm Bồ Tát, Mạnjusri/ vn: Văn Thù Sư Lợi and Ananda/ vn: A Nan: Meditation on the **Hearing. The process can be analyzed as follows, starting with**

- Summary 25 Arahats and Bodhisattva used many different methods to reach the level of Original Mind.EM of Enlightenment. Before that, they had spent many previous spiritual practices before reaching enlightenment. Thus, disciplines and methods are not as important as the determination in the diligent practice to come back to EM. According to the Surangama Sutra, the enlightenment process is gradual and not immediate.

H. MEDITATION ON THE SOUND. (QUAN YIN) 327

This type of Meditation is not very popular. However, Buddha and Mansjuri advise that this method should be used at the time of Ending Dharma and, particularly, for Ananda. The technique is used by Bodhisattva **Avalokitesvara,** Bodhisattva Manjusri, Buddha **Avalokitesvara,** San Mat Satsang sect, and Eckankar nowadays. The principle is to focus alternatively on the Light and the sound to limit or stop the Mind from wandering. At the same time, the Mind is to focus on the third eye, an imaginary point in the forehead, slightly above the eyebrow line, and in the midline of the face..

The Significance of the sutra can be summarized as follows:
When attaining the void, the Creation vanishes, giving way to the state of Nirvàõa. The first was in accord with the fundamental Profound Enlightened Mind of all the Buddhas and possessed the same merciful power as the Tathàgata. The second was in sympathy with all living beings in the six realms of existence and shared with them the same plea for compassion

In the Surangama, Bodhisattva Avalokitesvara (Bodhisattva of compassion) describes his method as follows:

> Buddha said in the Surangama Sutra/Volume 3/Six Entrances:
>
> Ananda! For example, if someone uses two fingers to tightly block his two ears, due to the tiredness of his ears, there will be sounds in his head; the ears hearing that tiredness are illusional manifestations of the Bodi nature. Due to the two kinds of false characters, Movement and Stillness, hearing appears and reflects the object of hearing, called the nature of hearing; this nature of hearing is separate from movement and Stillness process and has no self-nature (note: Thich Duy Luc's translation: hearing in both ears. Le Dinh Tham's translation: hearing in the head top)
> - Ananda, you should know! This hearing does not come from movement and stillness, nor from the ear, nor from emptiness. Why? If hearing comes from stillness, then when moving, hearing follows stillness and extinction. It should not be able to hear movement; if it comes from movement, then when stillness, hearing follows movement and extinction, it should not know stillness. If it comes from the ear, there is no movement and stillness, so know that the hearing

> *originally has no self-nature. If it comes from emptiness, emptiness has become the nature of the hearing, then it is not emptiness, and emptiness itself hears; what does that have to do with your entry of hearing? So know that Ear Entry is false, it is originally not the nature of causes and conditions, nor is it the nature of self-nature (from Inner Consciousness).*

Meaning: The hearing does not come from the the Nature of Hearing of the original Mind which is distinct from the object or the ear. Why?, because if it come from the movement, the hearing ceases with the stillness,, and so on for the hearing from the stillness, ear or the Air (representing Emptiness). As a result Buddha exclude the movement, stillness, ear, Air as the self Nature of hearing

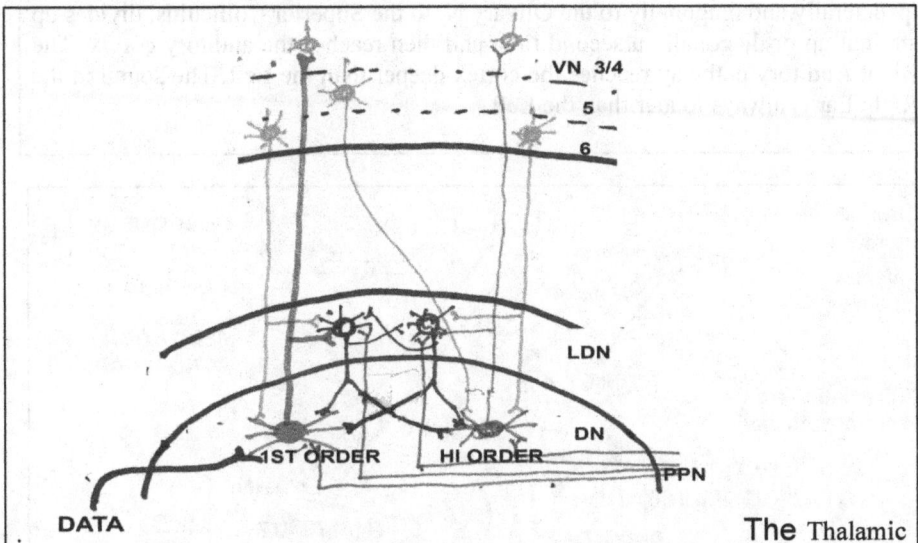

Fig 5.1A: Reticular Network/TRN, which acts as a filter in the Thalamus-Cortex connection (Sherman 2016). Pedunculopontine nucleus (PPN)
Information to the Thalamus composed of the 1st order and High Order neurons are transferred to cortical layers 3 and 4. Cortical neurons in layers 5 and 6 transfer instructions to the TRN and Thalamus to control the SOUND FILTERING mechanism with the help of PPN cells

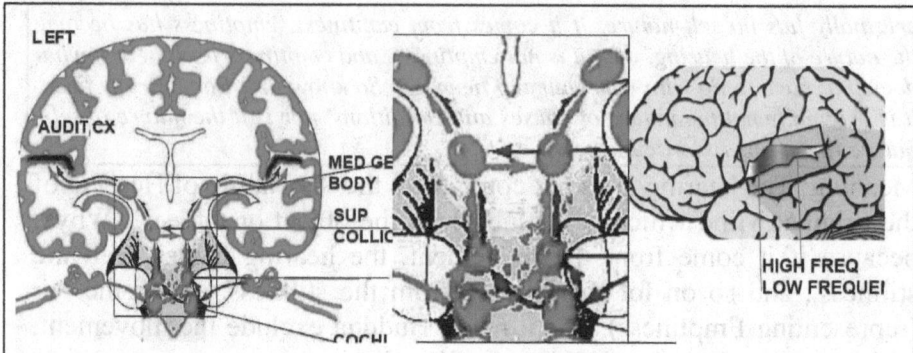

H5.3 The auditory nerve pathway to the Cochlear nucleus N. divides up ipsilaterally and diagonally to the Olivary N. to the Superior Colliculus, divides up straight up or diagonally a second time and then reaches the auditory cortex. The Right Auditory pathway reaches the cortex deeper than the Left. The sound in the Right Ear is always louder than the Left.

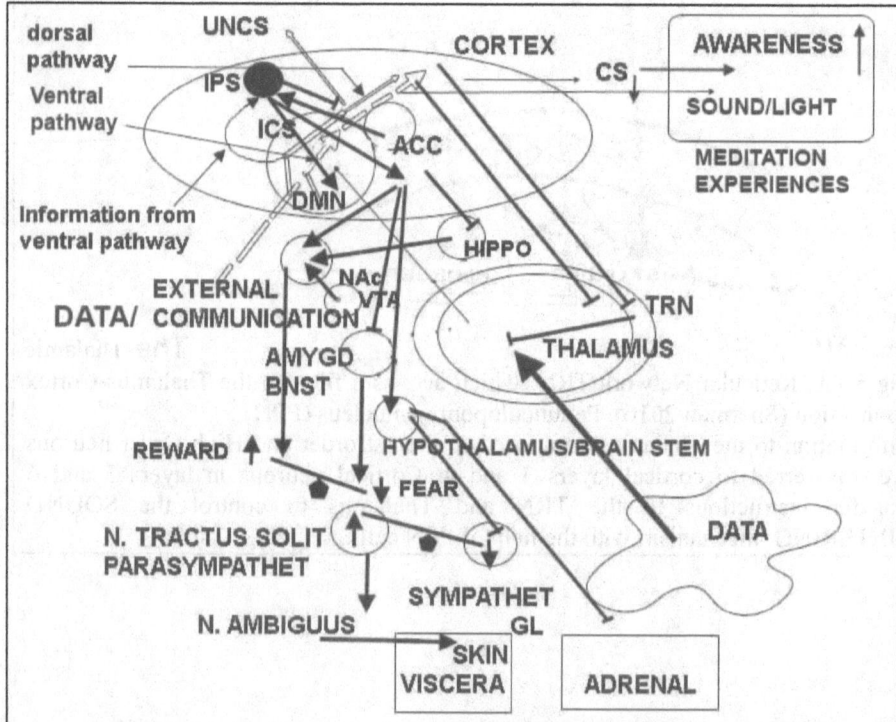

HI: High Order for sound filtering. H5.1B: Diagram: Meditation (black arrow) includes Information; the five sensory Consciousness/CS, and the wandering Mind are reduced or blocked when reaching Thalamus, and only a part of the Information reaches the Cerebral Cortex (sensory CS). Cortex sends inhibitory messages to the Thalamus, Amygdala, BNST /Bed Nucleus of Striata Terminalis (Connecting substance: GABA) (Emotion, Fear), facilitator input to Nucleus Accumbens/NAc (Joy, pleasure), HIPPO (memory) Hypothalamus-Pituitary-

Adremal axis (HPA, causing changes in endocrine system) and Sympathetic ganglia/Sympathetic. The inhibited Sympathetic System increases the activity of the Parasympathetic System (Sherman 2016). The open arrow is only activated in Meditation: The Inner Consciousness has reduced peripheral Information and Knowledge/CS so it reacts, sending Information back to the Thalamus and then to the cortex (in the Soul-in-body mechanism). The vmPFC of MMD can receive Information from the Soul outside the body. (Note: In the case of NOT meditating, peripheral Information always needs to be compared with Information in the inner consciousness to make CS).In summary, the Amygdala and the Sympathetic Ganglia, Hypothalamus, Pituitary gland, and Adrenal Gland are inhibited or have reduced activity. The DMN reduces activity but the Endocrine System reacts in the opposite way, so the Endocrine System and Cerebral Cortex increase activity.

1. THE BRAIN MECHANISM OF ATTENTION AND MEDITATION330

Buddhist scriptures are very voluminous and precise regarding the meaning and core of Tao and Zen. But when Buddha spoke, it was at a time when Science had not yet developed. The neuroscience mechanism of learning could not be understood by people of Buddha's time, so Buddha could not preach. Therefore, when preaching about the Original Mind, it is necessary to understand the formation of CS in the Surangama and the Flower Sutra; meditators must understand the mechanism of the mind/emotions, the external manifestation of the Mind and Celesika (the inner Mind). However, even though it has been studied and analyzed, the lack of understanding of Neuroscience also makes the separation challenging to understand, not thorough and can lead to mistakes.

Meditation includes two important mechanisms: Attention and Mindfulness

a). In Attention, information coming to the Thalamus is filtered by the TRN system to filter out information that does not need to be disturbed. The TRN neurons perform the filtration task; neurons from the 5th level of the cortex are stimulated by information coming from the primary order neutrons (unfiltered original information) and the Pedunculopontine nucleus (PPN) neurons from the brainstem. Finally, the HI/high-order neurons send the cortex more purified information. Thus, whether in Meditation or not, information is no longer original as it is

b) Sound Light in Meditation: information from the five senses is intentionally blocked due to the Bodhi mind for

Meditation. ACC is the part of the brain that searches for information. Because of the lack of peripheral information, ACC, through the intermediary of IPS, searches for information in the Inner Consciousness/Inner CS (mainly the Default Network/MMD. IPS is also the cortex that helps find targets for Attention in the stages of Finding and Attachment. As a result, in Meditation, the MM in ICS is retrieved to the present. MM from ICS of this type is in the absence of Attention (knowing that in Meditation, Attention is used to focus on targets such as breathing, wisdom eyes, and respiratory muscles). Because ordinary creatures cannot pay Attention to two targets simultaneously, information from ICS in Meditation cannot generate the CS. Therefore, information from ICS in Meditation is only known through a physical state of Sound or Light.

-**when starting Meditation**. Sometimes many scenes are often blurry. The scenes come from ICS (on the left, the scenes have a familiar character, and ICS on the right gives a scene with an illusional character like in the imagination of a normal person). The intensity of the light increases with the depth of the Dhyana. Initial Dhyana is less bright than Near Dhyana.

- Sound: The information is not accompanied by any conscious features. The auditory inner and posterior part cortex of the Rolando groove (corresponding to the cortex of the Right ear and high frequency) is easily recalled, even when there is no attention. In Meditation, the information of the superficial (Left ear) and deep (Right ear), anterior (low/bass sound frequency), and posterior (high frequency/like cricket chirping) parts are all recalled, so in Meditation, the Sound is heard synthesized on the top of the head with a high and low pitch. The more attentive and focused the anterior part of Cortex corresponds to low frequency or musical Sound like a flute, the more music is heard compared to when there is no TD or concentration.

2. SURANGAMA SUTRA WITH QUAN YIN METHO

Ananda....was instructed by Buddha to practise Meditation by *means of the organ of hearing..At first by directing the organ of hearing into the stream of Meditation, this organ was detached from its object, and by wiping out (the concept of) both sound and stream-entry, both disturbance and stillness became clearly non-existent. Thus advancing step by step both hearing and its object ceased completely, but I did not stop where they ended. When the Awareness of this state and this state itself were realized as non-existent, both*

> *subject and object merged into the void, the Awareness of which became all embracing*

San Mat, Eckankar, and Eckankar (Meditation on Light and Sound) are mindfulness meditation methods based on the principles of attention to light and sound.

- The meditation with attention on the light at the Third eye (using the **Dorsal pathway**) is to reveal the sound (Sound is revealed as a component of AWARENESS, comining from the Ventral pathway).

- When the Sound is loud enough after 5-30min), then the attention is switched to the sound stream that become the target of attention (therefore **Dorsal pathway**), the LIGHT is revealed as component of AWARENESS from the Ventral pathway).

The Quan Yin meditation which was first accessed in the Buddha's Book 6 is considered the basic and core book of Buddha.

Listen to the Sound stream as follows:

(Note: chapters 7-10, Surangama sutra are about tht Quan Yin meditation and practice)

　　i. **Hearing = hearing with the Sensory Hearing Consciousness**. The first step in hearing is entering the stream (Hearing in the inner (non-external) sound stream)

When focusing the Mind on the Wisdom Eye or any point on the body, the sound stream belongs to the everyday material world, high frequency like the sound of birds and crickets chirping. Only hear this sound in the Right ear. Why is that?, because the sound in each ear is separated into two streams going through the Left and Right auditory cortex. In particular, the sound in the Right ear goes deeper Cortex than the Left Ear. (H5.4 p. 298). The deep and inner part is probably less affected by the outside world.

　　ii. **The ability to hear:** (The 6th Consciousness/Knowledge= CORTEX When the input (stream) is gone, then the two phenomenon of the outside world does not interfere with the hearing Thus, gradually adding the ability to hear (The 6th Consciousness/Knowledge=CORTEX) then the ability to hear and the object heard both end [NO need to use the ears or the Cerebral Cortex to still hear (hearing with the MIND obliterated with IGNORANCE=SOUL= CONSCIOUSNESS];

When continuing to pay attention to the Sound stream of the Sensory Hearing, there is awareness of a second Sound stream, deeper and with a wide spreading nature, making the meditator feel like he is listening to the OCEANIC SOUND. This sound will gradually become clearer and louder and is felt to come from the middle of the top of the head (meaning that the sound comes from both side R ight and Left). One proof that the Sound stream is the ability to hear (CEREBRAL CORTEX) is as follows:

- When listening to this Sound, one often feels an electric charge stimulated at the top of the head. This feeling is because the Default Network (PCC, see Fg F4.4 pge 195) is where the Memory for the Inner Consciousness/CS is stored. Meditation in this stage is looking at the Inner CS. When the Inner CS Memory is stimulated, the synapses/neural communication from a resting state become active (see the image of the Soul structure on pge 52)
- This Sound often stimulates the Cerebral Motor Cortex to move the head and neck, causing the meditator to shake his head and neck rhythmically. Knowing that the motor cortex of the head and neck is located at the top of the head.

→The end of the ability (CS and sensory hearing) does not abide. If there is still knowledge that does not abide, there is still the CS and the sensory hearing, so the CS must come to the END, then that Emptiness will be attained.

The Sound Stream continues to change (going up, after a period of 5-30 minutes or more, the phenomena of perception at the top of the head will decrease or disappear, the sound will be more wonderful and refined and can change tone. The phenomenon corresponds to the end of the Ability to hear, but the sensory hearing may still be heard, but it does not hinder hearing the wonderful sound flow of the world beyond

iii. When the ability and Sensory Hearing are terminated (heard with Original Mind/Buddha NATURE). When the realm of birth and death is overcome, then nirvana appears, suddenly transcending the world and leaving the world....

Next, the Buddha told Manjushri to look at the 25 Great Bodhisattvas and Arhats who each presented the Dharma of attaining enlightenment at the beginning. The Buddha wanted Manjushri to choose a method for Ananda to practice and attain enlightenment. Manjushri chose the Quan Yin Dharma door, so he said: "Future practitioners should follow this Dharma, I also attained this door,... Other means depend on the six senses logistic, and on e Buddha's divine power are not the right way to practice..."

Although the Dharma was especially praised by Buddha Muni, it was never described how to practice it in details. Everyone vaguely understood that it was just to practice to sound contemplation. Why? No one asked or was not asked and of course no one answered. There is no official Buddhist scripture that notes how to practice Quan Yin Meditation. Apart from the description of Guanyin Bodhisattva. Sound (or Light) is not obtained from the ears and eyes but from the Inner Consciousness. Depending on the level of Zen, the Light and Sound become more and more miraculous as the level of Zen increases.

Impressions: The target of attention is not as important as the realization of God/ Buddha (called Enlightenment of various levels of Meditation) resulting from mindfulness. The purpose of spiritual practice is the realization of the Original Mind. Therefore, the method/ techniques are less important than the Meditation volition.

In this method, there is no contemplation on the Mind and Dharma. The reason is that the meditators using this method no longer need to learn (skt : Asaiksa/ no longer learning/ beyond learning stage/ vn: hạng Vô học , like Arhatship)

In Quan Yin method by Master Ching Hai or in San Mat

The meditator first pays attention to the third eye; The light will appear in the Third eye area. After a certain period, when the sound stream is strong, attention is switched to the sound stream. The light in the third wye will become stronger and more extensive

3. MECHANISMS THAT CREATE SOUND/VOICE AND LIGHT/IMAGE

There are two mechanisms:
Attention, a crucial element in the process, is responsible for bringing Buddhahood /Original Mind back to information. It originates from the volition which starts from the Insula cortex. This active role of attention empowers the practitioner, making them responsible for their meditation journey.

In Meditation, when focusing on the Wisdom Eye. Information from the Inner CS, the realm of our thoughts and emotions, is pulled back to the present because the ACC, the anterior cingulate cortex is looking for the source of information.

Insula→IPS→dlPFC→ACC→Inner CS

This information is not accompanied by the attention because of the Insula-initiated attention is already used for tha attention. The intention to pay attention carries the Soul and the Original Mind. Without the Soul and the Original Mind, information from the Inner CS in Meditation cannot become CS, so Meditator can only hear the Sound or Light stream but there is no voice or image.

There is occasional image when first starting to meditate, or sometimes in Meditation. The event shows that there is no attention but only mere concentration, so the inner consciousness information can be seen with the image (due to the presence of low level of attention), which is usually unclear.

When there is attention, there may be vague images or sounds such as the sound of wind blowing, ocean waves, or elegant sounds, unlike the images or sounds when out of body, the images and sounds are clearer when in daily awareness. In Meditation, attention is directed to a simple, predetermined target such as breathing or the wisdom eye. So, the information from the inner consciousness is less noticed; therefore, there is no detail. But when attention is paid to the inner consciousness, it becomes thinking, a mental process that involves the manipulation of information as a result of cognitive processes. This is no longer considered Meditation. - From the Buddha nature of the mind, that is the true sound and light.

From the above event, as argued before, when the Buddha invited 25 Arahants Bodhisattvas to present the methods for enlightenment. All methods are the means to reach Emptiness. Likewise, in Quan Yin meditation (or Vipassana), light, sound, or tactile (proprioceptive) are all brought back to Emptiness. Because all six senses represent the temporary division of the data. The original information is a single whole package emanating from Emptiness

4. Inner Consciousness, Buddha Light, and Meditation experiences 335

In the previous paragraph, the restriction of mindfulness and subsequent development of restricted Consciousness allows enhanced and enlarged Awareness to be developed. Because Awareness is closely connected with the Original Mind/ Buddhahood, the radiant light and the stream sound experienced in mindfulness likely originate from these states of Mind/ Awareness.

In addition, the restriction of external information coming to the Thalamus is remarkably reduced. The brain is the box of prediction.

The central role in this prediction is played by the ACC and the Zona Incerta, which are known as detection of errors or identification of different or novel information.

Without incoming information from outside, the ACC and ZI search for data in the inner Consciousness. This phenomenon is comparable to smartphones in areas without the Internet, like on airplanes. One has to turn on airplane mode to turn off the search mode. As a result, the ICS is looked at, and information in the ICS is retrieved. The retrieval of information from the ICS likely accounts for the imagery in meditation experiences.

In most Meditation studies, the Default Mode Network (DMN) is equivalent to the component ICS involved in storing the implicit and semantic MM. This network displays low activity that is consistent with the low retrieval output. This is in contrast with the high activity of DMN when thinking. This low activity does not reflect the decreased Ego in Meditation. The ego is the barrier to enlightenment unless Meditation cleanses the ego.

Since, the data retrieved from the ICS in the meditation does not receive the critical attention, therefore visual or auditory data do not represent as comprehensive images or meaningful sound as in case of thinking with attention. They simply appear as simple light or sound of varying intensity, and rarely as fuzzy images or brigt moon or sun non characteristic wind of ocean sounds.

Another mechanism of Meditation experiences is the connection of spirit to the inner CS. The connection is likely similar to that of the embryo's ensoulment or religious mediumship or in a person known for ESP/ extra sensory perception capabilities, which is quite familiar in southeastern Asia, including Vietnam.

I. FIVE DHYANA Meditation 335
, in Samatha Meditation

The first stage of Five level Dhyana meditation is marked by attention, a crucial element that paves the way for Concentration. This attention is channeled through the Dorsal Pathway, a significant route that aids in keeping the Mind still .

i.   The Dorsal Pathway of attention to the breath (Proprioceptive/Tactile): pay attention to the inhalation and exhalation at the nose or a region/point of the body--use attention to the Mind.

ii.   [The Ventral Pathway involves the contemplation to : Body, Feeling (receiving pain, itching...) Mind (feeling emotions) and Dharma (mechanism)].

Theravada Buddhism points out five stages / called the

1. FIVE DHYANA MEDITATION

• Initial application (**Finding**) or Intensive step (paying attention to the target such as breathing, the Wisdom eye) (vn: TÀM),

• Sustained application (**Attachment**) or Reflection (the Mind begins to stay on the target without wandering), the first two stages of the Five Samatha Meditations. In these two stages, the role of Mindfulness is secondary because meditation is for Concentration, then there will be Awareness (right Mindfulness). Thanks to the Mind no longer wandering, the meditator feels comfortable

• **Joy/Pleasure**, then the Mind/Knowledge can begin to look into the Inner Consciousness, the realm of images, lights, colors and sounds. All of these create a feeling of joy, lifting the spirit of the meditator

• Happininess Prolonged Joy,

The final stage of the Five Dhyana Meditations is marked by **'Concentration/One-pointedness'**. This stage signifies a deep and unwavering concentration, where the Mind is fully absorbed in the object of meditation.

•   When the Inner Mind is looked into, it is the period when the Mental inner CS is modified, eliminating the impure Mind and increasing the Pure Mind and Immeasurable Mind. (There is the concept of Contemplation to cure Drowsiness, Contemplation to cure Doubt .

• **Correlation of Four and Five Dhyna Meditation**

	Initital	Upacāra,	First	Second	Third	Arahat
Initial application	X	X				
Sustained application	x	x	X			
Pleasure		x	X	X		
Happiness			X	X	X	X
One-pointdness			X	X	X	X

2. Samatha/Mindfullness, the Dorsal pathway 337

The Insula-ACC, salient network, under the command of the volition, initiates the dlPFC-IPS management network that triggers the dorsal pathway to command the IPS to use the motor control of the attention system. When the goal (in the form of information) is chosen (for example, the rhythm of respiration), the other information (coming from the Inner CS is discarded by the IPS. The discarded information without enough attention is not reconsolidated and can be
- Stored in the subconsciousness.
- Prone to be subjective to the process of neuroplasticity with the cleansing of the related Memories (karma).
This Dorsal pathway, combined with the Ventral pathway, is used to inhibit the Mind from wandering.
This technique is efficient in promoting the meditator high to the scale of Jhana,

The method of meditation tends to inhibit the input data from reaching the sensory cortex and the absence or decreased Activation of the ventral pathway. As a result, there is absent or decreased contemplation or Awareness. In comparison to the Contemplation method, in mindfulness meditation, the sensory cortices are deprived of the input but not inhibited. Therefore, Awareness can not develop

Note on
***Neither Thinking nor not Thinking/ vn: Phi Tưởng Phi Phi Tưởng**. The term should be coined as Neither Feeling nor not Feeling. The five Skandhas consists of Form (the body/sensory organs), Perception (receiving the data coming in the Brain), Feeling (immediate reaction of the Brain without consciousness), Formation (integration od data in the brain) and Consciousness (formation of knowledge. Therefore in Either Thinking nor not Thinking, the term thinking should be replaced by Feeling because at this stage of data input there no consciousness formation yet, as a result there is no thinking (formation of thought) yet. Therefore Neither Thinking nor not Thinking implies a state of the brain and the body of the meditator with nearly complete insensitivity to data input*:

- Contemplation and awareness, The ventral pathway

With Enlightenment and realization of Omniscience/ Buddhahood. Insight and Awareness mean training one to see things as they are. This goal is achieved through mindful contemplation of four previously mentioned areas. In this technique, the most contemplated are the mind and karma, which are used to realize selflessness, the four immeasurable minds of virtue, and to relinquish the evils of the mind, which include greed, anger, and ignorance. It is evident that many intellects can realize the above insight without Meditation, only through reading spiritual books and thinking

The farside effect of this technique is slowing down the path of mindfulness Meditation because contemplation eventually diminishes the attention in Mindfulness.

The IPS, in this case, does not suppress or discard the data from the Ventral pathway (that retrieves the Memories from the inner CS) but uses the data as the object of contemplation (such as Greed and joy...). In the state of Mindfulness, the data is clearer.

· As a result, in the Vipassana method, there are many options in mixing the above techniques with Mindfulness and contemplation in successive or alternative order. The technique exemplifies this, the so-called Lục Diệu Pháp Môn (vn)/ Six Wonderfull Dharma . (see the next paragraph **SIX WONDERFUL DHARMA DOORS)** Method supported by the Respected Venerable Thích Thanh Từ due to the flexibility of the technique.

The most frequent technique is the concentration on the rhythm of breathing in and out, as one breaths slowly without intention or effort. In the meantime, observation is directed to a second phenomenon, one of the four types of five Maras listed above.

(Satipatthana, Anapanasati

Buddha discovered the Dharma of Zen after practicing Samantha with two Non-Buddhists. Alara Kalama reached the Boundless Space Realm and the Boundless Consciousness Realm and Uddaka Ramaputta reached the Neither Perception Nor Non-Perception Realm. But because he had not yet attained full enlightenment, he practiced asceticism in the forest with the 6 brothers Kondanna for 6 years, at last, to practice the Middle Way. After 49 days of practice, he attained full enlightenment through the practice of Vipassana.MS Dharma is based on the following fundamentals:

-The past and the future cannot be grasped

-But the Present can be grasped with each PRESENT MOMENT when using Diligence, Mindfulness and Awakening. These three characteristics make AWARENESS manifest. Knowing that

AWARENESS is already present in the Buddha Nature/the original Mind of everyone but is obscured by Ignorance in the form of Afflictions of Wrong Views and the manifestation through the intermediate of the Five Aggregates or Name and Form (Form: Body, Name: Feeling, Perception, Volition, Consciousness). The way to remove Afflictions and other bonds is to focus the Mind on the activities of Name and Form. When Name and Form are not active, Afflictions cannot be seen so they cannot be removed. Name and Form are KARMA. When understood like this, Karma has the opportunity to be removed to see AWARENESS

1. MINDFULNESS MECHANISM IN INSIGHT PRACTICE TO DISCOVER AWARENESS 339 **(H5.3)**

Insight meditation, Awakening + Diligence + Mindfulness together with Concentration, keeping the Precepts are the mechanisms to remove Karma. Practice itself does not give rise to Awareness because Awareness is already present in Buddha Nature, just need to wipe away Karma to have Awareness.

Mindfulness here has a relative meaning, because it is a phenomenon that occurs in the world of delusion, is not permanent in the phenomenon of momentary birth and death. True Mindfulness only occurs in the first Creation when Buddha Nature / True Emptiness through Mindfulness creates the World. Delusion is the Physical world and Mindfulness creates Metaphysical Nirvana

In the concept of birth and death of Name and Form, Volition, Diligence, Awakening and Mindfulness come to the mechanism of Breathing, walking, feeling, thinking and contemplating. Creating Name and Form... creates the removal of Karma. From there, we can deduce that Reciting Buddha's Name or Contemplating the Sound of Sound also uses a similar mechanism, the problem is that Diligence, Awakening and Mindfulness are needed to reach the present phenomenon

The practice focuses on AWARENESS through Contemplation
- Form/Body (for people with much Desire/ use 24 topics/Kasinas about the body (Thanissaro) such as breathing, walking, standing, bending, limbs, corpse, 32 characteristics such as hair...),
- Feeling (for people with Awareness and Greed/ use 9 topics such as joy, suffering and neither joy nor suffering),

- Mind (people with Wrong Views, Arrogance/ use 16 topics, mental states such as greed, anger, ignorance, limited concentration...),
- Dharma (people with strong Memory/ use 5 topics about Dharma:

    -The INNER mind gives rise to Than San De, Doubt, confusion)

    -The operation of the Five Aggregates in the mind

    -The six external objects

    - The operation of the Seven Factors of Enlightenment

    - The operation of the Four Foundations of Mindfulness

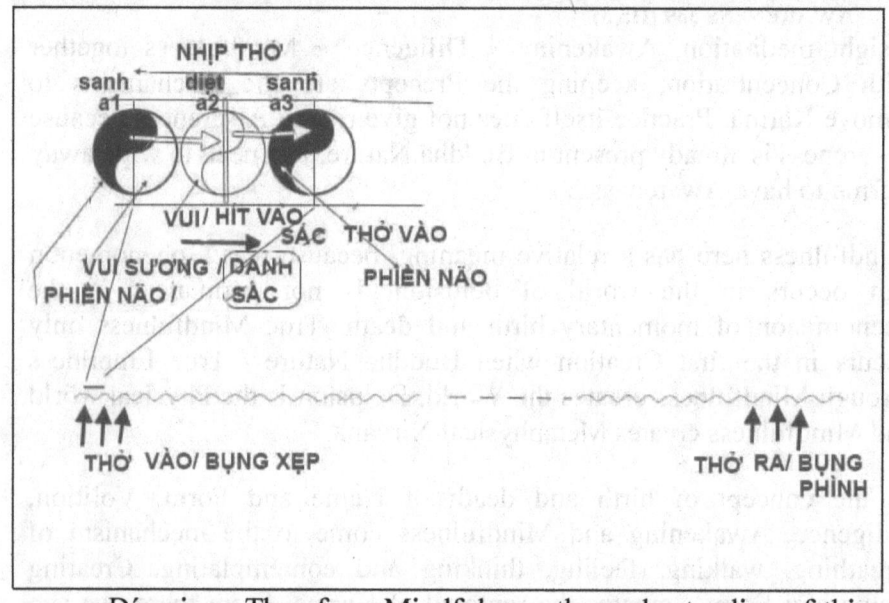

Dícusiom: Therefore, Mindfulness, the understanding of things as they are, that is EM. In the Surangama Sutra, where the Buddha invited 25 Arahats and Bodhisattvas to present the method of practice. All methods are ultimately based on this essence, EM.

Hypothesis on the phenomenon of NAME AND FORM arising and ceasing of living beings and the establishment of MINDFULNESS in Vipassana Meditation. Each inhalation (a1) and exhalation (a2) and then inhalation (a3) is two times of birth and death. Life has repeated moments of birth and ceasing of NAME AND FORM with many times of Affliction and Joy occurring, but people cannot know. Only when people, thanks to Diligence + Awareness + Mindfulness of the phenomenon of Exhalation/Bellied Belly 9 and other changing phenomena, feel the phenomenon, then the phenomenon of bloating (and joy arising) is recorded as the phenomenon of bloating belly (Kamma dissolving) is MINDFULNESS and joy is recorded in the inner consciousness

Note: Form is Karma and is Affliction. When Affliction is over, Joy and Happiness appear. Furthermore, when Form changes in response to movement, as we often know, movement helps to bring joy.
Friction is the Result of: 1) Birth (of the 5 Aggregates) 2) Aging (of the 5 Aggregates) 3) Death (of the 5 Aggregates) Suffering with Affliction: 4) Sorrow 5) Lamentation 6) Suffering 7) Worry 8) Anguish 9) Being close to people you don't like 10) Being apart from people you love 11) Wanting but not getting

2. THE SIXTEEN AWARENESS OF INSIGHT MANIFESTATION WITH CONTEMPLATION 241

AWARENESS consists of three types: Knowledge (by CONTEMPLATION), Inference (reasoning) and Abandonment (abandoning wrong views): recognizing whether Contemplation things are impermanent, egoless and suffering or not
1) Distinguishing Name and Form (Form: Body, objects of the five senses that can be known, Name and form that cannot be known/Soul/Feelings, Perceptions, Forms, Consciousness)
2) Awareness of grasping Dependent Origination: recognizing the root cause through the discriminating mind to point out the cause of knowing
3) Awareness of Investigation of the Three Characteristics/Impermanence, egolessness/suffering
Awareness of Destruction/The intermediate process between Birth and Death. Destruction is always followed by the practice of abandoning and correcting
4-11) Awareness of Fear Knowing Danger, depression, desire to escape, liberation, release
12) Awareness of Consciousness: surrender to the natural law of the Way
13) Conversion of Race to the Way
14) Awareness of the Way
15) Awareness of Results; feeling at peace, free from afflictions
16) Awareness of Reflection: returning to the mind in harmony with the world
BRAIN LEARNING: Recognizing the above event in the Four Foundations of Mindfulness in Meditation is the work of CONTEMPLATION using the Ventral Pathway, with the purpose of CONTEMPLATION meditation to purify the inner MIND. When the

inner mind is clean, that is, the veil of ignorance has been removed, there will be Awareness. Unlike other practices such as Quan Yin, in Vipassana, attention is important in both Contemplation and Mindfulness. But because Attention reduces AWARENESS, Attention needs to be balanced with Contemplation in Vipassana. For Quan Yin or Buddha Recitation, the more Attention (using the Dorsal pathway, the more AWARENESS (the lower Transmission path) there is.

In Samatha practice, only attention is needed and the Upper Brain pathway is used. In Samatha practice, the Lower Brain pathway is almost eliminated. Only the process of Form, Feeling, and Perception is used. Mental Formations and Consciousness are almost eliminated, so AWARENESS is not revealed. The information on this lower path will be suppressed and discarded by the IPS (inner sulcus).. This process of cleansing MM is unconscious because of lack of attention. However, the desirable effect is that the meditator can reach the fourth level of meditation or higher, such as Neither Perception Nor Non-Perception, a level of Samadhi without enlightenment and realization. The reasons are:

- The cleansing of Karma in the Inner Consciousness, as it stands, is inadequate. The information extracted by ACC is not only random and chaotic, also incomplete.
- Lack of subjective element of meditation or repentance that is contributory in the spiritual journey. It is through these practices that confession becomes more than a mere withdrawal of bad Karma. It becomes a process of arranging and understanding our actions in a more subjective and personal manner.
- The lack of understanding of Buddha nature is a significant gap in the knowledge. Without fully understanding Ignorance, the Three Dharma Seals, especially Non-Ego, the spiritual journey remains incomplete. It is crucial that striving to fill this gap is crucial.
-In Zen on Form, Feeling, Mind and Dharma. Information includes UNDERSTANDING and KNOWING to develop Mindfulness from there

In this case, the IPS cortex suppresses the information of this lower pathway and instead transfers a large part of this information to its attention. This process results in keeping the meditator on Concentration, and avoiding the Attention from unwanted information. But because there is still Concentration, the emotional information is clearer, and the evil Mind can be replaced by a good mind. For example: I know I am angry (while breathing in or breathing out), so I transform it into a good mind: I let go and become peaceful. This transformation is like turning a dark room into a bright one, where the darkness represents the negative emotions and the light represents the positive transformation.

In Samatha meditation, Attention is used to practice either Contemplation or Samadhi, or both are used sequentially or alternately with Samadhi and Contemplation in each meditation session. Attention is the ability to focus on a single task or object without being distracted to attain the Samadhi, while Contemplation is the practice of being aware of your thoughts, feelings, and sensations in the present moment. This method of practice is the method of the SIX WONDERFUL DHARMA DOOR, praised by Venerable Thich Thanh Tu because this method is flexible.

Focus on the Breath (top-down system) and observe the changes in the Breath (bottom-up system) to gradually improve the ability to focus the Mind and strengthen the CS to become mindful. Among the 25 Masters invited by the Buddha to present the practice to attain enlightenment, two of them used the method of Breath Contemplation (Chu Li Pan Te Ca, Sun Da La Nan Da). The focus uses the Breath to enter the door of the Mind to enter the Original Mind / True Emptiness. Sun Da La Nan Da also described the Light seen when practicing the Contemplation on the Breath, proving that he entered Samadhi to see the Light from the Inner Mind (also called Buddha light). In Samadhi, the Inner Consciousness is gradually purified over time, then Buddha Nature/Original Mind will appear, and Awareness can be obtained. When there is no Samadhi, practicing contemplation is difficult and inadequate. CS is a veil of Ignorance; Mindfulness increases with Attention +Samadhi and decreases without Attention +Samadhi.

Without Buddha nature, there can be absolutely no Knowledge and Mindfulness. The Majjhima Nikaya/Mindfulness of Breathing records the Buddha's words:

> "*He abides thus mindful, contemplating, considering, investigating that Dharma with wisdom. Bhikkhus, while dwelling thus mindful, the bhikkhu with wisdom thinks, contemplates, and investigates that Dharma, at that time the enlightenment factor of investigation of Dharma arises in him. At that time the bhikkhu develops the enlightenment factor of investigation of Dharma. At that time the enlightenment factor of investigation of Dharma is developed by the bhikkhu to perfection.*"

The sutra emphasizes the transformative power of meditation. It teaches that when the Mind is still, thoughts arise. Through Attention, we can distinguish RIGHT FROM WRONG when contemplating the Dharma in practice to achieve perfection. Even when the Mind is distracted by the impact of Feeling, Mind, and Dharma, the meditator can prevent it from running wandering by using the Ventral pathway to IPS to pull the Mind back to the Up-Down/Dorsal Attention System as mentioned above.

The method requires patience and persistent practice to make the two systems, Dorsal and Ventral, into a habit. This familiarization is just a form to express the connection of the nerve cells of each of the above systems with the predetermined goal. Daily practice strengthens the connection and gradually makes it automatic, inhibiting the Ego in the Mind of Attention.

The two systems above can connect and restrain each other, creating a harmonious balance. The Dorsal system actively inhibits the Ventral system in Dhyana and vice versa in Contemplation. This mutual inhibition creates a balance that eliminates the subjective nature of the Dorsal system. The Ventral system has the significant effect of eliminating the False Ego

Contemplation is different from Awareness. Contemplation uses the volition in committing to the meditation to perceive the true nature of events. We do things unconsciously without paying attention and are often misled by prejudices. So, the meaning is to renew and awaken, as well as remove past wrong prejudices. That is to build Mindfulness, including the four types mentioned above. Remember the MIND and good DHARMAS, such as NO- Greed, NOAnger, NO-Ignorance, Peace, Joy, Compassion, tolearance. Use these Mental Dharmas to remove the Evil Mental Dharma from the Inner Consciousness during Concentration. Only when concentrated is inner consciousness revealed as well as evil CS removed. At other times, replacing the good Mind and evict the Evil Mind is very difficult.

AWARENESS is the connection with Omniscience without much interference by CS in DMN/Inner, usually obtained when in Mindfulness. As a result, Awareness is not the acquisition in thr Mindfulness but a revelation in the Mindfulc ness when Samadhi or Comtemplation remove the evil MM.

In the Surangama Sutra, when enlightenment is obtained, Awareness does not come directly from Contemplation but is a direct result from the cleansing of Karma:

> *Ananda, you should know, you sit in the Bodhimanda, if the Delusional Thoughts (CS) are gone, then right at that moment of leaving the thought, everything is clear, movement and stillness do not change, remembering and forgetting are as one, you should stay here and enter the right concentration. Like a person with bright eyes in a dark place, the true nature is pure, in the Mind there is no light yet, this is called the scope of FORM WARMTH. If the eyes are clear, then the ten directions open, there is no more darkness, called the Delusional Thoughts are gone, then*
> *that person has transcended the Worldly life. But observing the cause is due to the Persistent contemplation up to the root.*
> *Ananda, while studying the Wonderful Revelation, forgetting all the four elements, suddenly the Form goes in and out of all materials without any hindrance, that is called the clarity that overflows before the eyes. That event is only a temporary function like that, not the attainment of Sainthood; do not consider yourself a Saint, call it a good realm, if you consider yourself a Saint, you will immediately fall into evil realm*

Similarly, when Ignorance obscures Feeling, Perception, Mental Formation, and Consciousness, it is removed by TD. This process leads to Buddha becoming a fully enlightened person.

So what is the role of Contemplation? Contemplation requires attention, so it also aids Samadhi and Samatha. By contemplating events in the body and mental life, it helps the Brain recover MM/Karma to cleanse away evil Karma. The mechanism is similar to repenting, and confessing. However, in Meditation, the recovery of Karma to the present can be much deeper. Thus, it can be said that Quan Yin or Vipassana is no different in terms of mechanism to have Awareness.

3. SAMATHA, SAMADHI/MINDFULNESS, CONTEMPLATION AND AWARENESS

Relationship of wakefulness/ Mindfulness and Awareness
Wakefulness (accounted by Acetylcholine), Consciousness (Norepinephrine), mindfulness (attention), and Awareness (ventral pathway) are four different states in Meditation:
Mindfulness is necessarily associated with wakefulness with a generation of Consciousness
Awareness develops as the result of narrowing down Consciousness / mindfulness. Therefore Awareness is original, "natural," and effortless.

Cognition and Awareness are mutually inhibitory, as shown in the Figure. Cognition is the product of the brain, after the Creation, False thought / Big Bang. Awareness is near the Original Mind, close to the Tathagata.

a) Concentration and Attention (Samatha) is to stop the Mind Consciousness or, more correctly, the Mana Consciousness and the whole CS. Information stops at the stage of Feeling, corresponding to the Dorsal Pathway. The Ventral Pathway is blocked, so there is no CS. For example, when a strange object flies into the face, the eyes close, the head dodges but does not know what it is. As a result there is no Mental Formation and the ventral Pathaway is involved. In Samatha, the concentration makes individual possible to become insensitive to the environment: water, fire, rocks, walls... are all the same. Likewise, space gradually becomes insensitive to the Mind so that the meditator can gradually reach the states of the Four Formless Dhyanas.

Only in Samatha/Silence is there little questioning of MM about the present, so there is little Awareness. The reason is that ACC-IPS is less activated, like Brain during Sleep

From the above principle, Samatha can help to reach the following four levels of FOUR FORMLESS DHYANA

- The Realm of Supreme Infinite Space: seeing the space in front as infinitely vast. Practice meditation to lose the sense of discrimination in that space
- The Realm of Infinite Consciousness: This type of meditation will go beyond Consciousness about the space that is being discriminated, In this Realm of Infinite Space, there is still Consciousness that can discriminate. The first teacher of Buddha achieved this level of meditation.
- The Realm of Nothingness: In the Realm of Infinite Consciousness CS is still known. In this Realm of Infinite Consciousness reveal that there is nothing to be reveal. When Consciousness is eliminated, there is also no Action, no more CS about Space
- Neither Perception Nor Non-Perception completely surpasses Consciousness and Perception to reflect, but CS is comlptely eliminated.

b) Samatha is not synonymous with Samadhi
- Concentration/Samatha:
    In Samatha: only pay Attention to the inhalation and exhalation,

In Quan Yin Methode: only pay attention Attention to the Wisdom Eye to see the Light, but do not pay Attention to the Sound that is ringing in the Lower Path

- Samadhi/Mindfulnes:
In Vipassana, to pay Attention to Breathing with reducing the Attention in the Upper Path, so the Lower Path has someAttention to create Mindfulness/Mindfulness. When concentrating, the Zen practitioner can see the Light or hear the Sound while contemplating the breathing. But the hearing and seeing above is rarely recorded in most books of writing in Google; perhaps the phenomenon is rare. On the contrary, in Quan Yin method, Awareness is not enough in extent, when compared to people who practice Vipassana method. For those who achieve the high level of Samadhi, like the Kiều Trần Như brothers, only need to listen to the Four Noble Truths to attain the Arhat fruit. Meanwhile, Anada who is enlightened, so he only needed one week of meditation to attain Arhatship.

In the Quan Yin method, is to pay Attention to the Sound that resonates in the Lower and Upper transmission paths, which also creates Mindfulness and Wisdom.

- Samadhi/Mindfulness) (sama: equal dh: "consciousness) Concentration is when there is participation in the Ventral Pathway, so there is Awareness. Mindfulness corresponds to the process of seeing and feeling things as they are. The memory is recorded in the Inner Consciousness, eliminating evil Inner Consciousness of the same category. Mindfulness is often used to treat nervous anxiety because it blocks the Manna Consciousness/thinking, and calms the Mind.
Vipassana/ Vi=Variations, passana=Awareness) is the process of having Awareness as the result of reducing or clearing away the veil of Ignorance. The removal of CS (veil of Ignorance) by Samadhi to have the view as it is, which is Mindfulness, is equivalent to thinning and lifting aside the veil of Ignorance to reveal AWARENES, which is the closest form of Original Mind/Omniscience. The method is Concentration, which requires effort to keep the Mind from wandering. Like a glass of muddy water that is not stirred, the dirt and dregs settle down, leaving only clear water. AWARENESS is not developed by effort. Because AWARENESS is already there, there is no need to try to develop. Trying to devekop AWRARENESS only makes people stupid, like a computer with artificial knowledge. Or makes the body stiff and numb like in a person practicing Qigong, equivalent to the state of Knowledge (Qigong) entering the body.

- CONTEMPLATION: when Contemplating Body, Feelings, Mind, and Dharma
AWARENESS appears, but Awareness may not be enough to cleanse Karma. When concentrated, the brain loses peripheral information, as a result the ACC start looking fot the data fron the Inner Consciousness. The inner CS will then be cleansed of false information. But when performing Samadhi or Mindfulness does not mean that Ignorance will be purified accordingly. So Samadhi and Mindfulness are not always associated with corresponding Awareness. Samadhi and Mindfulness , therefore, sometimes do not go together with Awareness. The reason is that Awareness needs Contemplation as well as Samadhi and Mindfulness while Samadhi and Mindfulness need attention. But with too much attention, Contemplation becomes TR, Attention with too little Samadhi and Mindfulness , Contemplation becomes learning, no longer meditation.

Understanding the importance of balance in meditation is crucial. Just enough Samadhi and Mindfulness, when balanced with Contemplation and Awareness, is the most effective approach. This concept guides you in navigating the complex interplay between these components, ensuring a harmonious and fruitful meditation practice.

Focus on the Breath (top-down system) and observe the changes of the Breath (bottom-up system) to gradually improve the ability to focus the Mind and strengthen the CS to become Mindfulness. Among the 25 Masters invited by the Buddha to present the practice to attain enlightenment, two of them used the method of Breath Contemplation (Chu Li Pan Te Ca, Sun Da La Nan Da). The focus uses the breath to enter the door of the Mind to enter the Original Mind / True Emptiness. Sun Da La Nan Da alone also described the Light seen when practicing the Contemplation of Breath, proving that he entered Samadhi to see the Light from the Inner Mind (also called Buddha light). When concentration and inner Consciousness are gradually purified over time, wisdom from Buddha's nature will appear, and mindfulness can be obtained. When there is no Concentration, practicing Mindfulness is difficult and inadequate. Knowledge is a veil of Ignorance, Mindfulness increases with Concentration and decreases without Concentration.

   Mindfulness (sama: equal dh: "Consciousness), is often used to treat nervous anxiety because it stops Manna consciousness/thinking and calms the mind. Awareness (Vipassana/ Vi=Variations, passana=Awareness, Knowing) is the process of having Awareness which is the result of reducing or cleasing away all evil MM. Normally, attention is achieved, Awareness automatically appears, but Awareness may not be revealed enough due to insufficient cleasing of Karma. When concentrated, the inner Consciousness loses peripheral information and becomes activated.

c) MINDFULNES (ĐỊNH) AND AWARENESS (TUỆ) DISTINCTION: SAMATHA, SAMADHI/MINDFULNESS AWARENESS IN VIPASSANA AND QUAN YIN MEDITATIONS.

* Samatha	VIPASSANA*	QUAN YIN
Mindfulness; Low Awareness Low Karma Cleasning: low Concentration should be Peaceful, single-minded attention to the goal	Mindlness moderare Awareness High Karma Cleasning modréate Right Mindfulness with revelation of Awareness through Contemplation of Weak Samadhi	Mindfulness high Awareness variable / moderate Wareness is revealed by Mindfulness, Karma Cleasning high Awareness may precede Mindfulness
Ignorance is less eliminated	Ignorance is eliminated at its root, but this elemination is not large in extent	Samadhi deeper than in MS Ignorance is eliminated according to the level of Awareness, but this elimination is more extensive
Dorsal Pathway of the Brain Not liberating because of lack of AWARENESS, can only be compared to Brahma	Both Dorasl (for stillness) and Ventral pathways. (for Awareness) Ending Samsara dependingon awareness and cleansing of karma.	Both Dorasl (for stillness) and Ventral pathways. (for Awareness) Ending Samsara according to the level of Awareness ang cleasing of karma
* The six wonderful Dharma doors combine Samadhi-Contemplation to increase Concentration so it takes more time		
*Samatha and Samadhi have many similarities in cutting off external information, so the internal information is retrieved back to the present and Karma is cleansed. However, in Samatha, the body becomes insensitive to all information and external circumstances. Because Samatha only uses the upper transmission path and the Five Aggregates process stops at Feeling stage. Inner consciousness can be removed, less when compared to Samadhi much less to conteplation		
*Samadhi/Mindfulness . Using Dorsal and Ventral paths, the Inner consciousness is cleansed much more. Therefore, Awareness can be discovered		
*Mindfulness is similar to Samadhi with more Awereness Kảm cleáing dêpnd on Sâmdhi and development ò Âêns		

Dícusion According to the imperfect law of the dualistic world, there can be no complete method of practice. It's a delicate balance-too much AWARENESS will lose Samadhi. The Four Foundations of Mindfulness need more Samadhi, so we must use the Six Wonderful

Dharma Doors method. The Quan Yin method of practice is good for those who already have AWARENESS; AWARENESS can be further supplemented by listening to the Dharma like the six brothers Kieu Tran Nhu, who listened to Dharma preached by Buddha attain Arhahat level or to learn the sutras to understand the creation completely and can still be supplemented when contemplating the Body, Feelings, Mind, and Dharma.

The Inner Consciousness will then be cleansed of evil information.
The realization (Cleasing away many veils of Ignorance) happens quickly or slowly depending on the practice method (Pure Land, Four Foundations of Mindfulness...), the practice method and the amount of Karma washed away. The role of diligence in this process empowers the individual and instills a sense of responsibility. Karma in this life can only be clensed away by oneself through practice with diligence.
It's important to note that when practicing meditation with the wrong method, there may be an imbalance between Samatha/Samadhi/Mindfulness and Awareness. The Buddha, for instance, reached the 8th level of meditation through an external path, but due to a lack of Awareness, he had to abandon this method. This issue will be further discussed later.

The removal of Knowledge by Mindfulness is equivalent to thinning and lifting aside the veil of Ignorance to reveal KNOWLEDGE, which is the closest form of Prajna Awareness. The method is to concentrate the mind and need Will to keep the mind from wandering. Like a glass of muddy water that is not stirred, the dirt and dregs settle down, and the clear water remains. AWARENESS is not developed by Will. Will is only used to keep the mind from wandering. Because AWARENESS is already there, there is no need to try to acquirre. Trying to acquire only incease the CS and makes people stupid like a computer added with artificial Intelligence. Or makes the body stiff and numb like in a person practicing qigong, equivalent to the state of qigong entering the body. Therefore, in Vipassana/Vi=Variations, passana=Awareness, Knowing), AWARENESS is natural (like when KNOWING the belly is bulging and deflating without trying to understand. In Quan Yin meditation, light comes naturally, and one cannot hope for it.

Mindfulness/ Cognition and the Awareness
Awareness (Fig 8.2)
In Meditation, the dorsal pathway is associated with mindfulness/ cognition, and the ventral pathway is associated with Awareness

Krishnamurti : Meditation is not a practice; it is not the cultivation of habit; Meditation is heightened Awareness

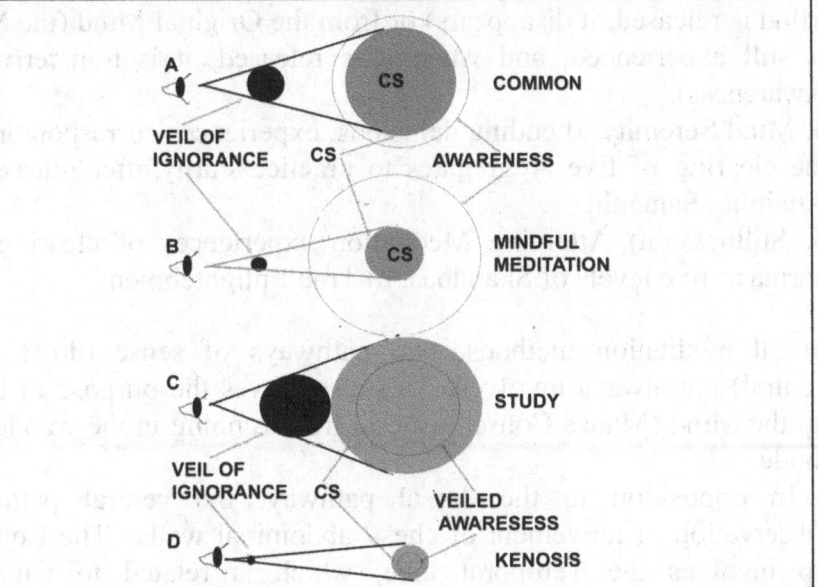

Fig 8.2 Clear circle: Awareness. Black circle: CS, Gray Circle : Understanding
A: Common
B: Meditation: attention to the restricted target: the understanding is narrower than usual and less detailed. The Awareness is uncovered. If the meditator changes the attention to the areas of Awareness, the Awareness immediately disappears due to the CS involved
C: Learning: increases understanding, obliterated Awareness, and ignorance develops.
D: Kenosis lack of proportional development of Awareness with narrowing awareness

4) THE SIX WONDERFUL DHARMA DOORS (vn: Lục Diệu Pháp Môn)

consists of the alternation of the Samantha and Contemplation leading to the realization of the Truth. The cycle can be repeated as per intention

Includes six processes
1. Counting the breaths counting the breaths,
2. Following the breathing rythm
3. Samntha: keeping the mind still a) relaxing, the mind is quiet, b) Mindfullnes
4. Contemplation, the Four Foundation of Mindfulness technique,

5. Looking into oneself (Inner CS): a) the question of whether the experience is from the Mind (Inner consciousness: so when the Mind is released, it disappears) or from the Original Mind (the Mind is still experienced, and when it is released, it is transferred to Awareness).
5. Mind Sereinity a) ending delusions, experiences corresponding to the clearing of five Aggregates to practice Purity/after release, b) Attaining Samadhi
6. Stillness: a) Attaining Meditation experiences of clearing the karma at five levels of Skandhas, b) True Enlightebment

In all meditation methods, two pathways of sense (dorsal and ventral) are always involved. This also serves the purpose of tying up the Mind (Manas Consciousness) from running in the wandering mode.
· In opposition to the dorsal pathway, the ventral pathway: Observation of movement of chest/ abdominal walls. The Bottom-up involves the Temporal lobe, which is related to emotion, cognition,/ Consciousness (Zanesco 2013, Raffone 2010, Tăng 2015, Bernajee 2019, This latter pathway provides information on the overall body. In this process, the thinking/ creativity/ Manas Consciousness are inhibited by events. Using minimal attention, this is associated with the Awareness view, which is objective and reveals the look close to the Tathagata true view.
The respective pathway and Consciousness/ Awareness are mutually inhibitory. The higher the attention, the more restricted the area of focus, with accompanying a narrower consciousness. The limited Consciousness results in an enlarged and clear Awareness. The two pathways inhibit each other for the attention

In all methods of mindfulness, the Inner CS is exposed because there is no external input; data (MM) in the Inner CS is retrieved. Therefore, removing bad MM or inserting good MM into the Inner CS is easier. This process is executed through contemplation (good and true Daharmas). In no meditative period, the above removal and insertion are possible but difficult and not efficient.

K. FACTORS AFFECTING MEDITATION 354
Because meditation is a process of the mind, besides the meditation method (this book focuses on Vipassana and AQuan Yin meditation), the inner factors, Bodhi determination, external environment, and the

place where meditation is performed play a very important role, lack or mistake can waste time and ruin the effort of meditation.

1. Morality
If the purpose of Meditation/MD is to cure illness, the belief in MD to calm the mind and change the body's reaction is important and also needs to be accompanied by keeping the precepts of practice. If MD aims to eliminate suffering and end the cycle of reincarnation, then the precepts are very important. Nirvana is not a place to welcome people who have many merits in MD but lack morality and especially do not understand the Dharrma such as about Buddhism, equality, self, eternal uniqueness, and the omniscience of Creator/EM. Human morality is knowing oneself as Selflessness/Anatta and Impermanence. Accompanying this is Right Livelihood (Life), which is practicing Anatta, so one does not have an inflated ego, is not jealous or competitive, has right speech and right action, avoids thinking of evil, and does not do evil, Compassion, Joy, and Socialization. Compassion, Joy, and Socialization, in addition to love, keeping the soul always happy with everyone, reality, being ready to let go of all ties, Social Life, and Family are great obstacles for MD because they know that people live in this world to correct mistakes, not to enjoy. Enjoy just enough to support learning. Afflictions must be considered a remedy to practice liberation. Do not consider afflictions as obstacles to despair and discouragement.

2. Mindfulness is the recognition of the Memory
. The meditator can have the right thoughts (thinking) and views (concepts of life) thanks to Right Mindfulness. Studying the scriptures helps us clearly understand Nirvana, Emptiness, and Dharma. Mindfulness is the highest level and the meaning of the teachings of the Buddha. Understanding Dharma is an important factor for effective meditation. The highest level of meditation, especially Mindfulness

3. Faith in the Dharma and the determination
Faith in the Dharma and the determination in the spiritual journey are not just necessary, but they should be continuous and intense throughout life, especially during meditation. Developing the Bodhi mind and Diligence requires a significant amount of mental energy, making it crucial to allow the mind to rest and regain energy to continue Diligence. Thanks to Diligence, the mind is always in a state of Awakening. Sleep, dreams, and drowsiness during meditation are not just a waste of time, but they also hinder the effort of meditation.

4. Contemplation topic

must be suitable for the contemplation of three human misconducts of Greed, Anger and Delusion (see pae 289)

Karma and the nature of our actions play a significant role in the speed of our progress in meditation. Heavy karma and a slow nature can significantly slow down our practice, making it important to be mindful of our actions and their consequences.

5. Non-nervous body part:
has many effects on TD. Sitting cross-legged and half-cross-legged, hands clasped together, back straight is the factor to maintain the determination to practice. Sitting on a chair when MD requires mental factors and effort to balance with the comfort of the body. In Sâmdhi, along with breathing and the body's stillness, is a feeling of perfect balance, as if the whole body is covered by a pure environment consisting of Soul and non-neuronic (Qi Gong)

6. Role Of Divine Blessings And Relationship Between Elightenment –Diligent Effort Paramita, Endeavor Of Self Realization

Achievement of the spiritual practice depends on many factors, including divine blessing/ vows made by Buddha/ personal karma/ veil of ignorance. In the Right Dharma Age (the first 500 years, including 49 years of the proclamation of Dharma by Buddha), a large number of Buddhists attained enlightenment to level of Arahat for a short time of practice.. For example:

Kauõóinya: attained level of Arahat in 5 days
Mahà-Maudgalyàyana   5 days
Sàriputra   15 days
Yasa   7 days

Buddha explained that although the initiation made by Buddha is immediate, the enlightenment is progressive (depends on the clearing of the veil of ignorance), as highlighted in the following Sermon in Lankavatara Sutra.

> In order to the stream of perceptions of his own Mind, Mahamati Bodhisattva once more asked the Buddha, "Bhagavan, how is the stream of perceptions of beings' minds purified? By degrees or all at once?" The Buddha told Mahamati, "By degrees and not all at once. Like the gooseberry, which ripens by degrees and not all at once, thus do tathagatas purify the stream of perceptions of beings' minds by degrees and not all at once.

Buddha replies to Bodhisattsava Mahamatti's question: The Mind of disciples is purified gradually , not at once, but Buddha illuminates at once not by degree. Therefore in the technique of Sudden Enlightenment, the sudden enlightenment only occurs in practitioners with a long history of practice, such as a case of the first five disciples of Buddha

6. ALTERNATIVE METHOD OF CLEARING THE MIND TO ACHIEVE THE RIGHT VIEW, MINDFULNESS AND THOUGHT OF THE EIGHT FOLD PATH.

In Meditation (and likely worldly life), the virtuous mind should be activated to suppress the evil Mind (like Greed). In Neuroscience, the suppression of Memory suppresses not only the Memory of concern but also other similar types of Memories. The other method is to detach from the evil Memory as, like water, wind leaves no trace after it passes over.

Vitakkasaṇṭhāna Sutta The Removal of Distracting Thoughts 1. THUS HAVE I HEARD. 238 On one occasion the Blessed One was living at Sāvatthī in Jeta's Grove, Anāthapiṇḍika's Park. There he addressed the bhikkhus thus: "Bhikkhus."—"Venerable sir," [119] they replied. The Blessed One said this: 2. "Bhikkhus, when a bhikkhu is pursuing the higher Mind, from time to time he should give attention to five signs. 239 What are the five? 3. (i) "Here, bhikkhus, when a bhikkhu is giving attention to some sign, and owing to that sign there arise in him evil unwholesome thoughts connected with desire, with hate, and with delusion, then he should give attention to some other sign connected with what is wholesome. 240 When he gives attention to some other sign connected with what is wholesome, then any evil unwholesome thoughts connected with desire, with hate, and with delusion are abandoned in him and subside. With the abandoning of them his Mind becomes steadied internally, quieted, brought to singleness, and concentrated. Just as **a skilled carpenter or his apprentice might knock out, remove, and extract a coarse peg by means of a fine one, so too...when a bhikkhu gives attention to some other sign connected with what is wholesome...his Min**d becomes steadied internally, quieted, brought to singleness, and concentrated. 4. (ii) "If, while he is giving attention to some other sign connected with what is wholesome, there still arise in him evil unwholesome thoughts connected with desire, with hate, and with delusion, then he should examine the danger in those thoughts thus: 'These thoughts are unwholesome, they are reprehensible, they result in suffering.' 241 When he examines the danger in those thoughts, then any evil unwholesome thoughts connected with desire, with hate, and with delusion are abandoned in him and subside

....

. "Also, Rāhula, while you are doing an action by speech... (complete as in §10, substituting "speech" for "body") [418]...you may continue in such an action by speech. 14. "Also, Rāhula, after you have done an action by speech... (complete as in §11, substituting "speech" for "body")...you can abide happy and glad, training day and night in wholesome states. 15. "Rāhula, when you wish to do an action by Mind...(complete as in §9 , substituting "Mind" for "body") [419]...you may do such an action by Mind. 16. "Also, Rāhula, while you are doing an action by Mind... (complete as in §10, substituting "Mind" for "body")...you may continue in such a mental action. 17. "Also, Rāhula, after you have done an action by Mind... (complete as in §11, substituting "Mind" for "body" 639)...you can abide happy and glad, training day and night in wholesome states. [420] 18. "Rāhula, whatever recluses and brahmins in the past purified their bodily action, their verbal action, and their mental action, all did so by repeatedly reflecting thus. Whatever

> recluses and brahmins in the future will purify their bodily action, their verbal action, and their mental action, all will do so by repeatedly reflecting thus. Whatever recluses and brahmins in the present are purifying their bodily action, their verbal action, and their mental action, all are doing so by repeatedly reflecting thus. Therefore, Rāhula, you should train thus: 'We will purify our bodily action, our verbal action, and our mental action by repeatedly reflecting upon them.'" That is what the Blessed One said. The venerable Rāhula was satisfied and delighted in the Blessed One's words.

As discussed above, and in Vitakkasaṇṭhāna Sutta, The Removal of Distracting Thoughts, evil unwholesome thoughts connected with desire, with hate, and with delusion, Buddha advises that disciples should give attention to some other signs connected with what is wholesome just as a skilled carpenter might knock out, a coarse peg by means of a fine one.

If, this method is not efficient, then he should contemplate the danger in those thoughts thus: they result in suffering

If this method is not efficient again, he should try to forget those thoughts by not paying attention to them and abandoning them.

If, this method is again not efficient, he should try very hard *"with his teeth clenched and his tongue pressed against the roof of his mouth, he should beat down, constrain, and crush Mind with Mind"*. *He will think whatever thought he wishes to think, and he will not think any thought that he does not wish to think*

In *Ambalaṭṭhikārāhulovāda Sutta Advice to Rāhula at Ambalaṭṭhikā,* Buddha teaches Rahula about the beneficial action of the physical part that should be done to self and to others

K. KOAN/ Kung-an and KOAN Meditation

Koan consists of a question, problem, or statement with or without an irrational answer or action given to the student. In other cases, Koan is the inappropriate or illogical replies or acts of the master to the student often posed by the master to the student. Since all world problems originate from the Emptiness with arisen False thoughts, all unanswered questions or acts must go to the Original Mind. Therefore, when Koan is posed, the student who couldn't find the answers must have a look inside the Original Mind, where the student finds out the Awareness instantaneously.

In Koan Meditation, the Koan is usually posed when the student is in a state of mindfulness, according to **Genki Takabayashi Roshi Master of** The Dōgen Institute. Koan is the high-level component that opens the lid covering the underlying Awareness.

Koan posed in the early stage of mindfulness only causes the Mind to be activated in an erratic state.

Likewise, when Buddha held up a rose for 1250 disciples to see, no one understood anything; only Venerable Kasyapa smiled then, and DP smiled back and transmitted the Dharma. The meaning is that Venerable Kasyapa had a moment of mindfulness with enlightenment at the moment as it is.

J. Method of Recitation of Namo Amitabha Buddha of Pure Western Land (of Ultimate Bliss) Doctrine

Namo Amitabha Buddha is the first word in the intercommunication in Buddhist temples or meetings between devoted Buddhists. This method of spiritual practice is the most common in Vietnamese communities because of the simplicity and the promising and attractive outcome, according to the Larger Sukhavativyuha Sutra or The Sutra On The Buddha Of Eternal Life, *Amitabha Sutra*.

> *The Peace Land is characterized by extended longevity, morality, beautiful physical appearance, Bodhi Mind, persistent Mind in the spiritual pursuit, reproduction by the transformation (not due to Male and Female interaction), no Samsara, with six miraculous powers, Sammā diṭṭhi, justified/ correct knowledge view, land consisting of gold, silver, pearl, peaceful.*

Recitation with mindfulness may enable one to be reborn in the Western and blissful land, free of reincarnation. This land is divided into upper, Middle, and Lower levels. Each level is divided into three sublevels: upper-upper, upper-middle…lower-lower. This lowest level is for persons with bad Karma but not for persons defaming Buddha. Persons with bad Karma who recite ten times Namo Amitabha Buddha before death can still be reborn in the Western Land but stay in the plant of lotus for ten Maha kalpa (1,3 billion years) (Kalpa;aeon/vn: kiếp- 3,6 million years)

> **Further reading**
> **Amitabha/ Amita Buddha/** Buddha of Unlimited Light. /Buddha of Boundless Light/ Buddha of Irresistable Ligh is very popular in China, Vietnam, Japan and Korea. He is the chief of the Western Pure Land, belonging to another system of worlds distinct from this system of world for innumerable thousands of year. The Amitaba Pure land is obtained thanks to innumerable buddha's merit
> He was a king named Dharmākara who was impressed by a high court official under his rule becoming Buddha.. With vows to give up his throne if he can be offered a land with 48 aspirations that any being in any Universe desiring to be reborn into the Western **pure land** (vn: tịnh độ) and having faith and sincerity and reciting it for even only ten times, will be guaranteed rebirth Amitaba's land or

> may I not gain enlightenment, The other vow promises is that he and his bodhisattvas (Amitaba Triad: with two assistant bodhisattvas, usually **Avalokiteśvara** (vn: Quan Thế Âm Bồ Tát) on the right and **Mahāsthāmaprāpta** [vn: Đại Thế Chí Bồ Tát] on the left.) will appear before those who, at the moment of death, call upon him.
>
> The Amitaba Pure land is full of splendor with infinite light, sound, precious gems, gold, happiness, joy with long lasting life and are free of worry and illness

Different Disfavored and Favored Opinions.

For some Buddhist devotees, Amita Buddha and related Sutras were not mentioned in Nikayas Sutra. Therefore the **recitation Namo Amitabha Buddha method was created by Mahayana Buddhists, particularly Patriarchs in the Northern Spread Buddhism in China, Vietnam, Korea and Japan. Furthermore, the simple and easy access to the Westen Pure Land made the technique doubtful regarding the logic and the authenticity.**

In fact, Amitaba-related Sutras consist of at least the Amitabh Sutra (Longer Sukhavativyuha Sutra—The Longer Amitabha Sutra and The smaller text of Sukhavati-vyuha is a summary of the larger one) and the Amitayur-dhyana Sutra Sutra/ The Great Infinite Life Sutra. The Sutras represent the preaching of Sakyamuni to Queen Vaidehi, who was imprisoned by the prince heir-apparent of Rajagriha, who revolted against his father, King Bimbisara. In view of the inversion of this world, Buddha Aakyayana, showed the Queen the best place to be reborn. The sutras were transcribed by Ananda after Buddha said at the end of the preaching: "Oh Ananda! Remember this sermon and rehearse it to the assembly. By this sermon, I mean the name of Amitabha." (Please see page 394)

In addition, in contrast to the disfavored opinions that not a single tine in the Four Recitation of Buddhist Hyayana have ever mentioned the name of Amitaba, numerous time Amitaba have been cited in Mahayana sutras and literature,. Example

Flower Adornment Sutra, Laṅkāvatāra, Surangama, Lotus, Diamond Vimalakìrti sutras and in nāgārjuna's *mahāprajñāpāramitopadeśa*

In Surangama Sutra, Akàtagarbha Bodhisattva who practiced the Buddha recitation method revealed:

Akàtagarbha Bodhisattva then rose from his seat, prostrated himself with his head at the feet of the Buddha and declared: .When the Tathàgata and I were with Dãpaükara Buddha and realized our boundless bodies, I held in my hands four big precious gems which illumined all Buddha lands in the ten directions, as uncountable as dust, and transmuted them into the (absolute) void..

Furthermore, even in Nikaya Sutra, Land similar to WesternPure Land was revealed by Buddha Sakyana

> 3. *"Here, bhikkhus, a bhikkhu possesses faith, virtue, learning, generosity, and wisdom. He thinks: 'Oh, that on the dissolution of the body, after death, I might reappear in the company of well-to-do nobles!' He fixes his mind on that, resolves upon it, develops it. [100] These aspirations and this abiding of his, thus developed and cultivated, lead to his reappearance there. This, bhikkhus, is the path, the way that leads to reappearance there.11.........*
> ....
> *Just as an ornament of finest gold, very skilfully wrought in the furnace by a clever goldsmith, lying on red brocade, glows, radiates, and shines, ... 'Oh, that on the dissolution of the body, after death, I might reappear in the company of the Brahm of a Hundred Thousand!' He fixes his mind on that... This, bhikkhus, is the path, the way that leads to reappearance there.*

Meanings: Buddha said to his disciple that monks of virtue can be reborn at their aspiration. Buddha described the land similar to the Western Pure Land.

Discussion on methods of Meditation.

Meditation aims to cleanse karma so that the soul can return to its source, the Buddhahood, Original Mind. Knowing that the cleansing of the karma is essential in the path anf consists of a) no addition of new karma by following the Noble Eightfold Path, b) voluntary and persistent development of bodhicitta, aiming from the Original Mind, and c) Meditation. Correct Meditation methods can serve the purpose since the retrieval and subsequent abolition of the MM are independent of the form of Meditation. The respiration, sound, and light...are barriers to the realization of Buddhahood since they are illusional products of the Creation. However, they are necessary to serve as the target of the focus of attention in Mindfulness. Contemplation in mindfulness meditation is an effective way to achieve the Right View, Right thought, and Right Mindfulness. For intellects without meditation Right View, Right Thought, and Right Mindfulness can be realized. This will tremendously shorten the path of Meditation with contemplation Due to the individual variability and individual's environmental condition; the methods are designed to fit individual needs." With variable and succession in the order of Samatha/ Calmness, Samadhi/ Mindfulness **(sama: equal *dh:* "consciousness) and Awareness/ Omniscience.**

M. Recommended Duration and the Posture.

The duration of the Meditation can not be specified, depending on the volition, motivation, and progress. It may last
for 30 min for beginners and up to many weeks for experienced meditators, like in the case of kenosis, Western Pure land, with near complete extinction of sensation and thought (vn:.Diệt tận định). The duration usually varies, but in general the human body is hard-wired to certain programs, as seen in the sleep or the habits of the workers:
The sleep consists of fragments of NREM and REM sleep and the duration for the workers or sports players varies from 1.-2 hours.
Therefore, for the beginner, the optimal duration of each section is 1- 2 hours.

L. The Third Eye (Fig 8.3).
The wisdom eye has been addressed in many religions and is known as the wisdom eye or third eye. It is located in the midline of the forehead, just above the eyebrow line, and likely behind the skin about 1-2 cm. According to the Indians, this is also the site of Chakra 6 (Ajna). Chakra, which represents the spiritual energy Kundalini. Kundalini energy is conceived to originate at the end of the spine and gradually become purified as the source of energy ascends toward the summit of the head, the Chakra 9, where the spirit is located. Chakra 9, also known as Sahasrara, is the highest Chakra and is associated with pure consciousness and the divine.
Kundalini is distinct from Qi/ Chi in Chinese Medicine or Martial Arts. According to Ancient Chinese, Chi originates from the kidney and is responsible for vital energy. The kidney appears early in the embryonic development of the body. On the other hand, Kundalini is a spiritual energy that is believed to be coiled at the base of the spine and can be awakened through various spiritual practices.
It is believed that the Kudalini is associated with spiritual energy and enlightenment, and Chakra 6 represents the transformed Kudalini /Awareness into energy Qi. As a result, Chakra is not related to any acupuncture points in Chinese Medicine.
Some believe that the wisdom eye is linked to the Pineal gland, located behind the posterior to the corpus callosum. This gland secretes a small quantity of N. N-dimethyltryptamine, a substance associated with some effects experienced in Meditation, including hallucination. This has led to it being dubbed the 'spirit molecule'. However, according to David E Nichols, Chair in Pharmacology at Purdue University, this belief is unfounded. The Pineal gland is actually related to the sensitivity to the geomagnetic field, which is important for orientation in flying or moving animals. In conclusion, there is no proven relationship between

the Pineal gland and Consciousness, but its role in spiritual experiences remains a fascinating topic of debate.

In India, the third eye, known as the Bindi, is marked in red ink. The vibrant red color symbolizes dedication to God. Interestingly, married women traditionally stop wearing the Bindi, adding a layer of cultural significance to this spiritual symbol. While the Bindi may not hold any specific spiritual significance, its cultural importance is undeniable.
In Hinduism, the third eye is not just a physical feature, but a symbol of profound spiritual significance. It represents the eye of Shiva, the Supreme Being who creates, protects, and transforms the Universe. In Shaivism, the third eye is a powerful symbol of Awareness, a concept that is central to many spiritual practices. Understanding the spiritual significance of the third eye in Hinduism can be a truly enlightening experience.

In Islam, the third eye is referred to as ala Khafi, consisting of all senses and Mind, an amazing gift from Allah for connection with the spiritual world. In Islamic mysticism, the concept of the third eye is often associated with spiritual insight and enlightenment, allowing the believer to perceive the divine reality beyond the physical world. In Christianity, the significant saying in the King James Version of the English Bible, the text reads:

In Christianity, the significant saying in the King James Version of the English Bible, the text reads:
The light of the body is the eye:
if, therefore thine eye be single,
thy whole body shall be full of light.
Jesus said (St John 9:5):"
As long as I am in the world, I am the light of the world

"Therefore, listen to the *sound stream and the light, from the wisdom eye is a significant progress in the spiritual practice*. The spiritual pathway is not the broad boulevard, but a through a narrow entrance as pointed out by Mathew (Mathew 7:13-14):"

Enter through the narrow gate. For wide is the gate and broad is the road that leads to destruction, and many enter through it.
But small is the gate and narrow the road that leads to life, and only a few find it".

In Buddhism, numerous times, infinite beams of light radiate from between the eyebrows accompanied by innumerable rays of light, illuminating all worlds in all ten directions, stopping all evils and sufferings, eclipsing all realms of demons, illuminating the circles of countless Buddhas, revealing the features, occult powers, and mystic transfigurations in the inconceivable realm of the Buddhas, illuminating the enlightening beings empowered to teach in the circles of all Buddhas in all worlds in the ten directions; having shown the inconceivable mystical power of the Buddha, ... (Sutra Flower adornment, page 701) .

The celestral eye is able to see anywhere any time ,(past and future), and is free of obstacle.
Flesh Eye is unable to freely see
Dharma Eye is able to see all forms and rules of the world
Wisdom eye like the celestral sun uniformly illuminates all Dharma Realm without exception of any worlds
(vn: *Thiên nhãn thông phi ngại*
Nhục nhãn ngại phi thông
Pháp nhãn duy quán tục
Phật nhãn như thiên nhựt,
Chiếu dị thể hoàn đồng Viên minh pháp giới cảnh, Vô xứ bất hàm dung).

In Meditation, the light of varying intensity is usually seen in the area of the third eye, which may explain Buddha's saying, **"Be your own lamp, seek no other refuge but yourself, Let truth be your light."**.

The DMN (including vmPFC; third eye) is involved in the disorder of the CS (Xu 2016, Poerlo 2017, Linder 2007), depressive anxiety and hyperactivity, schizophrenia (Broyd 2009, Khadka 2013, Chang 2014, Repovs 2010). . vmPFC is not involved in the Intelligence Quotient (Zhang 2014, Du 2016,Pankow 2015). In PTSD, the vmPFC shows decreased connection and NPY (Schreiber 2016). In addition, vmPFC sends links to dmPFC, dlPFC involving the execution of
orders from vmPFC. vmPFC connection to PCC, the store of autobiographic Memories, vmpFC, PCC, and other cortices of the DMN. These cortices are not active in Meditation since the retrieval of Memories is much lower in Meditation than in thinking
 (Brewer 2011, Weng 2013, Hölzel 2011, Harenski 2006, Beauregard 2001, Schaefer 2002, Creswell 2007, Raichle 2015, Lee 2018).

N. Kenosis.
Kenosis is the Greek term used for the doctrine of Christ's incarnation by emptying " ? non-specified component" to take the form of a "Servant": *Philipians 2:7, ESV) but emptied himself by taking the form of a servant, being born in the likeness of men.* In

Bible, after emptying, the Son of God, Jesus, remained God with spiritual power and Awareness.

In Islam and Hinduism, Kenosis is a concept deeply rooted in mindfulness. The term likely originated from the Greek 'Kenosis' to designate a state of mindfulness characterized by the extinction of six CS and thinking/ the seventh/ Manas CS in the Form Realm before attaining the state of neither thinking nor not thinking. In the state of Kenosis, the Mind is separated from the realm of form (therefore, in the Formless Realm/ vn: Vô Sắc Giới),
consists of:

- Realm of near infinite Space/ first stage Jhana of Formless realm: toward void and vast CS. Longevity: 20.000 kalpas.

-Realm of infinite space with infinite CS.

-Realm of Void Space, no more CS, longevity 60.000 kalpas.

-Realmn of neither thinking nor no thinking: no C=Mana CS and no Dharma for CS formation, longevity 80.000 kalpas.

It was said that Buddha reached the realm of neither thinking nor no thinking during the period of practicing Meditation focusing an Sammadhi, with involvement ò the contemplation component, with externalists. But Buddha finally relinquished this method of Meditation due to the lack of development of Awareness/ Omniscience. As a result, it is obvious that in Kenosis, mindfulness can lead the meditator to the state of extinction of seventh CS, but there is non-proportional/insufficient development of Awareness/ Omniscience.. This is supported by the phenomena of Fakir/ Farquir/ also called mendicant dervish, who claims the need of God is self–sufficient. The Fakirs are empowered with miracles like walking on fire and lying on sharp nails... The body of a person with developed Kenosis can become mummified.

The underlying mechanism is the non-proportional development of Awareness, likely secondary to inhibition of the process of integration of input data in the sensory cortices. As a result, there is decreased integration of data in the cortex and subsequently there no formation of CS or awareness. Complete enlightenment will not occur.

Kenosis, in its extreme form, is a practice that not only slows down the activity of the Form, Perception, and Feeling (the first three of the five Maras) but also slows down the Formation and CS due to the lack of data reaching the primary sensory cortices. This extreme nature is further highlighted by the near to full stop of activity of the cellular component of the other parts of the body outside the brain, including the vascular circulation. The fact that there is no bleeding when cutting the skin or oral mucosa in this state is a testament to the wonder of Kenosis. It is important to note that Kenosis does not likely develop in the contemplative method like Vipassana of Quan Yin, further adding to its intrigue.

A mummified body, supposedly that of (?) Huineng, the sixth Zen Patriarch, is kept in Nanhua Temple in Shaoguan (northern Guangdong). (according to Jesuit Matteo Ricci, 1589).

In Kenosis, information gradually decreases to the Cerebral Cortex, so it no longer causes sensation (no more stimulation of the upper and lower transmission pathways, so there are no reflexes). Therefore, CS and emotions cannot occur. The above process eventually slows down cell metabolism. The slowing down and eventually complete cessation of brain activity are followed by the cells of the entire body. (Thus, contemplation in meditation, a practice that involves focused attention and mental stillness, is necessary. It helps the cerebral cortex to continue working, and consciousness does not stop functioning as in Kenosis). The phenomenon of sterility in Kenosis has not been scientifically studied. However, it is not mysterious and unimaginable because the phenomenon is similar to a tree branch separating from a tree. If in a dry place, the branch dries up and retains its original shape like Kenosis. In terms of biology, plant and animal cells have many similarities. When the metabolism of animal cells gradually slows down, as in Kenosis, all metabolism slows down and then stops completely. Catalytic enzymes are contained in small vesicles in the cell that do not break, so bacteria and enzymes do not work to cause the putrefaction

TANTRIC BUDDHISM, Vajrayana

Kenosis, the phenomenon of a body that does not decay over time, is a unique feature of Vajrayana Tantra in Tibetan Buddhism. After death,

Bodhisattvas are reborn in other individuals, bypassing the Form Realm to enter Nirvana. Vajrayana, a blend of Hinayana and Mahayana, is influenced by the Madhyamaka of Nagarjuna Dharmakirti. His Holiness the Dalai Lama, the head of the Tibetan Buddhist sect, is believed to be the incarnation of Avalokiteshvara Bodhisattva, while Panchen, the second in Vajrayana, is the incarnation of Amitabha Buddha.

TANTRIC BUDDHISM, Vajrayana is distinct from Quan Yin Contemplation,. The esoteric School's practice is characterized by its strictness. It often involves 3 years of continuous retreat and asceticism, enduring the harsh cold of winter without heating. This rigorous practice and training in scriptures form the essence of Vajrayana. The Vajrayana Sutra of Mahayana focuses on the absolute and inconceivable Prajnaparamita nature of Emptiness, which differs from the reality of the Dharma in the Sarvastivada School of Hinayana Buddhism. This absolute Emptiness may not yet be seen in the Esoteric School. In Tibet, Buddhism is still somewhat influenced by the Main Religion, the old religion of Tibet from the 6th century when Buddhism began to be taught.

The precept against killing was not strictly enforced, perhaps because Tibet had little vegetarian food and also because the concept of the precept against killing was different from Mahayana Buddhism. The karma of killing could be a factor that kept the Lama in the cycle of reincarnation. The 5th Dalai Lama built the Potala Palace in Lhasa in 1645 and demolished the old Red and White Palace built by the Dalai Lama in 637, in line with the later trend of building temples in Mahayana and Hinayana Buddhism, different from the time when the Buddha lived in the forest and not in the palace of his father King Suddhodana.

O. Phenomenon Of Illusions/ False States In The Meditation.

Sutra Flower Ornament, Chapter 1 2(Chief in Goodness)says: One can crush the power of all demons. If one can crush the power of all demons, One can get beyond the realm of the four demons . If one can get beyond the realm of the four demons, Then one can reach the stage of nonregression . If one can reach the stage of nonregression
The four demons are: Celestrial Demons (Demón in heaven,skt: Deva Mara)

Kamadhatu, with his innumerable host, constantly obstruct the Buddha-truth's and followers Idealistic people who disturb Buddhism

The Maya is sinful of love or desire, as he sends his daughters to seduce the saints.

Papiyan (skt) is the special Maya of the Sakyamuni period.
Death Demons, the death-causing interruption of earthly life
Moral Afflictions such as anxiety, depression...
Five Maras associated with skandhas,

In Surangama sutra, Buddha teaches that Consciousness is necessarily connected to the Original Mind. After showing Ananda and the Assembly different methods of Meditation, Buddha briefly gave a review of various states of illusions. There are five types of illusions corresponding to five maras/ aggregates:
- Form/ skt: Rupa,
- Perception, sensation, feeling/ skt: vedana,
- Feeling/ skt Samna
- Formation of Consciousness/ skt Samkara
- Consciousness

Recognition of the states appearing in the Meditation is illusional and false in existence is correct.
The experiences in Meditation represent the progression. If considering these states as true in reality, this is incorrect and will lead to a state of mental disorder with confusion and erroneous activity (wrong belief of being Saint...). This problem will be discussed in paragraph VI) MEDITATION EXPERIENCES of this chapter.

P. Dreams In Meditation.
Sleep in Meditation can be associated with dreams as in normal REM sleep. Other types of dreams, like Lucid dreams, may simulate experiences in Meditation.

Q. Hindrances in Meditation.
1. Illness, Pain.
. Common illnesses like viral/ bacterial infection, pain... that may or may not be related to Karma and may interfere with the effort, volition of Meditation.

ii. Decreased volition: the choice is between ignorance and volition.

iii. Numbness pain of the buttocks, hips, knees, and lower legs related to the inappropriate position of crossed-leg sitting when the Meditation exceeds one hour or more (Fig 8.3).

. The slight feeling may disappear after 10-20 min when the attention moves away from the areas of interest.

Strong feeling requires a change of position, like squatting. Spreading the extended legs is not advised.

The numb feeling is usually not caused by the compression of blood vessels or nerve trunks in the hip or knee,. The numb feeling is not distributed according to the nerve territory distribution but in a stock-like pattern. This distribution has been described in patients with stock-like loss of senses in patients with conversion disorder. It is proposed that the pathogenesis is related to the stagnation of Chi/Qi / embodied Soul. This type of stagnation, known as stagnation of Chi in the meridian channels, is well described in Chinese medicine.

Physical exercise of the legs (cycling or walking) has been proven to decrease the pain and numbness of the legs. The mechanism is likely related to the increased mobilization of the Qi (Soul of the body) after the exercise

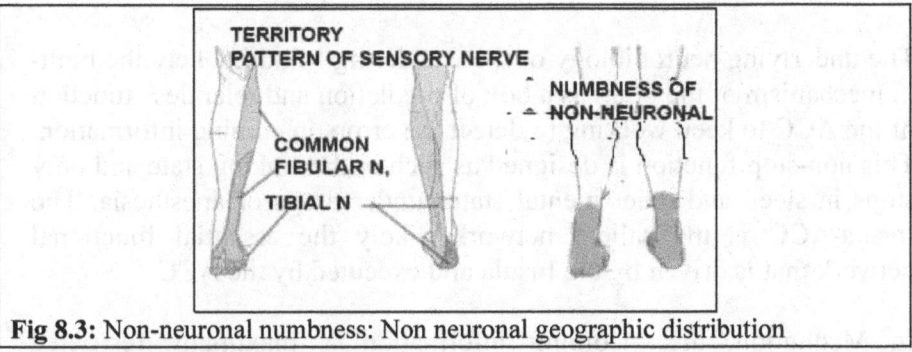

Fig 8.3: Non-neuronal numbness: Non neuronal geographic distribution

2. Somnolence and Sleep

Ranging from mild drowsiness to deep sleep with snoring and dreams. The sleep may last for one hour more, corresponding to one cycle of NREM or REM.

The eye closure will trigger the SCN to inhibit the DMH nucleus with a subsequent decrease of Orexin. Orexin is responsible for activating the wakefulness nuclei in the cerebral peduncles.

The wakefulness requires the activation of the wakefulness center, namely the LC (nor-adrenaline), TMN (Histamine), DR Serotonine, PAG (DOPA) LDT/ PPT and ForeBrain (acetylcholine) is necessary for Consciousness even in the absence of a wakeful state (as in the case of REM sleep).

Sleep and sleepiness in medication are caused by:

Lack of volition especially making up the Mind before each Meditation. This is the most essential factor. For sleep, humans need at least 5 hours for older people and 7 hours for young persons.

Lack of sleep, especially for a few days before the Meditation, is challenging to counterbalance with volition.

Sleepiness before the Meditation.

Illness with fatigue are removed when someone is enlightened

Medication inducing sleepiness.

Restless/ wandering Mind, Misleading/ False thoughts—
3 Transient thoughts Vitathavitakka
The Mind is like the monkey, and thinking is like the horse.
This is a most frequent hindrance to keeping the mediation from the goal of calming the Mind to enlightenment. The famous story is about Huike (The Second Master of Chinese Zen Buddhism) asking his master for help; Bodhidharma replied to his disciple's request by asking him to show the Mind. Of course, Huike couldn't find his Mind. The master said that he had already calmed the disciple's Mind.

The underlying neurobiology of the wandering Mind is likely the built-in mechanism of the brain as a box of prediction and relentless function of the ACC to keep working to detect the errors in coming information. This non-stop function is designed as such in the waking state and only stops in sleep and other mental states under drugs or anesthesia. The Insula-ACC is the salient network, likely the essential functional network that is driven by the Insula and executed by the ACC,

In Meditation, the incoming information is maximally restricted because of the dorsal stream of information transmission. ACC searches the information that is only available in the ICS; as a result, the retrieval of MM results in:
- Cleansing bad MM (karma)
- Wandering Mind

Therefore to control Mind wandering, the likely efficient technique is to develop the Awareness obtained by the Ventral Stream of the pathway of transmission in the brain to provide the data for the prediction of the ACC. The Awareness represents the Meditation experiences described below.

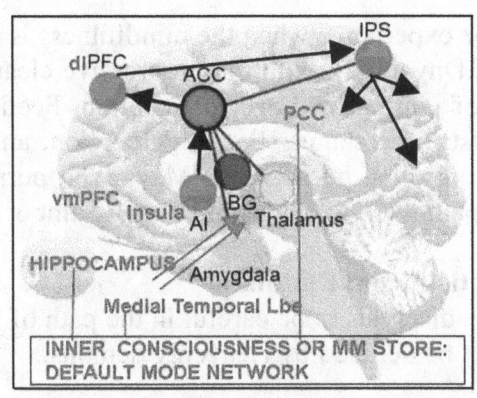

In Mind wandering and Karma cleansing, IPS plays a critical role in determining the focus of attention. Data retrieved from ICS often interfere with the designated attention focus. In this case, IPS can effectively block unwanted data (MM from ICS, itch, pain of the body) and prevent the Mind from wandering. Furthermore, the discarded MM from the ICS is eventually prone to synaptic neuroplasticity (destruction and repair), leading to karma cleansing. Please also page 161, 200 regarding role of IPS).

V. MEDITATION EXPERIENCES.

The experiences in Meditation are not the goal or purpose since Meditation is the natural way of Tao; therefore, it is the ideal way. They are phenomena that meditators encounter during Meditation. The beneficial effects like happiness, regulation of the physiologic function of the body, therapeutic effects on mental and bodily diseases.... are well known. The reader can refer to the internet for information.

In this section, spiritual experiences are described and discussed.
They result from the cleansing and discarding of the veil of ignorance/ stupidity (incomplete and deviated CS). As a result, the ultimate Omniscience/ Buddhahood is revealed. Buddhahood is associated with infinite sound, light, Omniscience, and other attributes like ingeniosity, stillness, and non-discrimination.

The basic experiences are the inner perception of sound and light (called inner sound and light) independent of the natural sound and light, believed to arise from the Original Mind. It is believed that when the Mind is relatively fixed (mindfully), the sound and the light will automatically appear (from the Original Mind).

The next step is the experience when the mindfulness is more advanced (from the stage of Dhyana 2), with the progressive clearing of the five types of the veil of ignorance: Form, Perception, Feeling, Formation, and CS. In the first three stages (Form, Perception, and Feeling), the mediator can be controlled by a Ghost/ Maya who purposely deviates from the spiritual path with delusions of being a Saint or Buddha.

Necessary instruction from Buddha:
Buddha warned his disciples to be careful in the path of Meditation and to be vigilant to not be deluded by Heavenly demons..

> **Bodhisattvas great Arhahat are unmoved when someone is enlightened, but demons and ghosts are upset and shocked to see their palaces break open.**
> Buddha said: *Though they rely on their supernatural powers, they are but externals and will only succeed in destroying you **if you, who own the five aggregates in your minds**, are deluded and let them do so.... On the other hand, if you fail to awaken and are thereby deluded by the five aggregates, then, Ananda, you will become a son of Màra and help the demons.*

VI. MEDITATION and KARMA CLEASNING:

THE PHENOMENON OF KARMA ASSOCIATED WITH THE FIVE SKANDHAS IN MEDITATION.

In the Lankavatara Sutra, after the Buddha pointed out how to practice becoming an Arhat to Ananda, he also highlighted the 'Mara obstacle' that practitioners may encounter during Meditation. This 'Mara obstacle' is not an external force, but rather the practitioner's own experiences that arise during Meditation. It is a crucial phase in the journey towards higher experiences. These experiences include self-seeing, self-hearing, and experiencing other physical sensations, all of which are distinct from inner Consciousness.

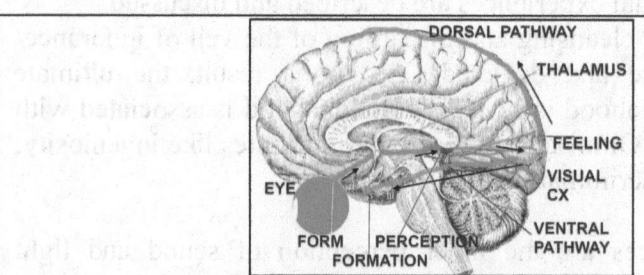

Pathway of vision: Retina→Thakamus→ Visual cortex :
→Dorsal Nonconditional reflex (ex: closing eye when seeing flying object)
→Ventral pathways; d Formation of CS

Depending on the realm, there are 5 types of Mara (Surangama Sutra, volumes 9,10). While having experiences in Meditation is beneficial, it can also lead to Delusion if one fails to recognize the true nature of their practice. This is known as the Mara obstacle, a sign of arrogance when one mistakenly believes they have reached the goal of their practice. However, this is also an opportunity for growth and deeper understanding. but still a Delusion, not considering oneself a Saint, if one gives rise to arrogance, it is called a Mara obstacle because when one has not yet reached the goal of cultivation, one thinks one has reached the goal. When Karma has begun to be cleansed away, meaning that the Karma's MM is significantly removed, the Zen practitioner has an individual experience. The mechanism is as follows The Soul consists of Dark Matter and Neutrino. In living beings, Neutrino is attached with electrons (with a charne of (-)), thanks to which the Soul sticks to the synapses with (+) charge, when the neuronal connection has just been removed from the synapse because it has just lost MM due to Meditation (Cleansing Karma, the change in voltage helps the Soul separate from the synapses. The Soul detachment is successive step by step and according to the principle of the order of Form, Feeling, Perception, Action, and Consciousness:

- The Form part corresponds to the synapses in the Retina, the Inner Ear... of the five senses that are detached first. When the Soul is attached to this part and is free (no longer attached to the Synapses, the Soul, due to the principle of being Non-Local, can contact the other worlds inside, manifested through the viewing of scenery, listening to the sound of Buddha and Bodhisattva preaching the Dharma...

- Next, the Perception part corresponds to the synapses in the Setnsory organs and Thalamus have been removed, so the experience is more wonderful and emotional, such as strong compassion, Awareness stronger than concentration, anxiety, joy, birth of self-life, lightness, abandonment of cause and effect, criticism, passionate love

- As for the Feeling Skandha, am, snapses in the Setnsory organs, Thalamus, and the primary snsory (primary) have been removed, so it is even more miraculous, thinking that he is Buddha, saints with wishes totravel and preach about lust, has the heavenly Knowledges, preaches that listeners will attain Nirvana immediately, accepts disciples, receives offerings, talks about previous lives, slanders and scolds disciples, eats little but still becomes fat, performs supernatural powers such as walking on water, denies reincarnation (no Awareness), wishes to live long

- As for the Mental Formation Skandha, the Setnsory organs and Thalamus primary sensort cortice and Temporal cortex have been removed, ending the merits of the Five Roots (like the 800 merits of the Eye Root), but has not yet attained Consciousness, so the understanding is not thorough. The experience of the nature of the CS, Seeing life as inhuman due to the deviated nature of the Bodhisattva (Human gives birth to Human, not the EM gives birth to humans, animals...). Likewise, misunderstanding the root of all things, not seeing Impermanence, humans are not born or destroyed, accepting the Ego: I give birth to everything, committing the wrong view, the Body and the Three of this person are crazy and crazy, misunderstanding Cause and Effect, Being - Non-being, being deluded into the world of desire is tasteless... (Tâm Điện Điển Luận) Regarding the Yin Consciousness, only then can it be replaced by the true nature when the Consciousness aggregates, living beings share the same root.

In the Lankavatara Sutra, after the Buddha pointed out how to practice becoming an Arhat to Ananda, he also highlighted the 'Mara obstacle' that practitioners may encounter during Meditation. This 'Mara obstacle' is not an external force, but rather the practitioner's own experiences that arise during Meditation. It is a crucial phase in the journey towards higher experiences. These experiences include self-seeing, self-hearing, and experiencing other physical sensations, all of which are distinct from inner Consciousness.

Depending on the realm, there are 5 types of Mara (Lankavatara Sutra, volumes 9,10). While having experiences in Meditation is beneficial, it can also lead to Delusion if one fails to recognize the true nature of their practice. This is known as the Mara obstacle, a sign of arrogance when one mistakenly believes they have reached the goal of their practice. However, this is also an opportunity for growth and deeper understanding. But still a Delusion, not considering oneself a Saint, if one gives rise to arrogance, it is called a Demonic obstacle because when one has not yet reached the goal of cultivation, one thinks one has reached the goal. When Karma has begun to be washed away, meaning that the Karma's MM is significantly removed, the Zen practitioner has an individual experience. The mechanism is as follows The Soul consists of Dark Matter and Neutrino. In living beings, Neutrino is attached with electrons (with an area of (-)), thanks to which the Soul sticks to the synapses with (+) charge, when the new Neuronal Synapses has just been removed from the Synapses because it has just lost MM due to Meditation (washing) Karma, the change in voltage helps the Soul separate from the synapses. According to the principle of the order of Form, Feeling, Perception, Action, and Consciousness:

- The Form part corresponds to the synapses in the Retina, the Inner Ear... of the five senses that are bound first. When the Hon is attached to this part and is free (no longer attached to the Synapses, the Soul, due to the principle of not being in place, can contact the other worlds inside, manifested through the change of scenery, listening to the sound of Buddha and Bodhisattva preaching the Dharma...

- Next, the Perception part corresponds to the addiction of the Form part, the Synapses in the Inner Ear and the Noisy Hill has been removed, so the experience is more wonderful and emotional, such as strong compassion, Awareness stronger than concentration, anxiety, joy, birth of self-life, lightness, abandonment of cause and effect, criticism, passionate love

- As for the Feeling Skandha, the part corresponding to the SYNAPSES in Sensory organs. Thalamus, and the primary cortex have been removed, so it is even more miraculous, thinking that he is Buddha, Indra, the Buddha is born to travel and preach about lust, has the Three Knowledges, preaches that listeners will attain Nirvana immediately, accepts disciples, receives offerings, talks about previous lives, slanders and scolds disciples, eats little but still becomes fat, performs supernatural powers such as walking on water, denies reincarnation (no Awareness), wishes to live long

- As for the Mental Formation Skandha, the corresponding parts, the SYNAPSES in Sensory organs,Thalamus and the primary côticés and Temporal Lobe have been removed, ending the merits of the Five organs (like the 800 merits of the Eye Root), but has not yet attained Consciousness, so the understanding is not thorough. The experience of the subtle nature of the CS. Seeing life as inhuman due to the wrong nature of the Bodhis natùe (Human gives birth to Human, not EM gives birth to humans, animals...). Likewise, míunderstanding the root of all things, not seeing Impermanence, humans are not born or destroyed, accepting the Ego: "I give birth to everything", committing the wrong view, the Body and the Mind of this person are crazy and crazy, misunderstanding Cause and Effect, Being - Non-being, being deluded into the world of desire, invéion

- Regarding the Consciousness skandha, only the true view of nature is established when the conception that living sentients share the same root from the Creator.

VI. MA Meditation

The Desire Realm, with its six paths of reincarnation, is an environment for learning and reformation with many challenges. As the meditator in Meditation gradually cleanses his Karma, his Soul can be separated from the Five Aggregates, starting with Form, Sensation, Perception.... These are the three stages in which the meditator is easily disturbed by illusions he himself creates. Seizing this opportunity, Heavenly Demons and other Souls often infiltrate to test him. These tests are not to be feared, but rather embraced as opportunities for growth. The Desire Realm encourages and flatters the meditator as the Supreme One to stay in the Desire Realm forever, but it's up to the individual to resist this temptation and avoid moral corruption. Surangama writes:

> " five supernatural powers, still fail to realize transcendental insight into the ending of the stream of transmigration for they have not broken their links with saüsàra; how can they let you destroy their dwellings? This is why they come to trouble and annoy you when you enter the state of samàdhi. However, in spite of their rage, these demons are there in your profound state of bodhi and are like people trying in vain to blow out sunlight and to cut water with a sword, while you are like boiling water that melts solid ice. Though they rely on their super natural powers, they are but externals and will only succeed in destroying you if you, who own the five aggregates in your minds, are deluded and let them do so. For these demons cannot harm you in your state of dhyàna if you are awakened and are not deluded. If you wipe out the (five) aggregates, you will enter the state of brightness where in all demons are but dark vapours. Since light destroys darkness, they will perish as soon as they approach you; .....
> ....
> . ânanda, now that the practiser is free from anxiety, after his receptiveness has vanished, he finds himself in the state of perfect dhyàna and likes its pure brightness. But he may be tempted to concentrate on the one thought of skil fully advancing, thus submitting to the heavenly demon who immediately possesses another man (to harm the meditater).251 This man, unaware that he is possessed will, as directed, preach the Dharma of the sàtras and think that he too has realized Supreme Nirvàôa. He will then come to the practisers place and take the high seat (reserved for reputable monks) to teach him the Dharma. To show his skill, he will appear either as a monk, Indra, a woman or a nun, and his body will send out rays of light that illumine the dark bedroom. The practiser will mistake him for a Bodhi sattva and will believe what he says; as a result, his mind will waver and he will break the rules and have desires. The man will speak of weal and woe, of a Buddha appearing at a certain place, of scorching fire in the kalpa of destruction and of future fighting and wars to frighten and ruin other people. This is the Strange Ghost who has become a 251. As the practisers mind is free from receptiveness, the demon is unable to influence it, so he uses another man to deceive and harm him. 292 demon in his old age and who now comes to trouble the practiser. When he is weary of his misdeeds, he will leave the possessed man. Then both teacher (the possessed man) and pupil (the practiser) will suffer all the

> *miseries inflicted by the royal law. You should first be clear about this temptation to avoid returning to saüsàra, but if you are deluded and do not recognize it, you will fall into the unintermittent hell*

379

Chapter IX: THE SLEEP
Summary.

Hypo is an essential part of the brain, regulating function of organs and structures that mainly work automatically and are not related much to the CS. Nucleus SCN/ Suprachiasmatic Nucleus is considered the body's central clock with a cycle of 24 hours (more precisely 24h 11 min) independent of the light and season. This clock is adjusted according to the light and season by the Dorsomedial Area of the Hypo and Pineal gland. SCN activates nuclei for wakefulness or inhibits nuclei of sleep by intermediate of MCH and lateral area of Hypo (secreting Orexin).

In addition, there are two groups of nuclei of the Hypo. The anterior group consisting of the ventrolateral preoptic (vlPO) and Median Pre-Optic nucleus (MnPO) are related to sleep. The posterior nuclei (for the wakefulness) consist of Locus Ceruleus (LC, epinephrine and norepinephrine), Dorsal Raphe (DR, serotonin), Peduncullopontine tegmental (PPT, Acetylcholine), Tuberomammillary (TMN) and Laterodorsal Tegmental (LDT, Acetylcholine, Ventral Tegmental Area (VTA, DOPA). Ventrolateral PeriAqueductal Gray

(vlPAG, DOPA). EEG subdivides sleep into NREM 1-2 ranging from slight to deep sleep, NREM 3-4 with sleep without CS, and REM sleep with Consciousness with dreams associated with muscle atonia.

Disorders of sleep consist of insomnia, which affects nearly half of the elderly population, and other disorders affecting the induction of sleep and REM disorders.

I. INTRODUCTION

Sleep is an essential part of life, the counterpart of wakefulness; both represent the principle of duality of the Creation. Therefore, it is unreasonable to ask "why we sleep". According to Greek mythology, Hypnos is the God of sleep, dwelling in the cave of River Lethe, in eternal darkness. His two other brothers are Thanatos (God of death) as twin brothers and Oneitrol (God of dreams). River Lethe is known for the power of its water that made any Soul forget the past life when drinking the water before being reborn. This story is similar to the legendary Vietnamese tale in that the Soul must eat the rice soup called Dummy Soup at the river border before crossing the bridge Nai Ha for reincarnation.

II. CHARACTERISTICS OF SLEEP
A. Electroenceplogram EEG and the Sleep .

1. NREM-REM.

EEG was discovered by Richard Carlos (1842–1926), was first recorded in laboratory animals 1912 and was recognized by The American EEG Society in 1947.

In addition, ElectroOccular gram EOG for the recording of ocular movement with two characteristic phases: Rapid Eye Movement (REM) and Non-REM (NREM), can subdivide sleep into 5 phases. The EEG registers the electrical voltage change of individual zones of cerebral cortices and underlying nuclei in the diencephalon, midbrain, and cerebellum using electrodes over different areas of the scalp. EEG has been useful in investigating different stages and disorders of sleep, including dreams and pathologic conditions of the brain.

2. The changes in EEG are categorized

-Low and rapid waves: in the wakeful state
-Slow and high waves in sleep, resting state, and mindful Meditation
-Delta(<4 Hz),
-Theta(4–7Hz),
-Alpha (8–12 Hz),
-Beta (13–25Hz
–Gamma 25Hz–200 Hz or even higher frequency

In Meditation, delta waves: relaxation, Theta: attention and mindfulness

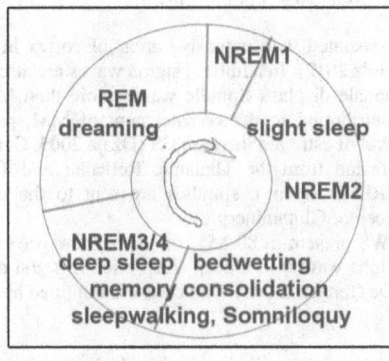

Fig 9.7 One cycle of sleep composed of five stages. **Five Stages of Sleep. Each cycle of sleep about 90-120min, 2-5 cycles per night.**

NREM 1, alpha wave + theta waves 5-10min, transitional from wakeful state to sleep; when waked up, one has the impression of not having been in sleep yet.

--NREM 2, 10-25 min. Bursts of rapid, rhythmic brain wave activity are known as sleep spindles. They are thought to be a feature of Memory consolidation—Light sleep.

-NREM 3 delta deep sleep, for refreshing and repairing the body with slow heart and breathing rates.

Bedwetting and sleepwalking and somniloquy.
- NREM4 is similar to NREM3, but deeper sleep than NREM 3-2 is repeated.
-REM, the first REM, about 10-20% in adults, compared to 50% in babies).
Dreaming: muscle atonia.

Alpha Sleepiness, REM sleep	8-12 Hz	Most common. Seen in Occipital cortex, not seen when openning the eyes **alpha wave in REM seen in Frontal cortex, semiarousal sleep**. Alpha wave intrusion seen with Delta in NREM in Fibromyalgia.
Beta Wakeful state	>13	Low amplitude, regular in the Anterior part of the Brain. When opening the eyes alpha waves change to Beta
Theta, Cx HIPPO CA1 Denta, Attention	3.5- 7.5	High amplitude in HIPPO, Lower in the cortex. In mice, seen with deep breathing during the attention focused on prey. In REM sleep HIPPO Theta related to MM uploading to vmPFC for MM consolidation. Theta in the cortex are not related to Theta wave from HIPPO.
Delta,SWS deep sleep NREM 34,	<3	due GABA from Thalamus , ARAS Occasionally associated with K complex . when Theta waves are present *cortex inhibited by GABA, Delta waves are seen , consisten with MM consolidation in vmPFC Decrease with aging, insomnia, Increase in Parkinson.
Gamma -In thinking -In deep meditation	>25	Faster than Beta waves, from Thalamus and ARAS (Urbano 2012)sent to all cortex. Disappear in Thalamic lesions, Depression, decreased in schizophrenia, late appearance with high amplitude in Alzheimer), seen in Dream REM Related to thinking (more frequently seen in Temporal Lobe), dreaming
K complex	Often<1	100 microV, (-) then (+) trong 0.5-1 sec, NREM stage 2 of sleep, From Frontal Lobe sent to Thalamus to oppose the inhibition, in thinking/MM retrieval.
Mu	7.5-12.5	Related to eye movement
Sigma Or Sleep spindle Necessary for MM		Associated with extensive areas of cortex in the consolidation of declarative (Holz,2012). In addition, sigma waves are also related to sensory input in sleep. Female displays Spindle waves more than Male 1.16 times due to Estrogen contributing to the enhancement of MM, particularly in menstruation (high level of estrogen increase CS (Dzaja 2005, Genzel, 2012, Manber 1999. Created from the Thalamic Reticular and Thalamic nuclei Network during NREM2 sleep), spindles are sent to the cortex (GABAergic and NMDA receptor Glutaminergic) SWS present in 80-85% of sleep. Slow waves (0.5-2Hz, lower amplitude than Alpha waves) in a deep sleep NREM ở giai đoạn 3 and 4 (with Delta waves). (De Gennaro 2008). The cortex is inhibited by GABA in sleep.
PGO waves or P waves	Pontogeniculo Occipital	From Peduncle →LGN→ Occipital cortex in REM, Modulatory involving Aminergic (Serotonin, dopamine and Norepinephrine), cholinergic nitroxergic và GABAergic neurons. PGO also involve vestibular, Amygdala suprachiasmatic (regulating REM sleep), auditory and Basal ganglion. In REM sleep PGO represents Dreaming state. (Gott 2017, Hutchindon 2015).

B. Suprachiasmatic Nucleus SCN/ and Molecular mechanism

Studies on Flies (Drosophila) reveal the clock genes PER and protein PER, and in mice, gene BMAL1 is similar to the clock gene in flies. In Flies, the gene is present in SCN and in many organs like GI, heart, and

Kidney... (Oishi 1998) Protein of gene CLOCK/MBAL1 attach to PER in mammal animal (Per1, Per2, Per3) and CYTOCHROME (Cry1, Cry2).

PER gene produces a protein that interacts with the sunlight to inhibit PER gene activity (gene repressor). As a result, a feedback mechanism develops (Cox 2019, Xie 2019).

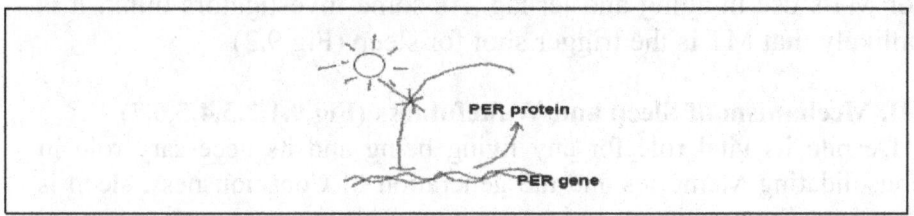

To control the below center of wakefulness.

SCN is the alarm clock switch between sleep and wakefulness. SCN activates the LHA, which, in turn, activates the wakefulness center through the hormone Orexin. SCN also activates the Dorsomedial Hypothalamus , then the Paraventricular nucleus/ PVH, which in turn stimulates the Superior Cervical Ganglion the, the Pineal gland
(Fig 9.2)

C. Melatonin and Pineal gland

The Pineal gland produces the hormone Melatonin/MT, which acts on receptors MT1 and 2 at SCN, CA2,3,4 of the HIPPO, cerebral cortex, and viscera. MT is produced when SCN is activated (darkness). SCN activates SCG, and then SCG activates the Pineal gland.

In addition, MT shows antioxidant, anti-, and proinflammatory effects, increases platelet production (thrombopoiesis), and lowers body temperature, blood pressure, and blood glucose. MT likely induces sleepiness, plays the role of autoimmune modulatory, and may contribute to treating Alzheimer's disease (due to the mild antioxidant activity), arterial hypertension, diabetes, bipolar disorder, COPD, and even Covid-19 infection (due to anti-inflammatory effect). MT is associated with a mild beneficial effect on immunity. MT inhibits enzymes involved in the degradation of cAMP in neurons, improving Memory.

More importantly, MT inhibits *Thyrotrophin Release Hormone (secreted by LHA to enhance the secretion of TSH/Thyroid Stimulating Hormone from the Pituitary gland). As a result, melatonin decreases the bal metabolism by lowering body temperature and blood glucose (along with lowering blood pressure, heartbeat, and respiratory rate)*

MT decreases with age when the Pineal gland degenerates with calcification of the gland. In aging, MT secretion decreases with a delay in peak secretion. As a result, MT's beneficial effect is inducing sleepiness (but not increasing sleep). This effect accounts for MT's use in aging and jet lag. As some investigators think, it is unlikely that MT is the trigger shot for sleep (Fig 9.2)

III. Mechanism of Sleep and Wakefulness (Fig 9.1,2,3,4,5,6,7)

Despite its vital role for any living being and its necessary role in consolidating Memories and the generation of Consciousness, sleep is automatic in its initiation and termination. This means that sleep control centers are not in the cerebral cortex but are all located in the hypothalamus and the brain stem.

The activities of the nervous system can be summarized as follows:

A. ASCENDING RETICULAR SYSTEM

WAKEFULNESS In the wakeful state, there are activities: wakefulness.

Consciousness and reflex reaction for survival.

- Nor-epinephrine for rapid reaction, anxiety and tumescence of the genital organ. Low epinephrine is associated with sleepiness, low level of attention.
- Acetylcholine for Consciousness.
- Histamine from TMN for itch sense.
- Serotonin from LDT for joy feeling increase is associated with fragmented sleep.
- DOPA from PAG for impulse feeling, muscle tonic contraction.

Further Reading

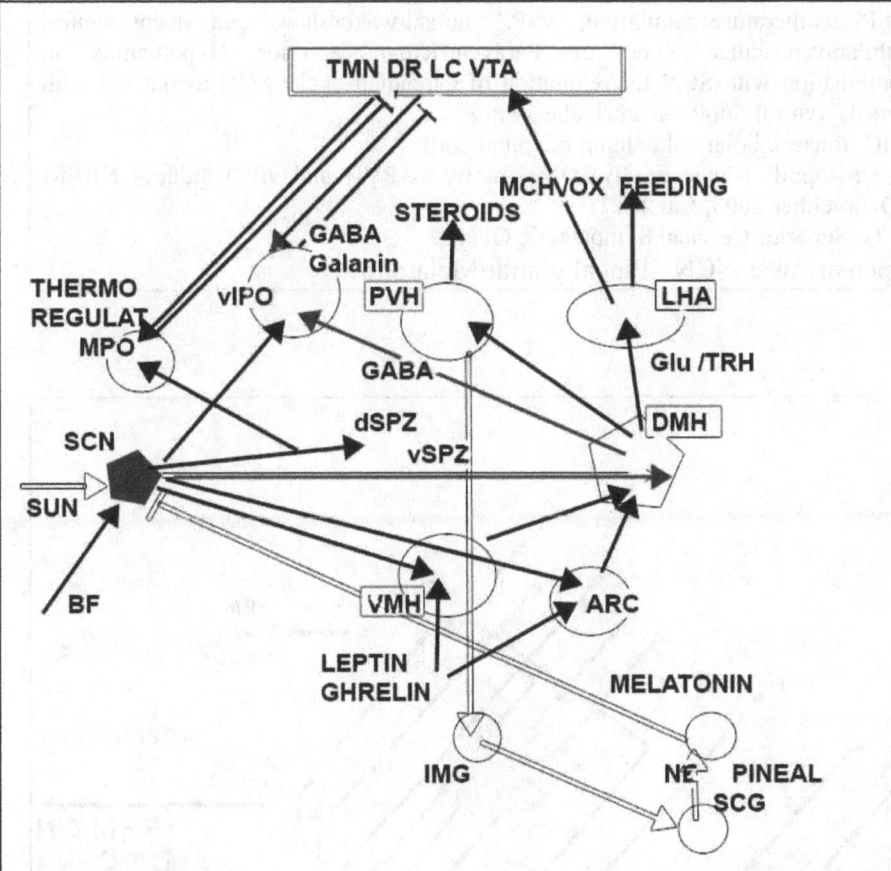

Fig 9.2: The diagram shows that upon receiving the light, SCN stimulates the center of wakefulness and eating and inhibits the center of sleep. Basal Forebrain and Tegmentum (Acetylcholine) also enable SCN Hypo to control eating, blood pressure, and sleep.

BF and Tegmen (Acetylchol) (Saper2005, Hut and Van Der Zee 2011 **O'Leary** 2014 Oakman 1995 Abrahamson EE 2001).
VMH: The ventromedial hypothalamus (VMH) causes fear, hunger, sexual activity, overeating, and obesity.
MnPO median preoptic nucleus: Homeostasis /sleep, temperature osmoregulation, regulation of other serum component.
DMH: Dorsomedial hypothalamic NPY (NPY mostly inARc) Energy and Glucose regulation
PVH: paraventricular nucleus (PVN, PVA, or PVH) eating regulated by the light , because of Orexin from LHA.
LHA/ Lateral Hypothalamic Area: Orexin and MCH
Tegmentum: The lower part of the midbrain comprises the red nucleus SN, PAG VTA and the upper part of the of ARN /Ascending Reticular Network

(dSPZ/temperature regulation., vSPZ: dorsal/wakefulness and sleep/ ventral SubParaventricular Zone of Paraventricular, anterior Hypothamus in combination with SCN for regulation of Circadian cycle/ SCN associated with dorsal / ventral SubParaventricular Zone)
IMC: Intermediolateral column of spinal cord
In Drosophila (domestic fly) DMH activates PVH and vlPO induces NREM (Deurveilher 2005, Liu 2017)
SCG: Superior Cervical Sympathetic Gland,
Open arrows: SCN –Pineal gland /Melatonin

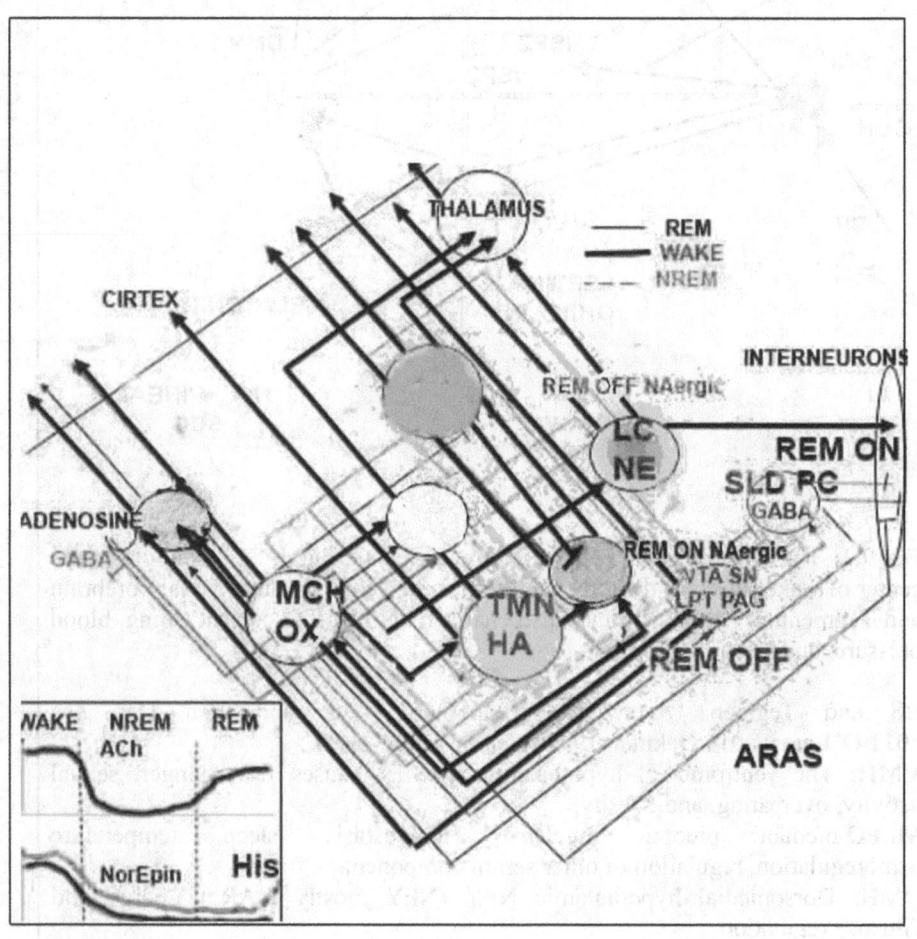

Further reading

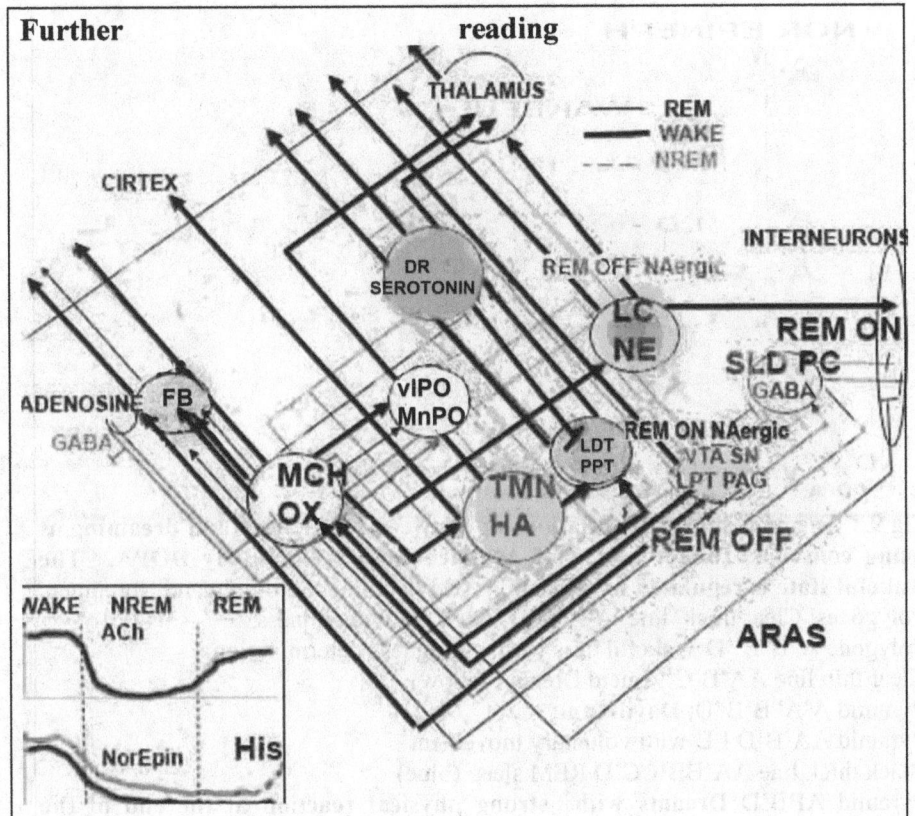

Fig 9.2 Diagram showing nuclei forming centers controlling the wakefulness and NREM/ REM sleep. Inset: Level of Acetylcholine (upper) and Norepin (lower) **(Becchetti 2016)**

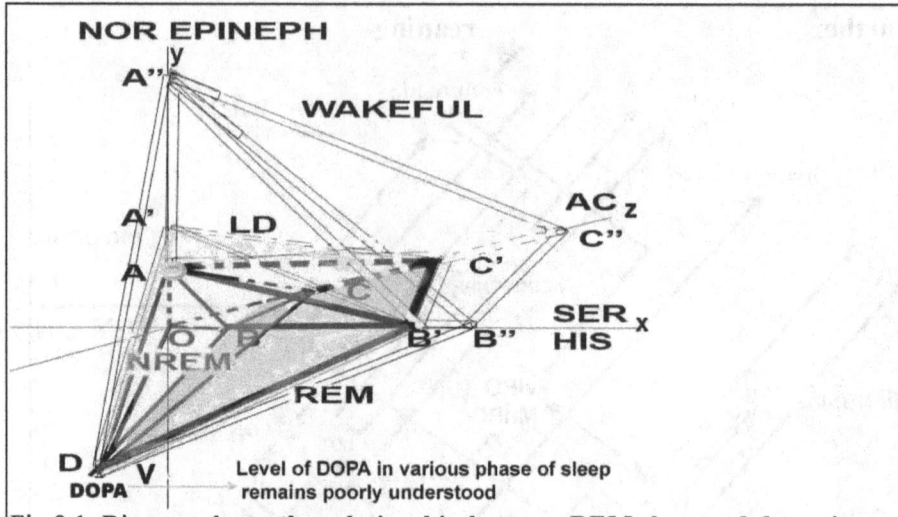

Fig 9.1: Diagram **shows the relationship between REM sleep and dreaming as being conscious (under action of Acetylcholine) inhibited by DOPA. The wakeful state is regulated by** Histamin Acetylcholine, Serotonin and Norepinep.
Polygones: Clear thick line: A'A"B'B"C'C"D: Wakefulness,
Polygone A"B'B"D Wakeful ness with willingness/determination
Clear thin line AA'B'C': Lucid Dream (yellow)
Pyramid A'A"B'B"D: **Daydream:**
Pyramid AA'B'D LD with voluntary movement
Black thick line AA'BB'CC'D REM sleep (blue)
Pyramid ABB'D **Dreams with strong physical reaction at the end of the dream**
Black thin line OABCD: NREM sleep (pink).
Pyramid: OABD Somniloquy
Note: Lucid LD cubic polygon is entirely within the Wakefulness cubic polygon. Positions of B' and C' vary according to the depth of REM and the Consciousness
Ox represents Serotonin/Histamine
Oy Nor epinephrine, Oz Acetylcholine,

- **SLEEP.**

Sleep cannot be voluntarily controlled since the control centers are located in the subcortical areas, the Diencephalon. However, sleep is almost always preceded by the state of sleepiness. Sleepiness is characterized by the lowering of metabolism to the basal metabolism with hypothermia, lowering blood pressure, heart rate, respiration rate, and blood sugar along with lowering the potential of increase of epinephrine, serotonin, DOPA, histamine, and Acetylcholine. As a result, inducing sleepiness is probably the efficient means to control insomnia.

The control of sleep is complicated. SCN initiates the event as follows: (Figure 9.2,3)

- Sleepiness: accumulation of Adenosines during the wakeful state and Melatonin secretion from Pineal gland under the influence the DMH (activated by SCN during daytime) with subsequent activation of PVH→IMG/interneuron in medullary cord→Superior Cervical gangion→ Pineal gland→ increased Melatonin to level enough to inhibit SCN.
- NREM sleep, no wakefulness, no Consciousness, and no dream: SCN activates sleep centers for sleep in the anterior Hypothalamus vlPO uses GABA to inhibit the wakefulness center of NREM sleep

Increased GABA and with subsequent decrease of Orexin, Epin, Histamin Serotonin

- REM sleeps: from <u>NREM to REM</u> : MCH in the lateral Hyothalamus inhibits wakefulness center, PnO Nucleus Pontis Oralis (Weber), activates BF LDT and PPT to secrete Acetylcholine) *(Maskos 2008)*

  - Orexin activate *S*LD-PC(SubLateral Dorsal) →GABA causing atonia
  - **Pn**O/ Pontine Reticular Formation/ *GABA (Watson 2007)*
  - Medullary Reticular Formation near the nucleus of the nerve VII may play a role in muscle atonia by inhibition

  the switch from NREM to REM with increase of Consciousness and dreams but no wakefulness. In animals that sleep with a half brain at the time, the hemisphere with NREM sleep is associated with lower levels of Acetylcholine than the other hemisphere

- Wakeful state, SCN activate DMH→LHA→ wakeful center in the posterior Hypothalmus

IV.OTHER MECHANISMS Error: Reference source not found

The BG and NAcare associated with movements of limbs and pleasure, pleasure promotes wakefulness. Mesolimbic pathway dopaminergic VTA, BG (motivation, desire for reward, pleasure, fear) also interfere with sleep.

DOPA: the role in the sleep is still poorly understood
- High DOPA: Happy, Motivated, Alert, Focused. Euphoric. Energized, high sex drive.
- Low DOPA Unmotivated, Unhappy. Memory loss. Mood swings. Sleep problems. Concentration problems. A low sex drive Having trouble sleeping. Having poor impulse control. Being more aggressive. <u>Attention deficit hyperactivity disorder ADHD).Parkinson's disease.</u>, Restless legs syndrome.

FURTHER READING

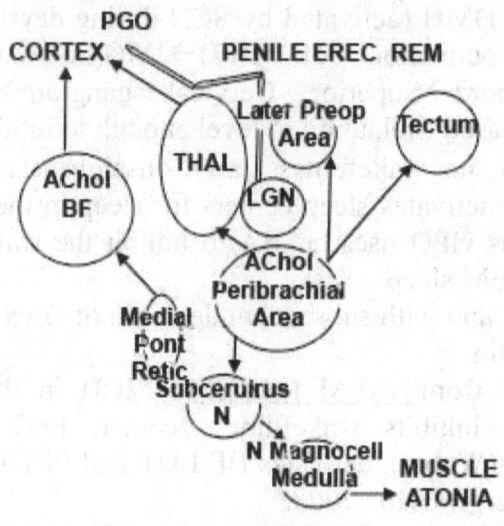

Fig 9.5

A REM: BF: Basal Forebrain, LGN: Lateral Genicular Nucleus
A. Mechanism of muscle atonia, Dream
with wave PGO, REM and congestion of genital organs in Male and Female. (consistent with the rise of Histamine at the wakefulness-up time in the morning .(Luthi 2016, Kovalzon 2016, Saper 2010 Gompf 2020) (Holland 2018)

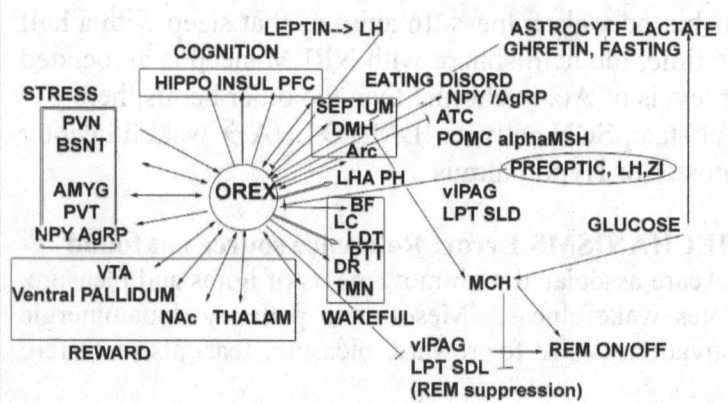

Diagram showing the Role of vlPO and Lateral HYPO (MCH/Orexin), Stress, Reward in Sleep, Eating disorder
(Anaclet 2012, Arrigoni 2019)

V. PHENOMENON IN SLEEP.
A. Muscle atonia: (under influence of GABA and Glycine)
Is important feature in REM, for the prevention of violent reactions or activities during dreaming.

Medullary Reticular Formation near the nucleus of the nerve VII may play a role in **muscle atonia.**

> **Mechanism:** The atonia is induced by GABA from SLD/ SublateroDorsal, activated by MCH from the HYPO. SLD is inhibited by LC/ nor epinephrine in the wakeful state. SLD inhibits the interneurons in the spinal cord that control muscle contraction. In addition, the glycinergic mechanism is also recently considered essential (Berger 2008, Arrigoni 2016

B. REM Intrusions.

In NREM sleep, REM may occur occasionally. In REM, the muscles are atonic, resulting in dozing sleep. Due to the volition to keep wakeful, in REM sleep, the sleep goes back to stage 3 and ends up with repeated flexing of the neck and head.

In addition, in REM intrusions in NREM sleep, dreaming may occur. Some researchers, dreaming of REM intrusion, may account for phenomena in OBE or NDE. REM intrusions may happen more frequently than believed. In near-death persons, REM is rarely demonstrated. As the REMs intrusion mechanism for OBE and NDE is not supported

C. RBD= REM Sleep Behavior Disorder.

Characterized by REM without associated atonia. As a result, dreaming (in REM) may be related to violent movement of extremities that may be harmful.

The disorders are often seen in men over 50 years old with degenerative neurological disorder, alcohol withdrawal syndrome, and patients with narcolepsy. The mechanism is unknown, probably secondary to the mechanism controlling the REM atonia.

> **Further reading**
> **D1. REMSD=REMS deprivation)**
> NAergic neurons (LC=REM On, LDT/ PPT= REM Off) decrease activity in NREM and are inactive in REM. LC is inhibited in LDT/ PTT in REM.. Achergic neurons of PPT associated Naergic neurons of LC and Somatostatin antagonist Acetylcholine in the regulation of REMS (McCarley 1975, Kumar 2012, Mehta 2017): Orexin decreases REMS, Ach antagonist or DOPA increases REMS. Serotonin plays no role in REMs (Arnaldi 2015).
>
> Histamine: H1 H3 receptors, Histamine decrease REMS and NREMS. TMN (Histamine), vlPO (GABA and Galanin) are ON/OFF buttons for Sleep Wakefulness cycle (Thakar 2011, Cheng 2020),
>
> Other nuclei regulating REM. In REMS deprivation insomnia, NAergic neurons are active, increasing NA. Insomnia *REM disorder*/REMSD is associated with

unstable mental state: bipolar disorder, Alzheimer, Parkinson, epilepsy (Narcolepsy), decreased CS, cardiovascular, respiratory, immune disorders, and fever.

D2. Local sleep.
It is one of the proposed mechanisms of Somnambulism (MD) in which a part of the brain is active.

The brain is composed of many parts acting as interconnected modules.
Communication between the modules is necessary for normal function. Therefore the concept of the locally inactive module, when other modules are active is feasible.

Conceived as such, NREM 3 and 4 may represent the phenomenon of locality. Slow Waves is another example. In Dolphins and other fishes, and birds, NREMS sleep only.
involves one side of the brain to accommodate their lifestyles in water and in the air (Mukhametov 1984, Oleksenko1992, Rattenborg 2017, Mascetti 2013).

Local sleep also represents a common phenomenon of the brain to accommodate Neuroplasticity (destruction and repair) for repair involving Interleukin 1, 6, and TNF (Kruger 2019, Mark 1995).

E. Sleepiness.
A physiologic condition preceding sleep, a state of falling asleep, characterized by a duration of a few minutes to hours, especially in insomniacs.

Causes: medications, adenosines, melatonin, fatigue, dullness, acute and chronic illness, psychiatric and psychologic conditions, and neurodegenerative disorders.
Endocrine disorders with low thyroxin in hypothyroidism, cortisol, low blood pressure, bradycardia, respiratory distress, and low blood glucose…. In physiologic conditions, aging and conditions like jetlag and melatonin effectively manage sleepiness.

F. Sleep Paralysis
Narcolepsy) and Cataplexy due to the loss of muscle tone
In REM sleep, bad dreams become panic. But due to atonia, one cannot react to the threat in the dream.
About 8% of the population for the entire life experience this problem

Narcolepsy) and Cataplexy due to the loss of muscle tone
Insomnia

Sleep is necessary for the brain and the body but not for the Soul. The mean duration is 7 hours per day, langer for children and shorter with aging, but the minimum time for the elderly is 5 hours.

G. Dozing
Is the sleep while standing or sitting. In the transition of NREM 1-4 and REM sleep, muscle atonia at the beginning of REM sleep causes the head drop. Since the volition of keeping awake in standing or sitting, the dropping appears to be repeated.

H. Insomnia.
1. Etiology.
sleep is controlled by the center of sleep, and the inhibition of wakeful centers in the ascending activating reticular network associated with Glutamate and hormones; ACh, NA, Ser, Hist, DA, and Orx.
- Melatonin secretion disorder.
- **Hypoactivity of sleep centers**: GABA to inhibit vlPO and MnPO.
- **Overactivity of** wakeful centers.

- LC secreting Epinephrine due to increased secretion, and lack of physical activity during daytime results in the brain accumulation of norepinephrine. Anxiety, work with too much attention (Yu 2018, Thakkar 2011, Arnali 2014, Vu 2009 yoshikawa 2021)
- Pain and itching (LHA, LC, DR, TMN PAG,...). (Moses 2003).. Histamine is low in the brain in the sleep NREM and REM. Histamine is increased in allergic conditions and at the time of waking up after sleep, even before the rising activity of norepinephrine.
- Excitement, pleasure increase serotonin. DOPA
- DOPA is increased with muscular exercise. High level of DOPA is associated with itch.
- DOPA is decreased as in Parkinson's disease
- **Sleep apnea, obesity.**
- Medication, caffeine.
- **Other states.**

vlPO: hyposecretion of orexin in Narcolepsy (daytime hypersomnia more REM than NREM, resulting in too many dreams causing fear) and cataplexy (sudden muscle atonia). Decreased orexin with increased GABA may play a role in insomnia in the elderly.
-Decrease and late in the secretion of melatonin, especially in the elderly, due to the degenerative and atrophic changes of the pineal gland.

Degenerative changes in the sleep and wakefulness centers, including Alzheimer's disease, and Parkinson's disease (Gong 2021).

2. Clinical implication usually associated with
- **Sleep fragmentation**; each fragment lasts 1-1.5 h., decreased recent.
- **Memory and Consciousness, and increased the risk** of cardiovascular disease hypercholesterolemia, hypertension and, obesity due to secondary interaction with a center of osmoregulation and eating in the Hypothalamus.
- **Increase of tumor necrotic Factor /TNF and interleukin** produced from glial cells and a decrease in melatonin and cortisol secretion (Jehann 2017) results in a high incidence of gastroduodenal ulcer, IBD, aphthous stomatitis, Type 2 Diabetes and cancer.

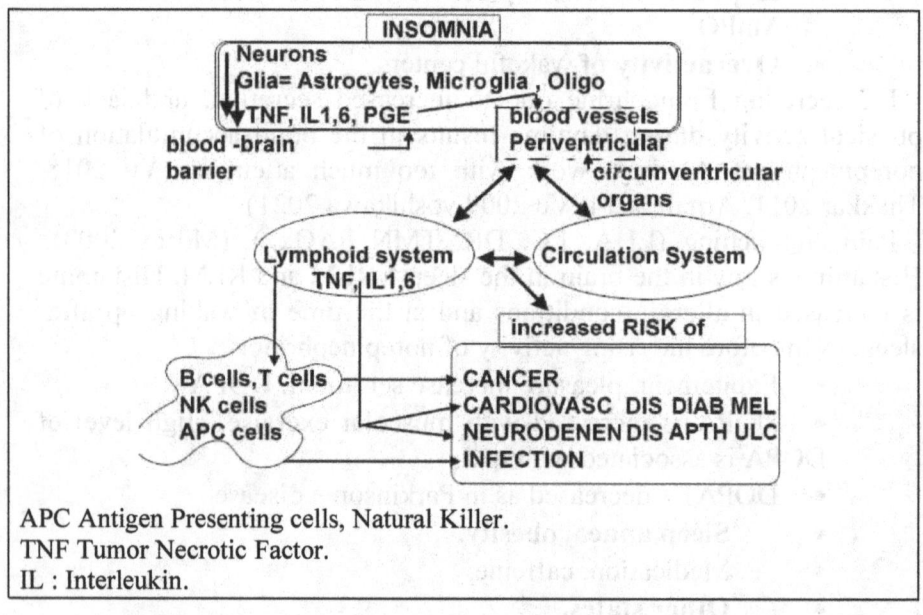

APC Antigen Presenting cells, Natural Killer.
TNF Tumor Necrotic Factor.
IL : Interleukin.

- **Decrease of the activity of PVH** (that typically activates the sympathetic cervical ganglion) results in low melatonin and cortisol.

- Other evident and immediate symptoms are tiredness, poor Memory, mood disruption,

3. Treatment.

Cognitive behavioral therapy is based on the understanding of the mechanism of wakefulness and sleep to facilitate the natural and automatic state.

Sedative and hypnotic drugs are helpful in the short term but are commonly accompanied by drug abuse because drugs interfere with automatic and natural living.

4. Resetting the sleep.

When experiencing long hours in bed in a wakeful state, some insomniacs may experience the benefit of getting up and doing some light activity for 5-10 minutes. When going back to bed, sleep eventually comes, especially after taking melatonin. There is no reasonable explanation for the mechanism. If one imagines that the sleep mechanism is more complicated than a machine, such as a computer. Resetting the device or the sleep mechanism is sometimes helpful.

I. Restless leg syndrome.

Restless leg is due DOPA secretion disorder, Cortico-Striatal-Thalamic-Cortical system with low sensitivity d DOPA receptor in the BG, **low Iron in the brain with a secondary increase** of glutamine.

J. Mouth Dryness

In NREM sleep, the Parasympathelic and Sympathetic systems automatically decrease in activity. In insomnia, there is a disproportional decrease resulting in high sympathetic as compared to parasympathelic activities. The low activity of the parasympathic is accompanied by the low secretion of its mediator, Acetylcholine superimposed by the poor hydration (insufficient consumption of water), mostly in older adults. Insomnia in elderly persons is often associated with the disproportional high level of sympathetic activity (by the intermediate of the Superior Cervical Ganglion). Drinking water before sleep usually does not improve the symptoms and is accompanied by excessive nocturia. The hypotonic solution in the foregut eventually triggers the hypersecretion of VIP (vasoactive intestinal peptide), activating the parasympathetic system of the hepatoportal area. This change stimulates the thirst/ osmolarity center in the Hypothalamus

Chapter X: THE DREAMS
*Dreams are true while they last,
and do we not live in dreams?'*
-Alfred Tennyson, Poet

Chuang Tsu once dreamt that he was a butterfly fluttering among the flowers. After waking up, he wondered whether Chuang Tsu was a man who dreamt of being a butterfly or was he a butterfly who dreamt of being a man

I. GENERALITIES

The Dream is an involuntary perceptive sensorial, primarily visual and emotional phenomenon that is conscious during sleep, especially in REM sleep. Dreams may last up to 30 min, many times a night, and become longer at the end of the sleep.

According to human cultures, from East to West, dreams represent messages or divine revelations from external entities, deities, and God. For Indians, life consists of three states: waking, sleeping, and dreaming. For monotheistic people, dreams are from the voice of God. The Old Testament frequently includes stories of dreams with supernatural inspiration. In the New Testament, Dreams may represent divine revelation, such as

Matthew: *"But as he considered these things, behold, an angel of the Lord appeared to him in a dream, saying, "Joseph, son of David, do not fear to take Mary as your wife, for that which is conceived in her is from the Holy Spirit."*

Or false message

John 4:1
Beloved, do not believe every spirit, but test the spirits to see whether they are from God, for many false prophets have gone out into the world.

In Chinese culture, the imperial court usually got advice from a royal dream interpreter for good or bad omens of the dreams.

Chuang Tzu is well known for his interpretation of life as a dream:
In a dream, Chuang dreamed that he was a butterfly, freely flying around from flower to flower. But when he awoke, he realized that he was Chuang Tzu. But Chuang Tzu asked himself: "Was I dreaming I was a butterfly, or am I really a butterfly dreaming that I am Chuang Tzu?"

Freud called dreams the interpretation of dreams, the royal road to the knowledge of the unconscious activities of the Mind. The id and libido are the automatic and primary sources of mental energy. Freud considers the libido involved not only in the sexual component but in high levels of mental energy like Creation and appreciation of art. The Ego is self-recognized in society, and the superego is the high level of morality acquired in society.

The Ego and super Ego constitute Consciousness, which also includes unconsciousness. The ego and Superego control, to some extent, the Id. As a result, Dream is the manifestation of id and subconsciousness during REM sleep.

Carl Jung includes the manifestation of the Mind in Dreams, suppressed and not suppressed in life. Therefore, Dream is individuation, part of life. Carl Jung emphasized not the libido but the significance of repeated images/ scenes in dreams. The personality represented in the Dream may represent a person in society as the phenomenon of transference.

II. MECHANISM.
A. Cortices of Dreaming.
EEG investigation in dreams demonstrated significant changes in the Parietal and Occipital lobes (Siclari 2017) and inner conscious areas that include:
Limbic and Paralimbic (piriform Cortex, entorhinal Cortex, and Para-Hippocampal Cortex), also called.
Anterior Paralimbic REM Activation *Area* (APRA) = bilateral confluent Paramedian zone,
Anterior Cingulate Cortex.
Anterior Insula,
Orbital Frontal Cortex, Temporal pole caudal ParaHippo, mPFC Amygdala/ Hạnh Nhân. (Mutz 2017).

The overall PFC is not activated when compared with the waken period despite the Consciousness being present in both states.

The connections between areas in the dreaming could be more cohesive and coordinated due to the lack or insufficient involvement of the PFC, especially the dlPFC and vmPFC. Other cortices that are not sufficiently activated are the inferior Parietal Cortex and TPJ.

B. Default Mode Network and Dreaming.
This network is mainly a site for storing Memories and Consciousness. Therefore, the DMN is variably activated in the dreaming.

C. Dreams and Neurohormones.

In NREM sleep, there is no Consciousness and no wakefulness. Neurons initiating PGO are located in the Pons near the superior cerebellar peduncles. In NREM sleep, these neurons trigger the slow waves (delta waves) in SW Sleep/ deep sleep (period of recovering resources of the body and the brain).

- Nor epinephrine (LC nucleus): "fight and flight" hormone accounts for wakefulness, alertness, anxiety.
- In Acetylcholine (Forebrain, LDT n REM, in which Acetylcholine is of high level with REM, the slow wave is replaced by the PGO.
- Serotonin (DR) and SSRI selective serotonin reuptake inhibitors in treating depressive neurosis promote dreaming. This phenomenon is due to the hallucinogenic effects of Serotonin and the general activation of the Serotonin of the Cortex (Pace-Schott, 2008DOI:10.1007/978-3-7643-8561-3_12)
- Histamine (TMN): wakefulness, mood, anxiety regulation, happiness, promoting food eating.
- DOPA, VTA, PAG also in TMN, DR for counterbalance and regulation of Histamine, Serotonin, muscle atonia.

D. PGO waves (Fig 10.1)
PGO waves are the electrical activity recorded in the Pons, Geniculate bodies of the Thalamus receiving the visual and auditory inputs.
PGOs are also recorded in animals like cats and dogs...in REM sleep (Morrison 2014, Revonsuo 2000):. The PGO may reach a level near that of epilepsy. PGO can be elicited by the stimulation of the brain stem (Caudolateral peribronchial area at Pons) with Acetylcholine

Therefore, PGO briefly precedes the dream, likely representing the electrical potential stimulating the somatosensation, visual cortices/ and even other cortical areas storing the autobiographic Memories. Due to the absence of the involvement of the PFC and the dlPFC areas for coordination, the dream manifests non-orderly images and stories that appear illogical in life.

Further Reading
Pons is inhibited by Serotonin (pleasure) from DR/ Raphe nucleus. (https://www.youtube.com/watch?v=D_y56PQvZZA) PGO induces imagery in dreaming. REM is related to Gamma waves (from Thalamus and ARAS to connect to cortices). The decrease of gamma waves in REM is consistent with the

decreased connection between vmPFC with Parietal and temporal cortices that are not activated in Consciousness formation) (Corsi-Caberra 2008).

Therefore, parietal and occipital cortice lesions are usually not associated with dreaming, but in midbrain lesions, PGO waves are still present and can generate dreams. This concept of dreams is quite different from that of Freud, who postulates that dreaming reflects suppressed unconsciousness/ subconsciousness such as Id, libido… (Morrison 2014, Revonsuo 2000 Paulson 2017). The PCC is the autobiographic store of the present life along with other cortices like the Precuneus for spatial Memory, RSC for storing the Memory of the previous lives, MTL, Lateral-inferior.

Parietal/ superior-Temporal region, and Temporo Parietal Cortex/ TPJ (Angular gyrus/ AG) zone of inter-action between vestibular, tactile, auditory and visual senses and entorhinal cortex. TPJ is known for its role in imagery (Eichenlaub 2014). Stimulation of TPJ or inferior Temporal Lobe in patients with brain surgery under local anesthesia can illicit dreams like the scenery outside the room

in a reported case. The connections with autobiographic Memories are usually limited to the past. Therefore a dream is more related to the current than to the past. In addition, Memories in the unconsciousness can be the source of input information in the dream, as suggested by Freud.

E. Weak, Non-coordinated Connections and Insufficient Activation of Consciousness center

Information in dreams mostly originates in the inner Consciousness. According to Freud, the dream is the royal road to subconsciousness.

The PGO waves initiate the retrieval of the information, and it is reasonable to believe that dream needs the presence of the Souls that ultimately is required for MM retrieval.

The lack of logic in the dreams is due to the weak, non-coordinated connection and non-activation of different centers of the inner Consciousness. The central network is not activated in dreaming, and the vmPFC is weakly activated. The involvement of the vmPFC suggests the possible participation of external Souls/ Consciousness in the dream. Altogether, given the random activation of the inner Consciousness by the PGO waves, the feasibility of intervention of external Souls, the interpretation of dreams must be based on mental status, the inner Consciousness, the unconsciousness, and the susceptibility to communication with external Consciousness, or power.

F. Dream Contents

According to Freud, dream interpretation is the 'royal road' to the unconscious. The PGO initiated from the Pons activates the sensory cortex, particularly the visual cortex. These waves likely arise either from the RT or LT side. As a result, one hemisphere is initially stimulated; however, due to the interhemispheric communication through the corpus callosum, the contralateral side is also involved in dreaming. Researchers commonly believe that the RT side is more involved in dreaming than the left side, and the RT side is more often connected with the LT side than when the LT side is initially activated. There are also case reports that in patients with LT brain lesions, patients stop having dreams. Studies in patients with callosotomy (separating two hemispheres) showed that the left hemisphere content in the RT-handed person is close to reality with the physical aspect as it is seen, **and images are mundane, simple.**

The RT hemisphere content of the LT-handed person is associated with the affective, conceptual, artistic, and even supernatural aspects of life associated with its LT-side counterpart. Therefore, the data in the RT hemisphere represents the analog to the sample of the LT hemisphere amplified with multi-dimensional, exponential power. The evidence is supported by dream contents from the RT hemisphere, which appear to be bizarre, fanciful, and grotesque, *with* vivid, *splendid, magnificent high imagery and expressive* and imaginative *states.*

As a result, the RT hemisphere content is an infinite asset for imagination. The two cerebral hemispheres play distinct, separate, complementary functions of the LT and RT hemispheres. The RT hemisphere amplifies by multiplying the data stored in the LT side with exponential power. For example, when receiving the image of a tall building, the LT hemisphere stores the image as such. The RT hemisphere multiplies the building in number, dimension, shape, color, and content.

G. Attention and Acetylcholine in dreaming

The role of the attention center (IPS/ Intraparietal Sulcus) and Acetylcholine is critical for the dream contents to be conscious. In LD, in addition to IPS, Acetylcholine and Norepinephrine likely account for lucidity in the dream, as LD can occur at the beginning or the end of the sleep (see Fig 9.1). However, due to the absence of the role of dlPFC, the content could be more consistent.

III. Categories of Dreams.

A. Mind wandering. Dreaming and imagination, when being in the wakeful state, could be related to creative activity, planning, and thinking. The Inner Counsciousness from the rigt hemisphere may play pivotal role
This is different from Lucid dreams and sleepwalking.

B. Somniloquy.
Usually occurs in persons 30-70 years during REM sleep with RBD disorders and NREM sleep. They speak in single words like "No," in short sentences that are correct in grammar, in low voice/ murmurs, in a loud voice, and even in foul language (10% of cases). The spoken words are difficult to understand because the phonation mechanism is poorly regulated.
The neurobiological mechanism needs to be better studied. If somniloquy occurs in NREM, REM intrusion likely plays a role because dream only occurs in REM. If the REM, RBD is the mechanism because in REM , voluntary muscles are atonic.

In NREM, there is no Consciousness; the somniloquy content is not remembered.Therefore, it likely that Somniloquy ocurs inNREM with high level of DOPA (pyramid OABD)

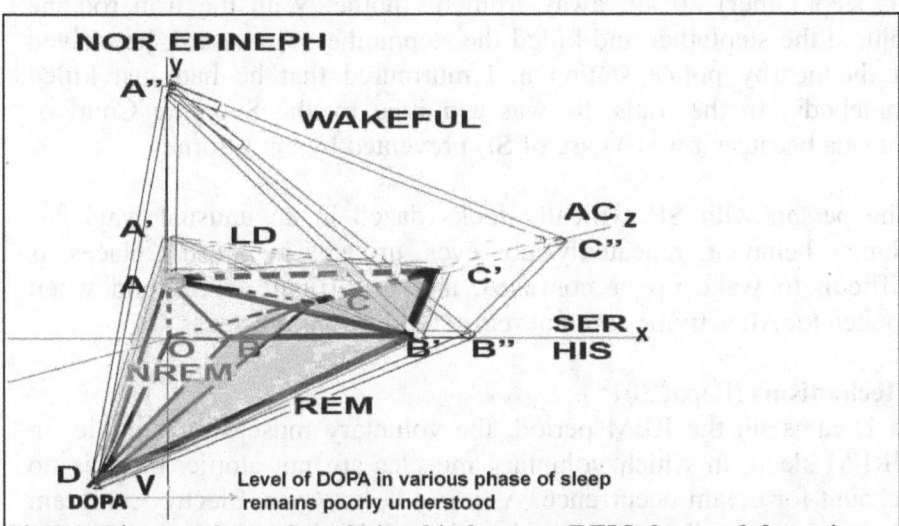

Fig 9.1: Diagram **shows the relationship between REM sleep and dreaming as being conscious (under action of Acetylcholine) inhibited by DOPA. The wakeful state is regulated by** Histamin Acetylcholine, Serotonin and Norepinep.
Polygones: Clear thick line: A'A"B'B"C'C"D: Wakefulness,
Polygone A"B'B"D Wakeful ness with willingness/determination
Pyramid A'A"B'B"D: **Daydream:**
Clear thin line AA'B'C': Lucid Dream (yellow)
Pyramid AA'B'D: LD with voluntary movement
Black thick line AB'CC'D REM sleep (blue)

Black thin line OABCD: NREM sleep (pink).
Pyramid ABB'D Dreams with strong physical reaction at the end of the dream
Pyramid: OABD Somniloquy

Note: Lucid LD cubic polygon is entirely within the Wakefulness cubic polygon.
Positions of B' and C' vary according to the depth of REM and the Consciousness
Ox represents Serotonin/Histamine
Oy Nor epinephrine, Oz Acetylcholine, Ov: DOPA

Dreams with strong physical reaction at the end of the dream likely occurs in REM with sudden increase of DOPA

Daydream: a stream of consciousness that detaches from current, external tasks when attention drifts to a more inner CS, resulting in a pleasant visionary and wishful creation of the imagination. Attention is displaced from the external world. The intentional drifting is likely associated with a rather increase in DOPA than epinephrine. DOPA accounts for the joyful impression (pyramid A'A"B'B"D)

C. Somnambulism/ SB. Sleepwalking.

An example: On the night of 23rd October 1987, Mr. KP woke up, went around the house, then drove the car to his stepfather (husband of his stepmother) 20 km away from his home. With the Iron rod, he injured the stepfather and killed the stepmother. Afterward, he arrived at the nearby police station and murmured that he had just killed somebody. In the trials, he was acquitted by the Supreme Court of Canada because it was a case of SB prevented by his Attorney.

The person with SB typically looks dazed in an unusual way, has clumsy behavior, repeatedly rubs eyes, urinates in unusual places, is difficult to wake up or confused, and is difficult to respond when spoken to. All activities are not registered in Consciousness.

Mechanisms (Popat 2015).
In Dreams, in the REM period, the voluntary muscles are atonic. In NREM sleep, in which voluntary muscles are not atonic, there is no account for dream occurrence. As a result, however, Electrooculgram has not been available to elucidate the mechanism of SB. As a result, the mechanism remains hypothetical
· Poor or absence of the involvement of the vmPFC that is responsible for the morality components of the behavior· Role of RBD
responsible for dreams in SB and muscle activity in NREM
· Role of REM intrusions (fro dream occurrence)
· dlPFC not working in SB

· questionable analogy to Charles Bonnet syndrome
-Local sleep in which only part of brain is inactive for sleep

All the above mechanisms can not account for phenomena in SB just because in a dream, the muscle is atonic and the highly unusual long duration of RBE or REM intrusions> Furthermore, these sequence of the events. Activities in SB appear to be more orderly and rational than in ordinary dreams. Furthermore, SB demonstrates some similarities with DID/double personality like:

- Implicit Memory like driving and eating not affected
- Amnesia
- Coordinated behavior and activities that are not related to the personality of everyday life
- History of mental neurotic disorders

Therefore SB may represent the DID only occurring at night during sleep. The person wakes up at night with the altered ID/ personality in a wakeful state

D. Peduncular Hallucinosis (Fig 10.3).

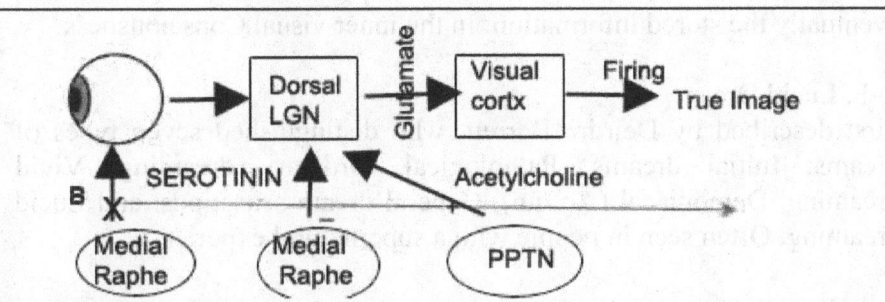

Fig 10.3 A Normally, the peduncle inhibits the pathway to make the pathway exhibit the image.
 B lesion of the peduncle causes the disturbance in the exhibition of an image resulting in hallucinosis."
The visual pathway to LGN is inhibited by Median Raphe and stimulated by PPT> V1.
In Hallucinosis, decreased Serotonin is unable to inhibit LGN.

The phenomenon occurs in the wakeful state, commonly at night, and represents the visual hallucination in which the images and scenes of concrete objects and living beings are colorful crispy vivid, smaller in size, and non-threatening. These features are characteristic and distinct from dreams and psychiatric hallucinations. The conditions are very rare, often self-limiting, and are usually attributed to lesions of the reticular system in the midbrain, thalamus, and basal ganglia. It is believed that the visual input is filtered and partially inhibited in normal conditions.

Patients are often associated with sleep disorders that are also a manifestation of the reticular ascending system (inhibition: serotonin and facilitation: Acetylcholine).

Patients of delirium tremens, narcolepsy-cataplexy syndrome, Lewy body dementia, Parkinson's disease, and temporal lobe epilepsy are prone to develop peduncular hallucinosis.[3] Peduncular hallucinosis is more common in patients with a long duration of Parkinson's disease.

E. **Charles Bonnet syndrome.**
The syndrome represents the visual illusion in a patient with a long history of visual loss due to optic nerve or visual cortex lesions. The illusions respond to the treatment with anticonvulsive drugs like **Phenytoin**.
. It is believed that the illusion develops as a compensatory mechanism for the loss of visual inputs. The retrieval of visual images or scenes represents the role of the brain as a box of prediction. Without an eternal information source, the brain keeps searching for input and, eventually the stored information. In the inner visual Consciousness.

F. Lucid Dream
First described by Deirdre Barrett, who distinguished seven types of dreams: Initial dreams, Pathological, Ordinary dreaming, Vivid dreaming, Demoniacal (Ác qui), General dream-sensations, and Lucid dreaming. Often seen in people with a supernatural experience.

In LD:
- Awareness of the dream state (orientation).
- Awareness of the capacity to make decisions.
- Awareness of Memory functions.
- Awareness of self.
- Awareness of the dream environment.
- Awareness of the meaning of the dream.
- Awareness of concentration and focus, subjective clarity of that state.

In 1992, Deirdre Barrett added more criteria:
- The dreamer is aware that he is dreaming.
- Objects disappear after waking.
- Physical laws need not apply in the dream.
- The dreamer has a clear Memory of the waking world.
- LD occurs in a wakeful state; therefore, the Memory of LD is intact.

LD usually happens after the age of 20 years old, at an incidence of 55% in the population, with more than once in 25% of the entire life, and lasts for 15 min.
LD often occurs when falling asleep in a lucid state; therefore, the LD dreamer can control or actively control many events in the dream. The duration of LD is equal to the time in life.
, In addition, LD can increase creativity, thinking.
during the LD because one can suppress unwanted stimuli in LD and focus only on the subject like writing, planning.

LD occurs in REM sleep, with beta waves at the parietal/ occipital lobe. These EEG changes suggest that Consciousness accompanies LD that involves the corpus callosum, AMGD, Parietal, Occipital Cortices, and dlPFC. HIPPO is not much activated n LD. Galatamine can induce LD.
Laberge thinks that LD is composed of episodes of micro awakenings with the presence of REM... For Hobson, LD consisted of two states: Sleep and wakefulness.
Gamma waves of 40Hz and decreased delta waves are recorded in the frontoparietal cortex and absent in non-LD REM sleep. Gamma waves are also recorded in CA1-3 of the HIPPO in non-LD REM sleep. Lesions of the above areas may cause LD.lesion of the

Thalamus are not associated with LD. However lesions of the intralaminar nuclei of the Thalamus or the activating ascending reticular system may result in LD

Drugs inducing LD are acetylcholine, galantamine (inhibitor of acetylcholinesterase.

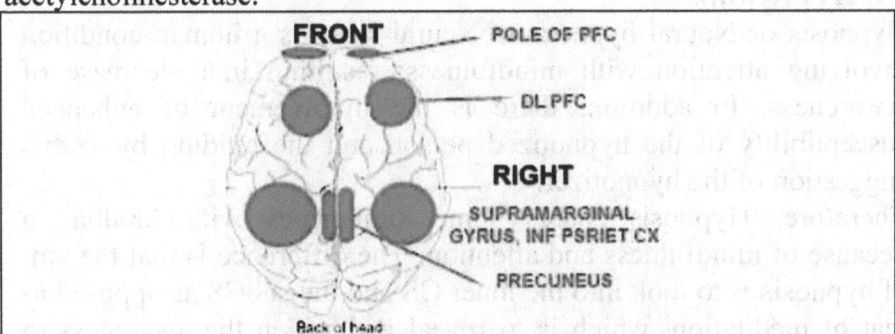

Fig 10.4 fMRI LD is associated with areas of the cortex, with hyperactivity bilateral PFC, and Precuneus and inferior Parietal lobules (angular and supramarginal gyri) Baird 2019, Christoff et al., 2003; McCaig et al., 2011, Dresler 2012)). Recently, some investigators have used LD to treat nightmares and design as one in the LD is aware of being in a dream and has the volition to modify the content. This is done because, in LD, the vmPFC and the dlPFC are

> equally activated, along with the TPJ and the Precuneus cortex. In patients with PTSD, therapy is effective. The quality of sleep is much improved with fewer nightmares https://www.ncbi.nlm.nih.gov/pmc/articles/PMC7471655/#

Fearful emotions and despair often accompany nightmares or bad/ unpleasant dreams in REM sleep in children and adults under stress PTSD, and high fever: inappropriate sleep position and certain drugs. The imagery mechanism plays a key tool. (Please see the chapter on Consciousness). In bad dreams, the dreamer usually does not wake up, unlike in nightmares. A night terror is distinguished from a nightmare since night terror occurs in NREM sleep, turned into early REM sleep, reaching the stage of atonia. Children with night terrors are awakened in a frightening state..

G. Other states of dream.
a) Hearing sounds in the dream.
b) Omen in dreams,
c) Vision of dead relatives.

H. Ghostly vision in a dream.
https://en.wikipedia.org/wiki/Greenbrier_Ghost
https://www.youtube.com/watch?v=44UosbwFRMs)

The phenomenon is often documented in newspapers rather than in medical or scientific journals. It is different from common types of dream because the content is not retrieved from the stored Memories in the inner CS but from an external source
Consciousness is inserted into the Memory system during REM sleep

I. HYPNOSIS.
Hypnosis or Neural hypnotism/ Neural sleep is a human condition involving attention with mindfulness, resulting in a decrease of Awareness. In addition, there is the involvement of enhanced susceptibility of the hypnotized person and the guiding by verbal suggestion of the hypnotizer.

Therefore, Hypnosis shares some similarities with Meditation because of mindfulness and attention. The difference is that the aim of hypnosis is to look into the inner CS and the subCS as opposed to that of meditation, which is to reveal and widen the awareness to approach the knowledge as close as possible to the Ultimate Omniscience.

In human history, hypnotism and mediumship are similar in their mysticism. According to Franz Meisner, a German physician,

hypnotism is popular (Hypnosis was known at the time as 'Mesmerism'). Hypnosis represents a mystical force that flows from the hypnotist to the person being hypnotized. Others suggest that "animal magnetism is involved (intangible or mysterious force).
The hypothesis was categorically rejected nowadays. Nevertheless, Hypnosis has been used in medicine by many researchers as behavioral therapy to treat acute and chronic disease like pain management, insomnia, DID, anxiety in IBD, and hot flashes.... other types of Hypnosis is stage hypnosis for entertainment, self-hypnosis (the person is hypnotized by oneself (treatment of stress, obesity by following a type of diet...), music hypnosis, brainwashing for committing crime, intelligence in military, forensic purpose...

B Techniques: Induction to induce the subject into a trance that depends on the susceptibility of the subject (classified into scale Subhypnotic -Full hypnotic -Hypnotic coma) induced by the hypnotizer's suggestion

The most common technique is Braidism, asking the subject to look at an object like Left fingers held at a distance. The suggestion aims to fix the Consciousness to a single idea or by verbal suggestion.
Therefore, the subject CS, i.e., is narrowed down to a simple object similar to meditation. As a result, the inner CS and SubCS are retrieved during Hypnosis.
In the long term, this will create a type of reflex that Braid called an ideomotor reflex response, ideo-dynamic, psycho-physiological" (Mind-body), or mono-ideodynamic **(based only on attention to a single idea).**

Neuroanatomy.
Increased activities and connectivities of ACC, PFC, dlPFC, IPS, Precuneus, PCC...The Thalamus is activated in Hypnosis. These cortices play the role of management, Inner CS for mental imageries.

Relationship with Sonambulism and Hypnosis.
The French Neurologist Jean-Martin Charcot played a critical role in legalizing Hypnosis in 1882, and Charles Richet, Nobel Award in 1913 has, coined the term meta psychism" for parapsychological research on Hypnosis:
"Metaphysics is not yet officially a science, recognized as such. But it is going to be. . . . At Edinburgh, I was able to affirm before 100 physiologists that our five senses are not our only means of knowledge and that a fragment of reality sometimes reaches the intelligence in otherways. .

Chapter XI: MORALITY, FOUR IMMEASURABLE MINDS OF VIRTUES, EMPATHY, ALTRUISM, THREE POISONS (GREED/ ANGER/ IGNORANCE).

I. MORALITY.

Represents Codes of virtual conduct according to Religion, Philosophy, and Culture/ Society. Morality represents the correct path of conduct for a person to follow. The Ultimate pathway is the Tao, which can not be expressed in terms, but the conduct to follow is possibly defined. Since the fundamental principle of the world is based on duality or multimodality, the codes vary with each group of human beings.

Therefore, Morality is not fixed. What's considered acceptable in one culture might not be permitted in other cultures in different geographical regions, religions, and families. In addition, they usually shift and change over time.

In Far Eastern culture (Vietnamese, Chinese...), the origin of the pathway is important for understanding the successive part of the path. Once the origin is well defined, the phenomenon that happened or is happening can be radically understood. Further, the phenomenon that will happen may be predicted. As a result, much emphasis is focused on the origin of the pathway known as the Tao. The term is commonly used in Lao Tsu Taoism. Unfortunately, Tao can not be expressed in terms. It is
represented by a state of Fuzziness in Lao Tsu's Taoism, formless Void and darkness upon the face of the deep in Christianity, Emptiness in Buddhism, and Fullness in Hinduism (Upanishads). These various terms are used to express the origin of the creation. The creator is also known as God, Holy Spirit, and Buddhahood. As characterized before, the origin is uncreated, unperishable, and permanent. God is omniscient, omnipotent, and ingenious in creation.

Morality is the expression of the Tao. Morality in this definition is unique in its phenomenal expression regardless of Christianity, Buddhism, or Taoism..., since this type of Morality is from the level of Omniscience (of the Creator)

On the other hand, the origin of the path of conduct is not as ultimate as in major religions, such as in the case of the path of conduct in a society or schools of thought. In this case, the codes of virtual conduct are created based on the principles of equality, fairness, and non-discrimination conceived by the group. The establishment of the codes is performed and varied according to:

- Time
- Space
- The status of consciousness, collectively from the group, plays a significant role in shaping codes of conduct. Consciousness, which varies from one individual to another, is a key factor influencing the establishment and variation of these codes. This understanding can enlighten us and provoke thoughtful consideration of the variations in conduct between individuals and between groups.

A. **Morality in major religions**, with the revelation of the Creation and Morality as the logistic corollary of the Creation regarding the concept and the fact.

1. BUDDHISM:

FOUR IMMEASURABLE MINDS OF VIRTUES, Appamanna, Brahmavihara (p)/Tứ Vô Lượng Tâm (vn)
Composed of:
-Loving-kindness/ skt: Mettri/ Metta/ vn Từ. Love dedicated to all sentient beings
- Compassion, pity/ skt:Karuna/ vn Bi, to save from suffering, the impermanence of life, Empathy, helpfulness, Donation,
- Inner joy/ skt Mudita/ vn Hỷ, sharing joy with other sentient beings,
- Detachment/ skt: Upeksha/ vn: Xả: concerning physical and emotional aspects of life Perfect equanimity immeasurable. No authority/ Egolessness, Forgiveness

The detachment is crucial and must include the self/ Egolessness. Human beings and the remaining Universe are born by the Supreme Level/ the Emptiness/ God/ Buddhahood with transcendental Awareness/Omniscience. Therefore, Humans are children of "God," similar to children in the family. A child is privileged to survive, including sheltering, feeding, clothing, and schooling proportional to their needs. He owns nothing in the family and has to follow his parents' instructions. The privilege is followed by duty and obligation of non-violation of other living sentient. As a result, Buddha taught the disciples about the Three Dharma Seals/ Three marks of existence, or three characteristics of all phenomenal existence.
- Impermanence/ Anicca (p) this Universe, and everything in it is created, therefore perishable.

- No-self—Egoless/ Anatta (p) Human beings are children of God. Only God I has the Self, but to teach his children, he selflessly acts.
- Suffering/ Dukkha (p) Suffering because of ignorance, mistakenly considers life and his belongings, including the physical body, as permanent and as parts of the owning

Therefore, human beings are egoless. All natural resources belong to God. As children of God, all profits of their work belong to God, who gives them a part proportional to his work. In a company or a family, the profit surplus, although at the company's or family's disposition, belongs to God. As a result, the profit surplus must be used for the benefit of the country, community, or other needy people, and giving out to other people does not represent donations by the donor but gifts on behalf of God. The feeling of agency (false Ego) and ownership develops along with the consciousness residing in the Default Mode Network (DMN), which tends to deviate living beings from the mind of virtue. Additional component is the vmPFC playing the role of Morality and Empathy. Therefore therefore for rich people, plenty of assets in surplus constitute a burden for life and may cause bad karma when the asset is inappropriately managed or abused

An example of detachment, egolessness:

Cranes fly in the sky, casting shadows in water Neither do the cranes intend to leave traces, Nor the water aims to keep them (Zen master Hương Hải, translated into English by Việt Phạm)
(vn: Nhạn quá trường không ảnh trầm hàn thủy, Nhạn vô di tích chi ý, thủy vô lưu ảnh chi tâm.

Buddha said, in Udan VIII.8:
those who have a hundred dear ones have a hundred sufferings. Those who have ninety dear ones have ninety sufferings. Those who have eighty... seventy... sixty... fifty... forty... thirty... twenty... ten... nine... eight... seven... six... five... four... three... two... Those who have one dear one have one suffering. For those with no dear ones, there are no sufferings. They are free from sorrow, free from stain, free from lamentation,

-- **Five precepts in Buddhism**: Abstain from 1) killing, 2) stealing, 3) sexual misconduct, 4) lying, 5) substances abuse.

The recommaded path of Morality is spiritual practice mainly represented by the Noble Eightfold Path: RIGHT

1. **Effort (self-motivated)** According to Buddha self initiation for Bohdi is crucial in the spiritual path. For those who is initiated secondary to a depenent cause, the path to attain enlightenment is for ever as long as many eons (in Surangama Sutra)

2. **Remembrance** (memorizing and retrieving good memory, The memory in the Inner Consciousness is essential in creating action speech thinking view...

3. **Livelihood (practice of Egolessness)**, Egolessness is the trademark of Buddhism. Egolessness is accompanied by No Ownership (the soul /mind, body, all assets that belong to the Creator

4. Thought (Thinking),

    Right Inner consciousness (memories) is prequisitwe for thinking, speech, action and view)

5. Speech,

6. Action (work).

7. View (understanding),

8. **Mindfulness meditation** that will lead to the Truth realization (enlightenment)

2. CHRISTIANITY

Morality is represented by the Atruism evidenced by the Charity. The term Charity comes from Caritas (carus) meaning immense love. Matthew 22:39:" *'Love your neighbor as yourself.'*
The 10th commandment [1) no other gods, 2) no idols. 3) not taking the name of the LORD in vain. 4) Remember the Sabbath day, 5) Honor your father and your mother. 6) not murder, 7) no adultery, 8) no stealing. 9) no false witness, 10) no covet (greed)].
wrote: Example Selflessness is extensive in Jesus teaching mostly highlined by his crucification as a symbol of salvation of the humanity sins. To apostles, Jesus said (*Lc 14,11*):
For everyone who exalts himself will be humbled, and he who humbles himself will be exalted."

3. **TAOISM** is characterize by the non-action, no fighting, non-suppression of the others

...The more it move the more it yields, more word count less
the sage stays behind, thus he is ahead, he is detached, thus at one with all, through selfless action, he attains fulfillment

(attention: Taoisn could be nor considered as a religion because the lack of the transcendental revelation and the absence of discussion of life after death.)

MORALITY IN SOCIETY, ETHICAL GROUPS: BASED ON THE CONSCIOUSNESS
1. NEUROSCIENCE and MORALITY

The ventromedial PFC plays a crucial role in the development of Morality, as exemplified by the case of Phineas Gage. He is a correct, reliable, and friendly construction worker in a railroad company. After being injured with an iron rod perforating the skull and the vmPFC, he became an immoral person in his conduct at work. Neuroscientific studies of the PFC reveal that vmPFC is connected with the dorsolateral PFC, playing the role of the overall activity of the PFC as an executive manager. dlPFC is connected with the intraparietal sulcus IPS, considered the central commander of emotion and motility, in particular in this case with Amydal (fear center) Nucleus Accumbens (joy center), Periaqueductal Gray that is related to the parasympathetic system (Figure below)

- Piaget theorized that moral development develops over time, in certain stages, as children learn to adopt certain moral behaviors.
- B .F. Skinner's behavioral theory: Environment forces shape an individual's development.

2. Morality and Philosophers

a) Philosophers used the most selective Consciousness for their theories, in contrast to major religious revelation that is transcendental and omniscient. The revelation is original and is not been obliterated or distorted by the human brain. The brain is a filter system for the original incoming data that needs to be connected with emotion and motility centers for communication with other sentients. Therefore, this filter is considered a veil of ignorance. Of note, ignorance is not non-intelligence but the deviation of the knowledge away from metaphysical areas or areas of interest

b) Morality developed by the duty. Duty is secondary to the privilege of survival, which the Creator gives. This survival privilege must be in harmony with the privilege of the other sentients. In an ethical group, the duties and privileges are governed by regulation, which constitutes the codes of Morality of the Group.

c) to compensate to this deficiency and bias, Emmanuel Kant proposed that, inaddition the the empirical knowledge, there is a priori knowledge (or experience that is possibly transcendental (or or influenced by karma of previous lives-according to this author). As a result, Morality may also be related to the metaphysical realm. Since philosophers are rarely referred to or are unaware of the concept of karma of previous lives, the commandment of Morality is not compulsory.

d) Confucianism
The concept of the descendent order of the King, Teacher, and Father is inappropriate nowadays as the King is no longer considered as the heavenly Son. The doctrine of rectification of names/ rectitude/righteousness means that conduct appropriate to one's own role to one's condition.

e)Free Will (see p)

f) Morality and Atheism
Atheists differ from philosophers in that they deny supernatural spiritual power or God. Morality codes derive from reality and are accessible to consciousness.
Morality and Legal system is based on complex copy-written codes.

4. Conscience is a popular concept. Conscience is a product of the brain of each individual. Therefore, conscience is different from Atruism-Charity and Four immeasurable Virtue. It is also different from philosopher-related Morality (created by the Consciousness of the philosopher or collectively in the ethical Group). Since Conscience is the product of the brain. It is also a special type of inner Consciousness that has a higher level of Morality. Inner Consciousness is a model of CS specific to each individual, and a referral model is used to compare with incoming data. Data similar to each type of memory in the IC is labeled according to the model memory. Identical to other memories in the IC, conscience is formed from birth and is continuously upgraded
According to Mencius, humans are born with perfect and pure character. It is true for primitive humans like Adam and Eva before

they committed sin by following the serpent's seduction and then acquired the discriminative mind. For newborn babies later in the Creation, the babies are always associated with karmas from previous lives therefore, their minds are not originally pure.

Morality with a common feeling of being good is the lowest level of Morality. For example, some people say that there is no need to practice spirituality. What to do is complete the assigned duty to get the pay and behave well to everybody. However, it is difficult to determine how much is complete or good.

5. Morality in Ethical Group. Each Group usually has its code of Morality according to the environment type works. The Captain of a ship or General, commander in chief of a city, sometimes satisfy themself when the ship or the city is lost. Suicide may contradict the code of survival of the Creator in major religions. Karma may play a role; however, in these situations of contradiction, karmic law is usually overrun by the morality law dictated by the Creator.

6. Morality and Science

Since science is known for its methodology, categorization, and accurate prediction of future events, it has been useful in establishing morality codes. However, science deals with the physical aspect of life and related theories, whereas Morality involves both the physical and metaphysical aspects. Furthermore, science can not go beyond the quantum level, representing no man's land. The advance of science is associated with the gradual distancing of humans from Morality and even the denial of the metaphysical realm. Krishnamurti said, a few decades ago:

The crisis is there. The crisis is not in the world, it is not the nuclear war, it is not the terrible divisions and the brutality that is going on. The crisis is in our Consciousness, the crisis is what we are, what we have become."

THE PSYCHOPATH is a person who lacks empathy and is often criminal, law-breaking, non-loyal, and violent in behavior with low activity vmPFC. AMYG's decreased activity causes low secretion of Oxytocin from HYPO. Children with Low AMYG activity become stubborn and are associated with poor academic achievement. The bravery is often associated with an increased volume of AMYG, BG, NAc, VTA (nuclei producing DOPA)

Heredity accounted for 40-70% of Psychopathy. Oxytocin and Vasopressin receptor genes are altered in Psychopathy.

B. Neurobiology (Fig 11.1).
Example of Phineas Gages: After the injury to vmPFC, PG is transformed from a dedicated model-conscious worker to an immoral, irresponsible intolerable worker.

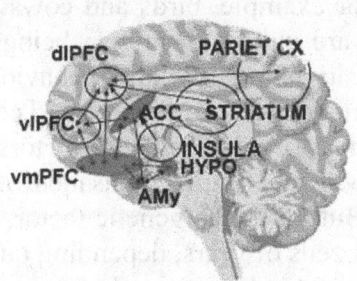

Fig 11.1: Neurobiological pathway of Morality consists of the connection of vmPFC and dlPFC other cortices and nuclei.
(reward, fear, Insula, BG, ACC, PCC (Marazzi 2013)
vlPFC: RT side to override or inhibit motor response, a component of the spatial attention system
Circle: lateral cortex

Morality is based on the concept of the Tao, the principle of Creation of the Universe that is governed by Consciousness, not by Transcendental knowledge/ or Spirituality. The highest center of Consciousness is the vmPFC, which is not usually the administrative center but significantly influences the administrative/ executive center, the dlPFC. The other centers are the Insula –ACC constituting the salient network for
impulse in making the will, the inner Consciousness, the store of MM of the present life, especially the PCC, the autobiographic Memory, ventral striatum NAc controlling the joy, and AMYG controlling the fear and anger.

Of interest, the Insula, located deep in the Sylvian fissure, is composed of three parts:
Dorsal Anterior connected to ACC : role in CS
Ventral Anterior to ACC : Emotion
    Posterior: connected to motor cortex: reaction to Taste Disgust (bad taste), feeling and feeling of internal organs.
Therefore, the insula plays the role of interface between the logic. CS is deep emotional feeling with an expression of the face, by controlling

eyes, nose, lips, forehead, cheek extremity muscles, secretory lacrymal glands of the eyes and nose to express the feeling.

Phylogenetically, the mechanism of Morality is one of the last development. It's safe to say that the gap between chimpanzees and human beings is marked by the development of Morality with the presence of the Orbitary PFC (OPFC that includes vmPFC). Animal studies have documented the development of emotions and love empathy in animals. For example, birds and cows show feelings, but Morality and creativity are unique in human beings. Damage of the OPFC causes hypersensitivity to immoral behavior with an adverse reaction like rage and strict concept of the violated codes of Morality.
In the development of Morality, genetic factors are essential as epigenetic factors. It will take many thousands of years for genetic factors to be acquired. But for the epigenetic factor, the timing is much shorter, about years to dozens of years, depending on the species.
In epigenetic adaptation, the histones play the supportive role of chromosomes and genes. Alterations of histone will make changes in gene expression. (Chen 2015, Rodgers 2015, Gapp 2014, Rassoulzadegan 2006, Sharma 2015, Han 2019, Hans 2019, Liu 2018, Dias 2016, Galton 2016, Bernacer2014).
The lateral and superior PFC, including the dlPFC, play the role of controlling and inhibiting the activity of using logistic reasoning.

Furthermore, it is noteworthy to mention that the regression of the codes of Morality inversely accompanies the progression in knowledge in science and technology. This is evidenced by many small and great wars in the past two centuries

C. Neurohormones in Morality
- Oxytocin made from HYPO from the mother is transported to the pituitary gland before being released into the bloodstream. Role in the milk secretion from the breast, contraction of the uterus, and enhancement of the Mother-baby love, feeling of generosity.
- Prolactin from both parents' pituitary glands. There is no action on the uterus, but Prolaction also exerts on the breast for milk production in the mother.
- Serotonins can not cross the Blood Brain Barrier. It is also produced by the brain's Dorsal Raphe nucleus and extracranial glands, mostly in the gastrointestinal tract. Effects: Joy, pleasure.

II. FALSE TAGGING THEORY(Asp 2013) (Fig 11.3).

As a neuroanatomical network/ homunculus, PFC plays a unique function in the belief and doubt (False tagging) in the conception and execution. In addition to the PFC, the working MM and the dlPFC are involved in the decision of belief or non-belief.

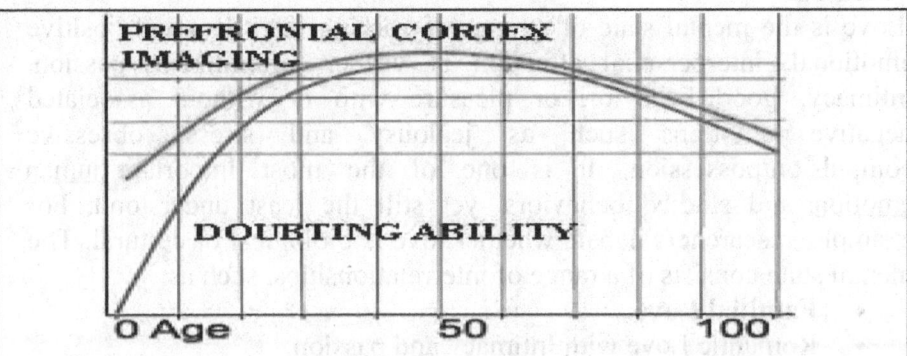

Fig 11.3
The doubting ability versus Prefrontal cortex degeneration varies with ages as highlighted in the diagram

III. SOMATIC MARKER THEORY.

Emotion usually plays a vital role in the decision-making. The sentiment is represented by the physiologic changes like respiration, heart rates, muscle tone, and endocrine secretion, recorded in vmPFC, HIPPO, NAc, and AMD as models for each type of emotion.

The decision-making of complicated problems often includes the Consciousness composed of motivation/ purpose, intention, and rationality, which are based on the management involving dlPFC and the emotion based on the vmPFC. The balance between Consciousness and emotion will result in the execution of the action paradigm.

IV FOUR IMMEASURABLE MINDS OF VIRTUES
See above, page 371
Zealous Atruism (Good Samaritans) (Sonne 2018).
Children 4-7 months start developing the theory of Mind, helping them to guess the Mind of other people and develop interrelationships with parents, family, and society. After the age of one, they learn to help other people.

In orphanages, children show poorer development of this capacity, with low activity of oxytocin receptors at Striatum, NA, VTA and Substantia Nigra (SN).

Narcissism, opposite to the mind of Virtue, associate with alteration of vmPFC and increased DMN activity

In summary, mind of virtue, zealous atruisn, narcissism represent the interplay between vmPFC, DMN and love hormones from NA, VTA and SN

V. LOVE.

Love is the mental state of strong (of varying intensity) and positive emotional, interpersonal affection of virtue, commitment, passion, intimacy, good behavior, or pleasure with or without associated negative emotions such as jealousy and stress, obsessive compulsion/possession. It is one of the most important human emotions and studied behaviors, yet still the least understood. For example, researchers debate whether love is biological or cultural. The mental state consists of a range of interrelationships, such as:

- **Familial Love.**
- Romantic Love with Intimacy and passion.
- Self- love.
- Divine love.
- **Friendship**: liking each other with varying degrees of intimacy.
- Infatuation often occurs early in a relationship with intense feelings of attraction without a sense of commitment.
- **Passionate love**: intense feelings of longing and attraction with physical closeness.
- Compassionate/ companionate love: This form of love is marked by trust, affection, intimacy, and commitment.
- Unrequited love: This form of love happens when one person loves another who does not return those feelings.
- Consummate love, xenia love....

Zick Rubin's Scales of Liking and Loving.
- **Attachment**: Needing to be with another person and desiring physical contact and approval.
- **Caring**: Valuing the other person's happiness and needs as much as your own.

Intimacy: Sharing private thoughts, feelings, and desires with the other person.

1. Romantic love.

Romantic love is one of the most frequent subjects in human literature and art. However, scientific study only began late in the second half of the 20th century. Ventral Pallidum Nucleus Accumbens,The Limbic system with Ventral Tegmental Area/VTA, Striatum, and nucleus accumbens uses DOPA, Vasopressin, and Oxytocin (Pituitary gland) as

neurotransmitters. The two later neurohormones also play a role in childbirth delivery and
lactation/ breastfeeding. In addition to these neurohormones, Serotonin (Dorsal Raphe), Dynorphin/ endorphin, and a mechanism of endogenous morphine + nitric oxyde Endorphin, and a mechanism of endogenous morphine + Nitric oxide are produced.

When there is strong emotion: The Hypothalmus-Adrenal Axis/ HAA with elevation of cortisone. They are often associated with survival, entertainment, feeding, reproduction, and the treatment of various illnesses. Addiction with decreased CS can be accompanied by chronic elevation of these neurohormones, particularly endorphin.
The insula connected with ACC likely plays an essential role in driving the love in controlling the PFC, Temporal, and Parietal cortices, Dorsal Striatum with related cortex, and Basal Ganglia Caudate Nucleus and Ventral Putamen.
 - Hippocampus related to "our love story".

b) Cortices with decreased activity
 - Dorsolateral Prefrontal cortex (dlPFC) and intra-parietal sulcus (IP Sulcus) involved in management and logistic thinking. The decrease is accompanied by a decrease of rational Consciousness, discrimination accompanied by the regardless consideration of the situation, time, and duty often encountered in love stories
 - Amygdala for fear.
 - Middle Temporal Lobe for emotion, romantic expression.
 - vmPFC (Theory of Mind).
 - Orbital PFC social interrelationship, morality.
 - TemporoParietal Junction/ TPJ mental imagery.

c) **Orgasm** is a particular state of altered CS but seldomly discussed in public and studied in neuroscience (with fMRI and PET). It is accompanied by a decrease in pain, fear, and impulse control. In some people, it can be triggered by just thinking about it.
Inability to experience or low sensitivity to orgasm is reported
Brain areas activated are similar in both sexes: related to sense touch (corresponding to the clitoris, followed by vagina and cervix in women; glans and penile shaft in men. Secondary cerebral areas involved are the Limbic system and PFC.

2. **Parental Love** (Fig 11.5).
Maternal (associated with delivery and lactation) and romantic loves share some similarities regarding neurohormones, subcortical nuclei,

and cortices. In Parental Love, the temporal cortex is more crucial in emotional expression, face recognition, and caring for the physical body of children. TPJ and the lateral PFC in empathy.
Hypo is less activated.

Paternal love differs slightly from maternal love because of the lack of delivery and breastfeeding. However, this lack does not make much difference because the neurohormone secretion is similar to Oxytocin, Vasopressin, and Prolactin.

Testosterone decreases when the father nurtures the child, and Prolactine increases during childbearing (by the mother's child). Although children are not created by parents, as pointed out, parents have the moral obligation of nurturing, loving, and guiding children.

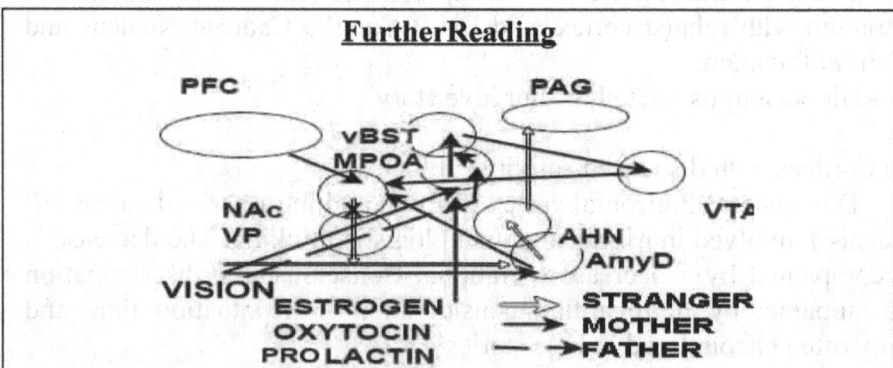

Fig 11.5 Mesolimbic Reward Pathway Nucleus Accumbens/ NAc: Nucleus Accumbens: VTA: Ventral Tegmental Area: joy
VP/ Ventral Pallidum: Joy, movement
MPOA/ medial preoptic area, related to optic nerve
PAG/ PeriAqueductal Gray, Indifference, AHN/ anterior hypothalamic nucleus.
vBST/ ventral bed nucleus of the stria terminalis, related to amygdala: fear, aggression

The diagram shows the different behavior of the Mother, Father, and Stranger.
   - Mother: PFC activates Nucleus Accumbens/ NAc and VP/Ventral Pallidum.
     Amyg activated
     Preoptic area and vBST/ ventral bed nucleus of the stria terminalis= similar to Amyg (activated by Oxytocin, Estrogen Prolactin, and the sight of the baby) activates VTA. VTA activates Nucleus Accumbens/NAc: Love and Joy.
   - Father: No mechanism of VTA, Nucleus Accumbens/NAc: Love and Joy
   - Stranger: Amyg activates AHT anterior Hypo and activates PAG indifference

3. Post Partum psychosis.

The disease occurs at a rate of 10% in mothers after childbirth. The clinical manifestations are depressive, obsessive-compulsive, anxiety neurosis/ psychosis. In animals, it is associated with the decrease of neuroplasticity (destruction and repair) in PFC, HIPPO, Nucleus Accumbens/ NAc, Amygdala, and disturbance of secretion of Norepinephrine, Serotonin, and Corticotropin-releasing hormone, GABA and Oxytocin.

Postpartum psychosis is different from non-postpartum (decrease of cognitive function of DLPFC, posterior cingulate, and precuneus/ cuneus) and increase pregenual PFC, ACC, ventromedial PFC, dorsal medial, thalamus, pulvinar, ventral pallidum/ putamen, ventral tegmental area (VTA), substantia nigra, tectum, and periaqueductal gray) (Alcaro 2010, Northoff 2011).

The disease also occurs in the father with a decrease of synaptic connection in the Orbitofrontal cortex, Posterior cingulate cortex, and Insula..

4. Filial Piety.
is considered an obligation in most cultures and religions, especially in Confucianism (551-479 BCE), because the lack of duty of children toward their parents is almost universal. Filial piety is also an essential component of Morality in familial relationship.

Birth and Nurturing gratitudes are almost the same (vn: Công Sanh thành và Dưỡng dục như nhau. Natural parents and nurturing parents only differ in the timing of nurturing. Natural parents nurture the intrauterine and eventually worldly life, and the Creator but not parents create the children. Foster parents only nurture the worldly life. However, there is a concordant wish between the Soul to be re-incarnated and the parents in the pathway of creating human beings.
  Contrary to parents, children mainly attach to parents based on intimate physical, mental, and moral relationships rather than the natural pathway as in parents.

EXAMPLE PARENTAL LOVE versus FILIAL LOVE
The story occurred at the time of Gautama Buddha, more than 2500 years ago. The Crown Prince Ajatasatru (skt) is son of King Bimbisara and Queen Vaidehi). He wished to overthrow his father to be the King. Knowing his son's intention, King Bimbisara abdicated to leave the throne to his son. After being King, Ajatasatru imprisoned his father and left him to starve to death, and only allowed Mother Queen Vaidehi to visit his father. He found out later that the father still survived because the Queen's mother pasted sugar molasses all over her body to

provide food. The mother was later imprisoned as well; in prison, the sufferance of Vaidehij was felt by Buddha. He appeared and showed them the Pure Western blissful Land. At the same time, the first son was born to King Ajatasatru, who adored his son so much. After death, King Bimbisara appeared to his son and advised him to repent of misdeeds and mental hindrances to Buddha. Later in his life, King Ajatasatru was a sincere Buddhist devotee who contributed much to Buddhist dharma. The story is the canon in the sutra of Amitaba Recitation in Buddhist meditation.

The example highlights the natural, almost universal, unconditional parental love (hardwired in the brain) versus filial love, which is often conditional and requires moral education and commandments.

5. The First Love.

From antiquity, the first love is one of the most impressive events in human life stories, evidenced in literature, music, and dramas.

In an experiment, the experimentee initially listens to a sound, then a second sound of a different frequency a moment later. The fMRI shows the first sound is associated with stronger changes than the second sound; even the second sound is stronger in intensity. The result means that the first sound exerts more influence on the brain than the subsequent sound. The mechanism of the phenomenon is that the brain is a box of prediction. It keeps the new information as reference data in the inner Consciousness. The subsequent closely similar events are compared with the first event in the process of CS formation.

As a result, sentient beings, particularly humans, are usually attached to their native place, mother, first song, first friend, first love...

6. Love and Beauty.

executive network dlPFC-IPS's role in controlling life's logic. Beauty is an essential factor in the relationship of romantic Love.. Love of the beauty is natural. This is highlighted in the following sermon in Nikaya, Long Discourse., 27 *(Agganna Sutta).*

(Aggañña Sutta: On Knowledge of Beginning) ...: *'There comes a time, Vasettha, when, sooner or later after a long period, this world contracts. At a time of contraction, beings are mostly born in the Abhassara Brahma world. And there they dwell, Mind-made, feeding on delight, self-luminous, moving through the air, glorious — and they stay like that for a very long time. But sooner or later, after a very long period, this world begins to expand again. At a time of*

expansion, the beings from the Abhassara Brahma world, [85] having passed away from there, are mostly reborn in this world.

Meaning: After the Creation of the Universe, sentient beings developed when eating the earth's products. Some sentients are good-looking, others are not. The discriminative Mind made the earth's products disappear....(A fact similar to the expulsion of Adam and Eva from Eden Garden after they developed the discriminative Mind.).

7.**Social Interaction** controlled by mPFC and consists of
- Social Influence,
- Conformity (change of opinion to go along with the majority in the group: informational confirmative and normative (to be likened by the group) and:
- Compliance with rules (set by the group).

8. Theory of Mind or mentalizing controlled by mPFC
Guessing the Mind of other people

VI. THREE POISONS: GREED, ANGER, AND IGNORANCE.

Greed/ Raga (skt) is the persistent attachment to the object associated with envy, passion, and ignorance. Greed is usually suppressed by morality. Greed manifested under Name (metaphysical part) and form (physical part). The form is composed of Killing, Stealing, and Sexual deviation.

Greed initially develops as an instinct of survival, a reaction to fear and anxiety. In an economic crisis, investors usually withdraw money to put in a safe account because of fear of losing more cash, in the principle of "Double coincidence of wants" (exchange of goods without the involvement of cash between seller and buyer). Like the generation of Consciousness, the brain must compare the new incoming information with inner Consciousness. For the investor there is a "Homo Economicus" characteristic for each investor. This network consists of Nucleus Accumbens/ NAc / DOPAmine (Reward system) and Hypothamus, VTA, and Amygdala as, in some cases, four immeasurable minds. Gambling, risky adventure, and cocaine amphetamine addiction also play a part in the process of Greed with activation of the NAc nucleus as seen in fMRI. (Kuhnen and Knutson (2005). Wickeret al. (2003), Wright et al. (2004). The concept is that testosterone may be a factor in the increased secretion of DOPA in Males and dictators.

Nuclei such as the Amygdala, NAc VTA, and striatum-producing DOPA are responsible for impulse/ compulsion and motivation. The decreased activity of mOPFC in the Greed is associated with activation of NAc /Ventral Striatum associated with joy (Seuntjens et al., 2015; Baumann and Odum, 2012; Lejuez et al., 2003; Pack et al., 2001; Zaleskiewicz, 2001(Barkley-Levenson et al., 2013; Kahneman and Tversky, 1979; Ko"bberling and Wakker, 2005; Tversky and Kahneman, 1992).

- Anger/ Aversion: acting coarsely or pitiless, creating sufferings with unimaginably destructive in cultivated path and conduct. Anger may have constructive effects on restoration and repair
Adrenaline and noradrenaline are accounted for most physical manifestations, including the heart, blood vessels, respiration, muscles

In anger, the top-down pathway cortex/ vmPFC-OPFC/ Morality, ACC/ detection of error-AMYG, Limbic system are connected to HYPO via stria Terminalis (Blair 2013). The mediodorsal of the Hypothalamus is activated with hormone secretion. The lateral HYPO may be accounted for a more violent reaction. The hormones are Serotonin, Catecholamin, Testosterone, Glutamate/ GABA (Jager 2017).

In patients with borderline personality, emotion is not inhibited by PFC (dorsolateral PFC, vmPFC, OPFC), Amygdala is overactivated, as evidenced by the decrease of connection between PFC and Amygdala (Volman 2016, Bertsch 2018). NAc connected to Medial Temporal Lobe and Posterior Cingulate Cortex/ autobiographic Memory (Sethi 2018, Kiehl, 2001; Maddock et al., 2003), social (Buckner, 2008; Vollm et al., 2006,(Greene et al.,2001; Harrison et al., 2008, Birstch 2018, Chen 2018)
- Ignorance/ Stupidity/ Delusion/ Unwillingness to accept Buddha- the Truth , skt: Mudhaya (p)—Moha (skt)—unintelligence. This is the most challenging component to recognize and constitutes the root of all deviation in life, including identifying the impermanence of the inflated Ego.

After the enlightenment, ignorance is identified and is considered as the beginning of the correction to implement the Four Immeasurable Minds of Virtue and eliminate the three poisons/evils in the path to walk the Tao.

VII FREE WILL AND KARMA.

1. Concept of Free Will and Libet's experiment (Fig 11.6).
Since antiquity, human beings have questioned whether they have the unrestricted right to act according to their will. The Orientals, under the influence of Hinduism and Buddhism, may have the opposite view of free will due to the concept of Karma and transmigration or samsara. This view of samsara is difficult to scientifically prove. Still, the concepts are very useful in understanding many spiritual phenomena. What happens now may reflect the results of past activities, including previous lives. Therefore, people reap what they sow, the activities partially control the benefits, and punishment can return to the remote past. Estimating how much a human being is free in daily mental or physical activities is difficult. Nguyen Binh Khiem, a highly respected Vietnamese Scholar (1491- 1585), said:" *Nine of ten wishes are not fulfilled.*
(vn: *"Thế nhân thập nguyện, cửu thường vi"*
Trong mười điều mong cầu, thì chín điều là không đạt).

Arthur Schopenhauer (1788 –1860), a German philosopher: Man can do what he wills, but he cannot will what he wills.",
The above quotes are for the theory of determinism.
All sentients, like the actors in the theater, live, act, and think according to the direction of the scriptwriter.

A significant number of Westerners, influenced by doctrines of existentialism and humanism, do not much believe in the existence of Karma and Samsara, decreased belief in religion and spiritual power (but the respect of spiritual faith and religion of others) usually hold the thinkings of the existence of the free will. Nevertheless, the infiltration of oriental wisdom has somewhat changed these individualistic views.

On the other hand, the experience of hardship in life and the spread of spiritual experiences by media also play a role in spiritual attitudes. Physicists, physicians, and neuroscientists considered the question of free will as controversial instead of denying it. In 1827, the first kind of meeting on Free Will, "The General Conference of Free Will Baptists," took place in Tennessee, U.S., and Seat Davidson County, Nashville. Until 2006, the official meeting on Free Will was organized in Sweden with the participation of neuroscientists and philosophers. A new specialty was created: the neurophilosophy.

2. In 1983, Libet performed a pivotal experiment that had a tremendous impact on the scientific community: The experiment was performed in vivo on a lucid patient who underwent neurosurgery under local anesthesia. The activities of the brain were recorded in the EEG. The experimentee was asked to perform a simple movement of the finger.

EEG detects the electrical changes in the brain, recording W (Will) moment of volition to act, M (Movement) moment of action, and RP (Readiness Potential) time of preparation. Period W-M=300ms, RP before M =1000msec, RP-W=700ms.

The results are that the changes in the brain precede the intention of movement (W), and the intention of movement precedes the movement (M). To Libet, the conclusion is obvious and straightforward: The will is preceded by another brain activity that is considered predeterminism. Predeterminism determines the intention of Free Will. Mark Hallett and Matsuhashi performed another experiment to determine the Free Won't. The Free Won't is also predetermined. Furthermore, the side (Right or Left) movement is also predetermined.

As a result, the impression of Free Will is illusional. Human beings have a false impression of Free Will and the permanence of lives, the world, happiness (free of sufferance), and the stillness of lives.

3. The neurobiological basis for the Free Will includes: (Fig 11.7).
· The Salient network comprises Insula involving volition and ACC involved in detecting an error in the incoming information.
The network receives inputs from the HIPPO, MTL, PCC (storing recent and semantic, autobiographic Memories).

· The central network specializes in the management, with hardwired connections of different areas of the brain, including areas for Memories, Consciousness, another mental status, and movement:
-vmPFC for morality, social interaction
- Basal ganglia including:
Dorsal Striatun, Dorsal Pallidus, Ventral Striatum= NAc and Ventral Pallidus. and VTA
Since the Libet experiment suggests that there is no Free Will, there have been so many discussions about the validity of the experiment's conclusion for the last four decades.

4. Opponents of No Free Will argue about the overall impression of Free Will in their actions, in the determinism of this macroscopic world, and the relativity in interpreting the result (Horst 2011, Roskies 2010).

5. The concept of Karma/ Samsara supports the proponents of supports the proponents of the No Free Will view. According to the Creation/ Genesis of the Universe in Buddhism, all the inorganic and organic materials, plants, and all sentients are created by the Creator, the personalized Ultimate or Original Mind, also called Ultimate or Transcendental Awareness/ Awareness. As a result, all sentients are children of God/ Creator. This concept is not different from the concept of Genesis in Christianity. Logically, as in a family, children depend on their parents at young ages. In the Universe, children of God are characterized by:
- Egolessness
- No ownership of any materials of the Universe,
- No Free will

But free to exercise the four immeasurable minds of virtue and set up Bodhi Mind to return home, the Nirvana.

All sentients have the conditional privilege of survival. In this conditional privilege of survival, they receive the benefits created in this word commensurable to their needs, including the development of creativity, the perfection of the Mind appropriate leisure time, and entertainment.

According to Nagarjuna in Mulamadhyamakakarika (Trung luận), the 14th patriarch of Mahayana Buddhism, when a sentient without self-nature (original nature. Note: Only The Emptiness/ Buddhahood has the original nature) acts without the guidance of the four immeasurable Mind is prone to create bad Karma. Simply because without self-nature, the sentient has no right for themselves. There is no right of action in this world of inversion/ upside down.

This phenomenon is similar to that in a family or in a company. The parents or head of the company have all privileges and power. Children or employees have to follow the guidance or the by-law. The benefits in the family or in the company are used for the member's benefit, commensurable to their needs or abilities. The surplus of the benefit will be served for development, research, and reserve funds.

One often witnesses the two components of the Consciousness involved in rivalry: the talent representing the manifestation of the Consciousness in the present life, and the destiny, representing the determinism dictated by the Karma or Consciousness of many previous lives. The famous Vietnamese tale of Kim Vân Kiều, written by the great Vietnamese poet, recites the life of Thúy Kiều, the beautiful and talented oldest sister in a family that includes a younger sister and youngest brother. Despite her beauty and talent, Thúy Kiều endures much suffering. This contrasts with Thúy Vân, who is inferior to her sister in appearance and talent but enjoys a good life. The author attributes this paradox to the destiny.

...In conceiving that everything is governed by the Creator
who attributes each human a form
To be doomed in the wind of dust, one must be in the dust
To be in a high seat, one will enjoy the prestigious place
No bias in favoring anyone
Despite much talent and destiny for sufferance
The talent can not be accounted
Since talent and accident often go along...
(vn:...Ngẫm hay muôn sự tại trời
Trời kia đã bắt làm người có thân
Bắt phong trần phải phong trần
Cho thanh cao mới được phần thanh cao
Có đâu thiên vị người nào
Chữ tài chữ mệnh dồi dào cả hai
Có tài mà cậy chi tài
Chữ tài liền với chữ tai một vần...)

(Note: Destiny represents the Karma of many thousands of life, while Talent only represents the good Karma of one life. Therefore Destiny often overrides the Talent.)
In this paradox of talent and destiny, morality contributes a part in the good future:
...When good seeds reside within the Mind
The heart outweighed much more than the talent...
(vn:...Thiện căn ở tại lòng ta
Chữ tâm kia mới bằng ba chữ tài)

It is important to note that acting according to Karma law will create successive chain of interminable Karma. On the other hand, acting according to Four Immeasurable, the Virtuous Mind is free of Karma because acting on the Eight fold Noble Path that is free of Karma.

The following story is an example: King Pasena of Kosala, a Buddhist devotee, wished to marry into clan of Sakya (of Buddha). To keep the Sakaya pure and clean, instead of sending a Sakaya Princess, a lovely girl of slave woman the King was sent to be made into The Queen.

The son of the slave woman was made into the crown Prince, Vitatubha... . At age 16, Prince Vitatubha visited Kapilavatthu, where his mother originally came from. The Prince was accidentally humiliated of his root from a slave woman, enraged, and wowed, declaring that one day, he would wipe out the whole clan of Sykayans. Eventually, after becoming King of Kosala, he marched with the army three times to fulfill the vow, but on three occasions, Buddha succeeded in restraining him by meditating on his route to the Sakayan clan. For the fourth time, Buddha looked into the Karma of the unending hatred. In the old times, the Sakayans killed all fish in a pond for food. The fish population becomes King and clan of Kosava people. Being aware of the return of Karma is not preventable, Buddha did not interrupt Vitatubha.

After the mass killing, , the King and the army returned home, On the way home, they camped on the river bed, and were killed by a sudden flood at night, ending the kingdon of Kosava.

Similarly, an over-excited defense reaction to a powerful and rude aggressor or an action of revenge are not advisable in most major religions. As Bible wrote (Matthew 5:38)

Eye for eye, and tooth for tooth.'[1]But I tell you, do not resist an evil person. *If anyone slaps you on the right cheek, turn to them the other cheek also.* And if anyone wants to sue you and take your shirt, hand over your coat as well. If anyone forces you to go one mile, go with them two miles. Give to the one who asks you, and do not turn away from the one who wants to borrow from you.
Love for Enemies:
You have heard that it was said, 'Love your neighbor[b] and hate your enemy.' 44But I tell you, love your enemies and pray for those who persecute you, **that you may be children of your Father in heaven**.

The action of war and non-conditional reflexes are not always permissible in morality since war is always followed by violence and immorality. Raising a white flag at an appropriate moment is a decent spiritual road. The judgment is always up to the highest level of Creation with law of Karma (eye for eye/tooth for tooth).

Upanishads wrote:
You are what your deep driving desire is. As your drive is, so is your will. As your will is so is your deed As your deed is, so is your destiny
(meaning: Destiny/Karmas→deed→will→drive→yourself/your action)

6. No Free Will in the Legal system.

The legal system does not often consider the principle of Free Will in criminal justice. Despite the judge being aware of this principle, the verdict at the court is not based on the concept of Karma, previous lives, because of the ignorance of humans in the current legal system.

7. Quantum Mechanics and Free Will.

In quantum mechanics, two photons or electrons belonging to the same system are shot in opposite directions in the thought experiment designed by Einstein, Podolsky, and Rosen called the EPR paradox. In this experiment, later confirmed in the laboratory by Alain Aspect (who won the Nobel Prize in 2022), the particles always spin in the same direction despite the distance.. Physicists call this phenomenon "entanglement" because the particles of the same system are interconnected with each other. Einstein used this evidence to argue with Niels Bohr that the quantum mechanics of the Copenhague interpretation is incomplete due to the hidden variable that governs the Entanglement of particles. In Middle school of Maha Buddhism by Nagarjuna, the Emptiness that represents the original Mind/ Buddhahood, Transcendental Awareness is characterized by the Eight negations:

i. No birth No death/ vn: Bất sanh Bất diệt)
ii. No end No permanence/ vn: Bất Thường Bất đoạn)
iii. No similarity, no difference/ vn: Bất đồng Bất dị
iv. No coming, No going away/ vn: Bất lai Bất khứ

The number iii Negations represents the entanglement phenomenon since the particles, despite being separate (No similarity) and different, are the same (No difference).

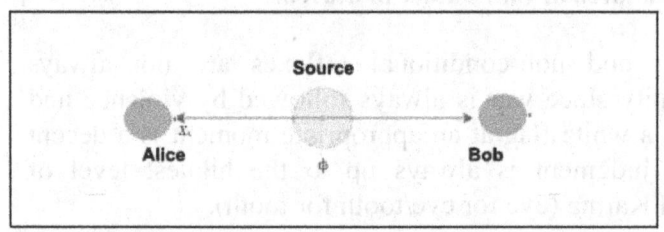

Fig 11.8A

This phenomenon of ERP paradox or entanglement in quantum mechanics illustrates the absence of "Free Will" since the particles are not free to rotate but follow the Emptiness status.

Therefore the EPR paradox can be interpreted as the photons being predetermined to spin according to a fixed spin. They do not have the choice to display the characteristic when observed by Alice and Bob.

When objecting to Copepenhague's quantum interpretation of quantum mechanics Einstein proposed a hidden variable accounts for the particle spin. However the hidden variable does not exist because. Emptiness can be accounted for the non-locality and interconnectedness. The Free Will theorem of <u>John H. Conway</u> and <u>Simon B. Kochen</u>la said: *If we have a certain amount of <u>free will</u>, then, subject to certain assumptions, so must some <u>elementary particles</u>. Of course, it is all right that humans have a certain right, that is, to only follow the instruction of the Creator/ Emptiness/ Buddhahood. Particles are free to rotate according to the initial determination of the experiment.*

In Summary, human beings are only able to act according to the four immeasurable minds of virtue and selflessness and are not free to act otherwise. The will and action based on the non-immeasurable minds of virtue will lead to the creation of Karma.

CHAPTER XII:
CURIOSITY, IMAGINATION AND CREATIVITY

Creativity is defined as the capability of forming something new and valuable according to an individual or group of people. Therefore, creativity is a personal expression and appreciation of the external world that may be inherent to each individual. The creation involves all aspects of every day ranging from art, music, singing, dancing, crafting (doodling, knitting, sewing, cooking, game playing, poetry or story, history and research writing, verbal communication skills, gardening, exercising, golfing taking yoga, performing a play reading meditating, daydreaming...). Creativity can be developed in even stereotyped activities, walking, swimming, and running... As a result, especially for seniors, there is a development of satisfaction, self-esteem enhancement, confidence, familial and social interaction, occasionally increased income and dexterity, lowering the rate of loneliness and depression, and encouraging a sense of humor. There is an overall increased quality of life.

A small study of 24 people aged 18-75, performed by Moncho Ouri in 2011, showed that older persons are as active as young people regarding creativity. However, this study has never been published in medical literature, which raises its medical validity.

In addition, there was a false claim that a study published in NEJM stating: most productive age of a person is from 60 to 70 years," that the "second most productive human stage is the age from 70 to 80 years old," and that the third "most productive" time is between the ages of 50 and 60

A further search in the literature found no peer-reviewed study supporting the claim that creativity peaks at the ages of 60 to 70 years. The ages of Nobel winners peak at the age of 40 or even younger for science.
Nevertheless, the ages of Nobel laureates appear to shift gradually toward the older group. Knowing that in the early 1900s, it appears that major scientific discoveries filled the vacuum of knowledge, which was opened by the discovery of the ultrastructure of atoms and is more open for scientists with *curiosity,*
imagination, and creativity than
experience. Since 1945, **experiences** proportional to science exposure have played an important role.
It has been observed that older people can maintain some level of creativity despite the everyday degeneration of neurons with thinning

of the cerebral cortex and eventual decline of the cognitive, visual, auditory, and locomotive functions..

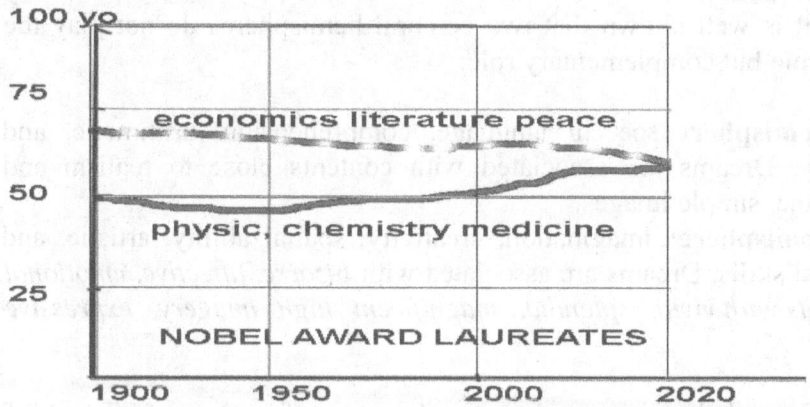

For insight into creativity, the following is an attempt to review the neuroscientific mechanism of curiosity, imagination, and creativity.

Curiosity is the instinct of living animals for survival, including the search for food, avoidance of danger, entertainment, and development. Curiosity decreases with age due to the enrichment of experience as one gets older.
Curiosity enriches the Inner Consciousness.

Imagination: contrary to curiosity, imagination is rooted in the information stored in the brain (the Inner Consciousness seated in the Hippocampus, Medial Temporal Lobe, Ventromedial Prefrontal Cortex and Posterior Cingulate Cortex). Depending on the nature of memory (MM or data), the MM in stored in the Left hemisphere or the Right side

Example:
1. In navigation, the overall map of navigation is stored in the RT hemisphere (allocentric navigation map, Allo= other/ not self), whereas the path of navigation (Egocentric map) is stored in the LT hemisphere (consisting Hippocamus for the map, Parahippocampus for landmark in the map, and reference frame in navigation)
2. In composing music, data is collected from the RT hemisphere, but for the experienced musician, the music composing is easier, and data mostly comes from the LT side.
3. The location and time when I first met my wife :
   Time X location T: in LT hemisphere.
   Significance, emotion, and aspiration of the meeting: in the RT hemisphere.

The reason is that each hemisphere does not store the same data but complementary data. Unlike paired organs like kidneys, thyroid lobes, adrenal glands, or liver (RT and LT lobes that play the compensatory role), it is well known that two cerebral hemispheres do not play the same role but complementary role:

LT hemisphere: speech/ language, comprehension, arithmetic, and writing. Dreams are associated with contents close to realism and mundane, simple images.
RT hemisphere: imagination, creativity, spatial ability, artistic, and musical skills. Dreams are associated with *bizarre, affective, emotional contents with vivid, splendid, magnificent high imagery, expressive states.*

Therefore, distinct separate but complementary functions of LT and RT hemispheres are a fact, not a myth as occasionally written on the internet.

In general, the RT hemisphere amplifies by multiplying the data stored in the LT side with exponential power. For example, when receiving the image of a tall building, the LT hemisphere stores the image as such. The RT hemisphere multiplies the building in number, dimension, shape, color, and content...... As a result, when a craft lands on the moon, the imagination retrieves data from the RT hemisphere with the following expectation of landing on the moon with a human being or landing on other distant planets.
Data retrieving from the right hemisphere usually requires attention and mindfulness. As a result, retreat and meditation are critical in the imagination and creativity.

Cao Bá Quát, a famous Vietnamese poet said:
Vắt tay nằm nghĩ chuyện đâu đâu,
Đem mộng sự đọ với chân thân thì cũng hệt
........Và
Kho trời chung mà vô tận của mình riêng
In conceiving everything with mindfulness
Dreams and reality are alike
and...
Universe store is common, but personal assets of awareness are infinite.

Creativity: as defined at the beginning of this writing. The mechanism of development consists of retrieving data, as in the case of imagination and experiences stored in the inner consciousness, and realising the data, converting it into reality with enrichment by curiosity. The experience's specificity is critical in converting the imagination into reality. Another source of data is spiritual through the interaction between the external mind and the medial PFC. Dreams are more likely from data of the internal CS from both RT and LT hemispheres than from external minds

Neuroscientific Wiring in the Brain for the Imagination and Creativity

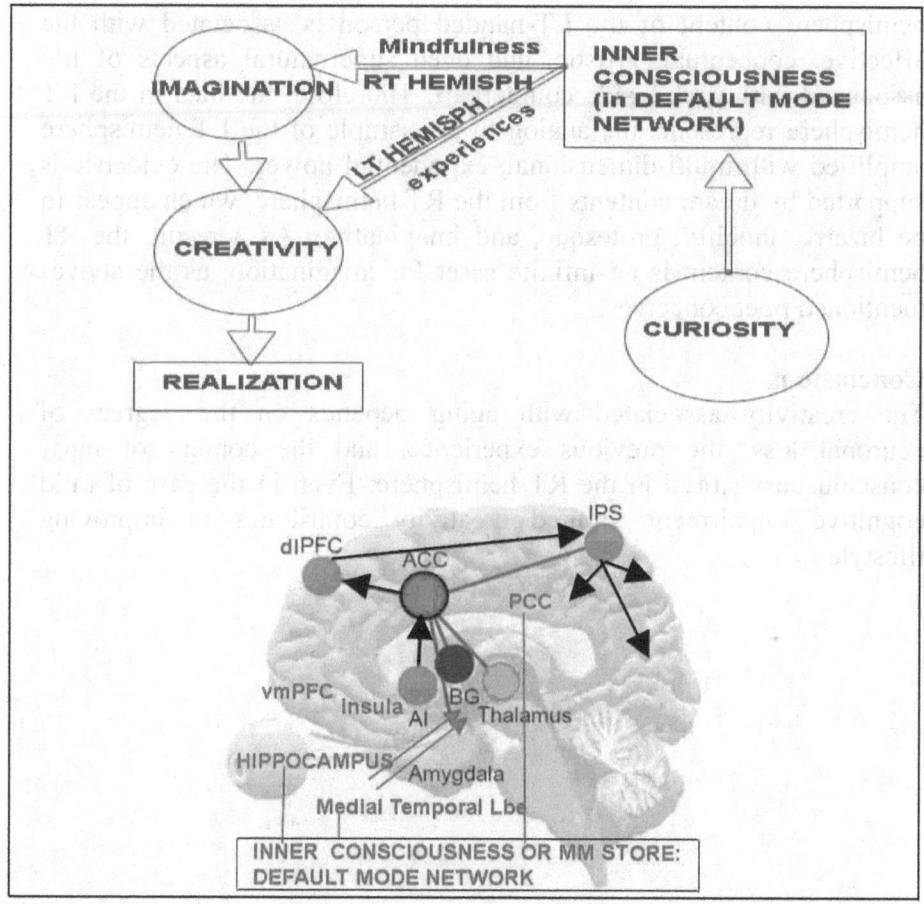

dLPFC-IPS (dorsolateral prefrontal cortex- intraparietal sulcus) plays the role of attention with IPS connected to various cortical areas of the Inner Consciousness for MM retrieval. The Salient network, Insula-ACC, controls DLPFC. The Insular cortex is considered the primary cortical site for volition in creativity.

Creativity and Aging
With aging, there is accumulated loss of neurons with thinning of the cerebral cortex, resulting in the loss of memories. On the other hand, experiences retained the MM subserve valuable assets to creativity.

In addition, MM stored the Inner Consciousness in the LT and RT hemispheres are not identical but remarkably different and rather supplementary than harmonious.

The difference in the MM contents in the RT and LT hemispheres resides in the inherent role of each hemisphere, as outlined in the previous paragraph. The LT hemisphere content in the RT-handed person is close to reality with the physical aspect as it is seen. The RT hemisphere content of the LT-handed person is associated with the affective, conceptual, artistic, and even supernatural aspects of life associated with its LT side counterpart. Therefore, the data in the RT hemisphere represents the analog to the sample of the LT hemisphere amplified with multi-dimensional, exponential power. The evidence is supported by dream contents from the RT hemisphere, which appear to be bizarre, fanciful, grotesque, and imaginative. As a result, the RT hemisphere content is an infinite asset for imagination, as the above-mentioned poet conceives.

Conclusions
The creativity associated with aging depends on the degrees of neuronal loss, the previous experience, and the content of inner consciousness stored in the RT hemisphere. Even in the case of mild cognitive impairment, limited creativity contributes to improving lifestyle

www.ingramcontent.com/pod-product-compliance
Lightning Source LLC
Chambersburg PA
CBHW010824070526
44583CB00022B/2920